T0271400

Agro-Climatology Advances and Challenges

NEW INDIA PUBLISHING AGENCY
New Delhi-110 034

About the Authors

Prof. Dr. T.N. Balasubramanian got his Bachelor degree in Agriculture from Annamalai University. Chidambaram, by 1966, followed by Post Graduate degree in MSc. (Agri.) in Agronomy and PhD. in Agronomy from Tamil Nadu Agricultural University (TNAU), Coimbatore, Tamil Nadu, India. He is trained in Agricultural Meteorology at the National level at IITM, Pune, CRIDA, Hyderabad and Division of Agricultural Physics, IARI, New Delhi and NARM, Hyderabad. He is trained in cropping systems at IRRI, Philippines and in extension education system at ETI, Hyderabad. He is the first Head of the Department of Agricultural Meteorology, TNAU, Coimbatore. He served for 38 years and six months at SDA and TNAU. After his retirement (31.12.2004), he served as consultant in MSSRF, Chennai, Agricultural Insurance Company, Chennai, RGR, Cell at TNAU, Coimbatore, ACRC, TNAU and Emami, Arruppukottai, Tami Nadu.

He has guided nine PhD students and seven MSc students including one foreign National student under common wealth fellowship during his academic period. He has published six books and six chapters in standard books, published in India. He has published more than 150 research papers both in India and abroad. He has attended 66 conferences and workshops both in India and abroad. He also organized many trainings, National conferences and workshops for the benefit of Indian Scientists and Development Workers. He was Academic Council member of TNAU, Coimbatore, Hospital Committee Member of TANUVAS, Chennai and member in Board of Studies of Faculty of Agriculture, Annamalai University, Chidambaram, Tamil Nadu. He conducted one ag MOOCs programme on Weather Forecast in Agriculture from the IIT, Kanpur, platform. He released one sesame variety for summer irrigated condition of Tamil Nadu along with co-scientists. His 10 agotechniques have been given for adoption to farmers of Tamil Nadu by TNAU, Coimbatore. He was PI for six research projects including one international project from Australia.

He was one among the high-level committee members for the Waste Land Development in Tamil Nadu and one among the witnesses to Government of Tamil Nadu for Cauvery Water Dispute Tribunal. He is a member in three professional societies and was office bearer and editor of the Journal, of Association of Agrometeorologists at Anand.

Prof. Dr. R. Jagannathan has obtained his Bachelor B.Sc. (Ag) Masters M.Sc. (Ag) and Ph.D. degrees from Tamil Nadu Agricultural University Coimbatore. He has completed nearly 36 years of service at Tamil Nadu Agricultural University in various capacities as Assistant Professor Associate Professor and Professor. Served as Head of the Department of Agricultural Meteorology at Agricultural College and Research Institute Coimbatore and as Dean at Vanavarayar Institute of Agriculture, Pollachi Post retirement from TNAU worked as Head of the Department and Dean of private Engineering Colleges in Coimbatore and Namakkal as technical administrator of the faculty of Agricultural Engineering. Currently serving as Adjunct Faculty of Tamil Nadu Agricultural University.

He received best Ph.D Thesis award in Agronomy and T. Nataraj Rolling Shield and Medal for Best Extension worker. He received best technical book award for his book titled Weed Management Principles from TNAU. He also received Life Time Achievement Award from the Association of Agro-meteorologist. He also obtained a Patent for coconut coirpith fertilizer briquette.

He visited 14 countries for various purposes like conferences, training etc. He has specialization in Crop Simulation Modeling, Weather Forecasting and Climate downscaling for agriculture impact studies through his trainings at National and International level. Developed knowledge base for automated weather-based agro-advisory system using weather windows developed for more than 100 crops grown in Tamilnadu. He has also developed weather scenarios for Maharashtra state for weather based agro-advisories. Mobilized funding for research programs to the tune of ₹3,750 lakhs from international, national and state level organizations.

He was involved in the preparation of curricula and syllabi for diploma courses of Tamil Nadu Agricultural University. Involved in development of Agricultural Meteorology syllabus for worldwide adoption as World Meteorological Organization nominated committee member. He was responsible for development of syllabus for B.E. Agriculture Engineering programme of an autonomous private engineering college.

He has published 43 research papers in peer reviewed journals and having 41 full papers and 19 abstracts in various publications of national and international conferences and seminars He also published 42 popular articles, 6 books, 15 practical manuals, 26 technical booklets and developed two interactive compact disks. Guided 14 Ph.D students, 19 Masters and two bachelors for their thesis research.

Dr. V. Geethalakshmi has doctoral degree (Ph.D. in Agronomy) from Tamil Nadu Agricultural University. Started her career during January 1989 and continuously serving in Tamil Nadu Agricultural University (TNAU) under various capacities and currently serving as the Director for Crop Management in Tamil Nadu Agricultural University. Her field of specialization is Agricultural Meteorology; Weather forecasting; Agro-advisory services; Climate change impact assessment and developing adaptation options.

In recognition of her scientific work, she was appointed as an International Expert of the CAgM (Commission on Agricultural Meteorology–a WMO body), in Open Programme Area Group -2. She has served as member in the expert committee for Science and Technology for women (2011–2013), DST and for the climate change programme under the Ministry of Environment and Forest. Currently serving as a Core committee member in the Expert Committee for Young Scientist in the area of Earth and Atmospheric Sciences in the Science and Engineering Research Board (SERB), Government of India.

Dr. V. Geethalakshmi has been nominated as National level coordinator for the NMSKCC programme and also serving as invited member of the expert committee (2016–2019). She has contributed for developing *State Action Plan on Climate Change for* Tamil Nadu State for agriculture sector. She is serving as Co-Chair for developing Science and Technology Innovation Policy 2020 of Government of India–for Agriculture Sector.

Dr. V. Geethalakshmi has obtained and executed research projects from International and national organizations to the tune of 30 crores and collaborated with Australia, Norway, UK, Africa, USA, Japan and many other countries. She has more than 100 research publications and guided more than 12 PhD scholars and about 25 M.Sc scholars.

Agro-Climatology Advances and Challenges

T. N. Balasubramanian
(Retired) Professor and Head
Department of Agricultural Meteorology
(Presently Agro Climatic Research Centre)
Tamil Nadu Agricultural University
Coimbatore 641 003, Tamil Nadu

R. Jagannathan
(Retired) Professor and Head
Department of Agricultural Meteorology
(Presently Agro Climatic Research Centre)
Tamil Nadu Agricultural University
Coimbatore 641 003, Tamil Nadu

V. Geethalakshmi
Director
Crop Management Studies
Tamil Nadu Agricultural University
Coimbatore 641 003, Tamil Nadu

CRC Press is an imprint of the
Taylor & Francis Group, an **informa** business

NEW INDIA PUBLISHING AGENCY
New Delhi-110 034

First published 2022
by CRC Press
4 Park Square, Milton Park, Abingdon, Oxon, OX14 4RN

and by CRC Press
6000 Broken Sound Parkway NW, Suite 300, Boca Raton, FL 33487-2742

CRC Press is an imprint of Taylor & Francis Group, an Informa business

British Library Cataloguing-in-Publication Data
A catalogue record for this book is available from the British Library

Library of Congress Cataloging-in-Publication Data
A catalog record has been requested

ISBN: 978-1-032-19837-8 (hbk)
ISBN: 978-1-003-26110-0 (ebk)

DOI: 10.1201/9781003261100

Be Weather wise—Otherwise—Not wise

Tamil Nadu Agricultural University
Lawley Rd, P N Pudur, Coimbatore-641003, Tamil Nadu

Prof. N. Kumar Ph.D., F.H.S.I
Vice-Chancellor

Foreword

I got an excellent opportunity to review the book on "Agro-Climatology: Advances and Challenges" written by three Professors of this University. From my personal observation, all the three authors have excellent knowledge on both practical and theoretical aspects of the subject, agro-climatology/agro-meteorology. I have gone through all the chapters of this book meticulously. Each chapter explains both theory and practical sides of the subject. Further, new topics on weather codes, astro-meteorology, livestock meteorology and installation of meteorological instruments in the surface observatory, have been dealt in this book which are most useful in the fields of agro-climatology/agro-meteorology. Further, the authors tried to include questions and answers for selected topic, definition for terminologies used in the field of agro-climatology/agro-meteorology and simple practical computation of meteorological problems which would kindle the readers and deserves appreciation.

I congratulate the authors for bringing out this book for study purpose and for use by the agricultural research and development managers for increasing the agricultural production and productivity.

I trust this book as an excellent publication to cater to the needs of agro-climatology education in the country and would definitely add knowledge to students, teachers, farmers and managers of the Government.

Date: 10.07.2020
Place: Coimbatore

–Sd– **N. Kumar**
Vice-Chancellor

Preface

The statement **"Be weather wise- Otherwise - Not wise** is the new jargon developed by the senior author of this book by 1998 and this indicates the importance of weather and climate to human society and other associates' life in the earth. This is an action-oriented jargon covering the lives of the earth from A to Z. In the earlier publications on agricultural meteorology and climatology, only theoretical parts have been covered elaborately. But in this publication, little part in theoretical side is covered, leaving major scope to cover under practical sides of the subject.

There are 10 chapters in this book covering crop -weather interaction and agro-met observatory, agro-climatic analysis, crop micro-meteorology, remote sensing, crop simulation models, weather codes and their management, integrated weather forecast and agro advisories, climate change, livestock climatology/meteorology and astro-meteorology. Hence this book becomes all in one publication.

Further selected and frequent asked questions from selected chapters have been answered in a simple way especially for under graduate students and other publics. To understand the text of the book, under terminology, simple details have been given for hard technical words. Further and above all, under practical tools, important computations and calculations have been given with example, which is the unique of this publication.

The authors feel that this publication would be very useful to under graduates, post graduates, research scholars, publics, teachers and also to the politicians to take policy decisions on the subject. The readers feedback is considered as very important to improve the quality of the present publication to a higher level in the forthcoming editions.

The authors are highly thankful to the Vice -Chancellor and Authorities of Tamil Nadu Agricultural University (TNAU) to use the institutional references for the chapters concerned. The authors are highly and sincerely thankful to the Publisher "New India Publishing Agency", New Delhi.110 034 for their motivation to write exclusive book on this title and also granting extra time to complete the book assignment during lock down period of COVID 19 from March 2020 in India.

Our sincere thanks to Dr. K. Bhuvaneswari, Dr.M.Dhasarathan, Dr. R. Gowtham, Dr. K. Senthilraja, Dr. P. Dhanya, Dr. N. Raja, N. S. Vidhya Priya, T. Sankar and D. Pugazhendhi technical staff of the Director (Crop Management), Tamil Nadu Agricultural University, Coimbatore for their full hearted support given to enhance the quality of the publication. The authors also registered their sincere thanks to the Professor and Head, and his technical team of Agro Climate Research Centre, TNAU, Coimbatore-3 for their support extended during the preparation of this publication.

Place: Coimbatore
Date: July, 2020

T.N. Balasubramanian
R. Jagannathan
V. Geethalakshmi

Contents

Terminologies

Absolute humidity: In a system of moist air, this is the actual quantity of water vapor by weight present in the total volume of the system.

Absolute scale: The lowest temperature theoretically possible is known as absolute zero. The zero of thermodynamic temperature is (-) 273.15 degree Celsius. The absolute scale is the temperature scale based on absolute zero

Accumulated temperature: This is the sum of the departures of temperature from a base temperature.

Actinometer: An instrument used to measure the intensity of radiant energy.

Adaptation: Adjustment in natural or human systems to a new or changing environment. Adaptation to climate change refers to adjustment in natural or human systems in response to actual or expected climatic stimuli or their effects, which moderates harm or exploits beneficial opportunities. Various types of adaptation can be distinguished, including anticipatory and reactive adaptation, private and public adaptation, and autonomous and planned adaptation.

Aerosol: In meteorology, it is an aggregate of minute particles (solid/liquid) suspended in the atmosphere.

Agricultural climatology: A branch of applied climatology to study the influence of climate on different processes of agriculture. Climatology is the science, which studies the totality of weather. This is in short form, can be called as agro-climatology.

Agro climatic Region: The grouping of different physical areas within a country into broadly homogenous zones based on climate and edaphic factors.

Agro ecological zones: A major area of land that is broadly homogenous in climate and edaphic factors, but not necessarily continuous, where a specific crop exhibits roughly the same biological expressions.

Agro eco system: The total complex of crops/animal composition in an area, together with over all environment and as modified by management practices.

Agro meteorology: The branch of meteorology relevant to problems of agriculture.

Agricultural drought: The soil moisture as a result of rainfall is well below the crop ET requirement during longer period of the crop duration.

Albedo: The ratio of the amount of visible light reflected to the amount incident on it.

Altitude: Vertical distance of a level, a point or an object considered as a point measured from mean sea level.

Altimeter: An instrument to measure the altitude from mean sea-level.

Alto-cumulus and Alto-stratus clouds: These are all middle level clouds in the troposphere of the atmosphere and the mean height is 2–7 km. Cumulus means heap, while stratus means layer or sheet.

Ambient Temperature: The hotness (temperature) of the surrounding atmospheric air.

Anemograph: This is an instrument to measure the speed and direction of the wind.

Anemometer: An instrument to measure wind speed only.

Aneroid barometer: One instrument to measure atmospheric pressure.

Anthropogenic emissions: Emission of greenhouse gases, green house gas precursors and aerosols associated with human activities. These include burning of fossil fuel for energy, deforestation and land use changes that result in net increase in emission.

Anticyclone: The atmospheric pressure distribution in which there is high central pressure relative to the surroundings.

Anti-trade winds: This is westerly air current. In both hemispheres, particularly in sub-tropical regions, these wind blow above the trade wind

Arid regions: Ecosystem with less than 250 mm of annual precipitation, where the annual Potential Evapotranspiration (PET) is > 1500 mm/year. The ratio between annual rainfall/precipitation and annual PET is between 0 and 0.3.

Aridity index: This indicates the degree of dryness of a climate, as a function of climatic factors like precipitation, temperature, heat, moisture, evapotranspiration, *etc*.,

Atmospheric pressure: The pressure exerted by the atmosphere as a consequence of the weight of the air lying directly above that point. At mean sea level, the atmospheric pressure is normally equal to 76 cm of mercury column. Standard sea-level pressure, by definition, equals 760 mm (29.92 inches) of mercury, 14.70 pounds per square inch, 1,013.25 × 10^3 dynes per square centimeter, 1,013.25 millibars, one standard atmosphere, or 101.325 kilopascals.

Atmometer: One instrument to measure the evaporation rate.

Available moisture: This is a moisture range unit or limit available to plants/crops to meet its evapotranspiration demand, which lies between field capacity and wilting point of the concerned soil type.

Azimuth: The angular distance from the north and south point of the horizon to the foot of the vertical circle through a heavenly body.

Barometer: An instrument used to measure the atmospheric pressure.

Bellani spherical pyranometer: An instrument to measure solar radiation falling on a spherical surface from the hemispherical sky and also that terrestrial radiation which is reflected from the ground.

Beufort wind scale: This scale was devised by Admiral Sir Franscis Beufort by 1805 to estimate and report wind speed. In the scale series of numbers are given from 1 to 12. The number on the scale can easily be calculated based on the effect of wind strength on surface features such as tress and sea state, *etc.*

Bio -climatology: This applied science studies the relationship between climate and bio organisms.

Blizzard: A weather abnormality characterized by low temperature, strong winds embedded with powdery snow or ice. In some cases, a storm may cause falling of greater amounts of snow or even the snow may be picked up from the ground. The wind speed generally associated with these storms are 15m/second or more during which the visibility is reduced to 150 m or even less.

Bolometer: One instrument to measure the intensity of radiant energy, with reference to the thermal component of wave band.

Bowen ratio: The ratio of vertical energy flux of sensible heat to latent heat at a given site in the same direction. The site may be a crop canopy or smaller area over sea or river or moist land etc., In any case the movement of sensible heat is proportional to the temperature gradient over any interval of height. Likewise, the latent heat transfer is proportional to water vapor pressure gradient. However, both the phenomena are dominated by eddy transfer.

Buoyancy: The lift experienced by a body by virtue of which it floats in another medium because of the density difference between the substances. Here, the lift means the net force exerted upward.

Campbell-strokes recorder: One instrument to measure the bright sun shine hours of a day. It is now recognized that during bright sun shine hours, the plant does photosynthesis.

Cardinal temperature: The minimum, optimum and maximum temperatures level computed for either the growth of an organ or the whole plant.

Capacity building: In the context of climate change, this refers to the process of developing the technical skills and institutional capability to enable them to participate in all aspects of adaptation. This goes up to mitigation process.

Ceilometer: An automatic instrument used to measure the height of the base of a cloud above the point of observation.

Celsius scale: In this scale, ice melts at zero degree and the boiling point of water is 100 degree. Temperature also expressed in kelvin scale, where one-degree kelvin is Celsius temperature + 273.16.

Cirrus: One high level cloud and the mean height is 5–13 km. The prefix refers to hair band like or hair lock. At sun rise and sun set these clouds will give marvelous colors and does not give rain.

Climatic normal: Optimal level of weather elements required for biological growth.

Climatology: It is derived from Greek language. This science deals with the factors that determine and control the distribution of climate over earth's surface or the science which deals with the totality of the weather.

Cloud: An aggregation of minute drops of water/water vapour over special or condensation nucleus in an indefinite unit with its base above ground level. The cloud forms when rising air gets cooled with lapse rate phenomenon to such an account that it can no longer hold all the water within it as water vapor.

Cloud burst: Unexpected torrential rain occurs all on a sudden with no stoppage and leads to flash floods and land slides.

Climate change: It is statistically significant variation in either the mean state of the climate or in its variability, persisting for an extended period (typically decades or longer). The United Nations Frame work Convention on Climate Change (UNFCCC), in its Article 1 defines climate change as a change of climate, which is attributed. to directly or indirectly to human activity that alters the composition of the global atmosphere and which is in addition to natural climate variability observed over comparable time periods. The UNFCCC thus makes a distinction between climate change (attributable to human activities altering atmospheric composition) and climate variability (attributed to natural causes).

Climate variability: Variation in the mean state and other statistics (such as standard deviation, the occurrence of extremes, *etc.*,) of the climate on all temporal and spatial scales beyond that of individual weather events.

Variability may be due to natural internal processes within the climate system (internal variability), or to variations in natural or anthropogenic external forcing (external variability).

Climate: Simply it is defined as average weather. Or more rigorously as the statistical description in terms of the mean and variability of relevant quantities over a period of time ranging from months to thousands or millions of years. The classical period is 30 years as per the standard given by the World Meteorological Organization (WMO). Climate in a wider sense is the state including a statistical description of the climate system.

Cloud seeding: Injecting the appropriate clouds with hygroscopic seeding materials like dry ice, silver iodide, sodium chloride etc., from an aircraft or by using ground generator for producing artificial rain.

Condensation: The process of formation of liquid from its vapor.

Cold front: A front advancing so that the cold air mass replaces the warm air mass. In general, upon the passage of a cold front different weather conditions exist.

Convection: Upward lift of relatively warm air.

Convective rain: Rain fall that is caused by the vertical motion of an ascending mass of air, which is warmer than its environment; the horizontal dimension of such an air mass is generally of the order of 15 km or less and forms a typical cumulonimbus cloud.

Crop calendar: A list of the standard crops of a region in the form of a calendar giving the date of sowing and various farm operations including harvesting date during the crop season in years of normal weather.

Crop efficient zone: A geographical area characterized by high spread of the concerned crop with its high productivity with minimum variation in its productivity from year to year.

Crop evapotranspiration: The total quantity of water lost in mm/day through crop transpiration and its surrounding soil evaporation of a disease-free crop growing in larger field under optimal soil condition and also under unlimited water condition.

Crop growth model: Crop growth is due to intermixing of physical, physiological and ecological processes and the representation of this system through simulation is called as crop growth model.

Crop season: The most favorable homogenous weather condition that exists for high crop's productivity from three to four months.

Crop yield index: A measure of comparison of the yields of all crops on a given farm with the average yield of these crops in the locality. This relation is expressed in percentage.

Cumulus: One low level cloud and can be seen at 0–2 km from the ground level. Cumulus means heap. The clouds develop vertically in the form of rising mounts, domes and the bulging upper parts resemble a cauliflower.

Cumulo-nimbus: Low level cloud and can be seen at 0–2 km. These are heavy and dense clouds in the form of mountain and associated with heavy rain, lightning and thunders. These are towering clouds and at times spreading out on top to form an anvil head.

Cup anemometer: Instrument to measure wind speed.

Cyclogenesis: Development of or strengthening of cyclonic circulation in the atmosphere.

Cyclone: An area of low atmospheric pressure in to which the air flow is deflected. The deflected air rises and blows around the low pressure in an anticlockwise direction at northern hemisphere and clockwise direction at southern hemisphere. In this synoptic weather system, a large storm develops in closed circulation over 1000 km area. Cyclone name is given to storm which develops at Indian ocean.

Cyclonic motion: Motion of the air in the counter clockwise direction around the centre in northern hemisphere.

Degree day: Calculation of the departure of daily mean temperature above base temperature of the concerned crop.

Depression: In meso scale and macro scale meteorology, an area of low pressure is commonly referred as depression. Certain times the trough is also called as depression.

Dew: Condensation of water vapour on a surface whose temperature is reduced by radiational cooling to below the dew point temperature of the air in contact with it.

Dew point temperature: At this temperature the saturated vapor pressure is equal to actual water pressure. This is the temperature at which a given parcel of unsaturated air must be cooled to produce a state of saturation at constant pressure and water vapor content.

Diabatic process: A thermodynamic change of a system involving transfer of heat across the boundaries of the same system.

Diffused radiation: Radiation reaching earth surface after passing or scattered through suspended solids/clouds/water vapor in the atmosphere.

Disdrometer: An instrument used to measure the size and diameter of the rain drops.

Diurnal: Daily cycle of 24 hours, where in the maximum and minimum weather values difference being evaluated for crop and animal performance.

Drizzle: Rainfall in general in which water droplets are very small ranging from 0.2–0.5 mm in diameter and it is not so continuous like rain.

Drosometer: Instrument to measure dew deposit and it is also called as Oetalies dew gauge.

Drought year: This is a year wherein the seasonal or annual rainfall is short (deficit) by more than twice the standard deviation of the concerned rainfall data series.

Drought: When precipitation is significantly below normal (< 75 per cent of normal) then this phenomenon starts operating. It causes serious hydrological imbalances that adversely affect land resource production system.

Dry land farming: The practice of crop production entirely with rainfall, wherein dry spell occurs during cropping season frequently causing crop stress for soil moisture. In some areas crop is raised with conserved soil moisture, where production is uncertain. This dry land farming is common in semi-arid climate. Crop may face mild to severe moisture stress during its life cycle.

Dry spell: A period of at least 15 consecutive days without rainfall.

Dust devil: The swirling dust observed in a hot day over the land.

Dust storm: This is an unusual weather phenomenon wherein, strong winds with dust filled air blows over an extensive area.

Eclipse: The passage of a non-luminous body in to the shadow of the another. The solar eclipse occurs when moon's shadow fall up on the earth, while the lunar eclipse occurs when the moon enters the earth's shadow.

Eddy: A translating and rotating air parcel is often called as eddy. The transient eddies move in space (*e.g.*, cyclone), while stationary eddies remain in fairly fixed locations (*e.g.*, anticyclone).

Effective temperature: The temperature above a certain minimum at which physiological processes such as growth of a crop plant, etc., are considered active. For many crop plants 5°C is considered to be effective temperature.

ELNino-Southern Oscillation (ENSO): EL Nino in its original sense refers to a warm water current that periodically flows along the coast of Ecuador and Peru, disrupting the local fishery. This oceanic event is associated with a fluctuation of the inter-tropical surface pressure pattern and its circulation in the Indian and Pacific Oceans, called southern oscillation. This coupled atmosphere-ocean phenomenon is collectively known as El Nino Southern Oscillation (ENSO). During an EL Nino event, the prevailing trade winds weaken and the equatorial counter current strengths, causing warm surface waters in the Indonesian area to flow eastward to overlie the cold waters of the Peru Current. This event has great impact on the wind, sea surface temperature and precipitation patterns in the tropical Pacific. It has climatic effects throughout the Pacific region and in many other parts of the world also. The opposite of an EL Nino event is called as La Nina.

Emission: The production and discharge of greenhouse gases from industry/ other sources and the effect of emission on atmosphere and bio organisms impact including human being.

Emissive Power: The total energy emitted from a unit area of a surface of a body per second. The total emissive power depends on the temperature of the body and nature of its surface.

Emissivity: This is the ratio of the emittance of a given surface at a specified wavelength and emitting temperature to the emittance of an ideal black body at the same wavelength and temperature.

Eppley pyrheliometer: Instrument to measure the solar radiation.

Equatorial: The wind/air originating near the equator is called as equatorial. The equator is the line that seperates the sphere of the earth in to two equal halves.

Evaporation: The process by which a liquid becomes gas or vapor.

Evaporation pan: The instrument to measure the evaporation from the open water body.

Evaporograph: Instrument to measure evaporation over a given period of time through graph.

Evapo-transpiration: The combined process of evaporation from the Earth's surface and transpiration from vegetation. Please refer crop evapo-transpiration also.

Evapotranspirometer: An instrument to measure evapotranspiration.

Extended forecast: The weather forecast is valid for five days from the date of issue. The main emphasis is given from prediction of change of weather type to weather hazards like strong winds, dry and wet spells.

Extra-tropical cyclone: A cyclone outside the tropics.

Extreme weather event: An event that is rare within its statistical reference distribution at a particular place. Extreme weather event may vary from place to place. It may be + of 75 to 90% of the normal or - of 75–90% of the normal.

Eye: In meteorological terms, the roughy circular area of calm or relatively light winds and comparatively fair weather at the centre of well-developed tropical cyclone.

Fahrenheit temperature scale: One temperature scale having freezing point of water at 32° and the boiling point equals to 212° at standard atmospheric pressure.

Fair weather: Fine sky. Appearance of cirrus or cumulus cover is in fair amount but not covering the sky totally.

Flood: The overflowing of rain water from a river, stream surface tank etc., from their respective boundaries of safe water holding level. This is mainly due to extreme rainfall event.

Fog: Obscurity in the surface layer of the atmosphere, which is caused by a suspension of water droplets, with or without smoke particle and which is defined as visibility is < one KM.

Forecast: This term was first used in the science of meteorology by Admiral Fitz Roy and this gives statement of anticipated condition for a temporal and spatial scales

Fortin's barometer: One mercury barometer to measure the atmospheric pressure, named after J. Fortin (1750-1831).

Freezing fog: Fogs form due to the cooling of moist air below its saturation point. If super cooled water droplets freeze on impact with any solid surface, then the fog is called freezing fog.

Front: An interface between two different air masses of different properties like sharp gradient of temperature, wind etc., A front is a mesoscale phenomenon in the lower troposphere and is considered as a region of most rapid weather changes.

Frost: The deposition of ice crystals on a cooler land surface or objects, by diffusion and sublimation. This occurs in the atmosphere when air temperature and dew point temperature are below freezing level.

Funnel cloud: A tornado cloud or vortex cloud extending downward from the parent cloud, but not reaching the ground. Technically this will not be brought under cloud.

Gale: The wind speed is between 32 and 63 miles/hour. In Beufort scale this is from 7 through 10.

General circulation model: General circulation is nothing but the average general distribution of wind in motion on the earth. If the general circulation is modelled in a three-dimensional procedure, then it is known as general circulation model.

Geostrophic wind: The wind moving parallel to straight iso-bars.

Glaze: An accretion of relatively clear ice.

Global radiation: Direct solar radiation from the sun + indirect diffused sky radiation received on a unit horizontal surface.

Greenwich Meridian Time: The time at the Greenwich Meridian is known as Greenwich Mean Time (GMT) or Universal Time Coordinator (UTC). The place Greenwich, where the royal observatory located at England is taken as meridian zero. The eastern time zone in the United States is designated as GMT minus five hour. When it is noon in the eastern time zone, it is 5 p.m. at the Greenwich observatory. In India if the meridian 82.5°E (Allahabad) has the time of 1730 hrs. and this means that, the time at Greenwich is 12 noon of the day.

Growing degree day: Sum of the difference between daily mean temperature and base temperature of the concerned crop in a growing season. Please see degree day also.

Growing season: Period of a year with prevailing homogenous weather favorable for crop growth and their productivity.

Greenhouse gases: A greenhouse gas absorbs and emits radiant energy within the thermal infrared range. The primary greenhouse gases in Earth's atmosphere are water vapor, carbon dioxide, methane, nitrous oxide, CFC and ozone. Accumulation of greenhouse gases in the atmosphere causes global warming and this leads to change in climate.

Gust: A sudden and short positive departure from the ten minutes average of wind speed.

Hadley cell: A mean thermally driven atmospheric circulation in which warm air rises and flows away from the equator at upper level and descends at the tropics or the wind moves from the tropics back to equator poleward.

Hail: One form of precipitation. These are all stone lumps of ice formed from protracted rimming in cumulonimbus updrafts.

Hair hygrometer and hair hygrograph: Instruments to measure relative humidity of the air.

Hail storm: A storm often too long and severe consisting of largely frozen rain drops ranging in diameter from 5–10 mm or still more.

Harvest index: It is the ratio between economic yield of crop and its dry matter.

Haze: It is a form of condensation. Haze is formed when water vapor is mixed with pollutants in the atmosphere. The visibility is reduced.

Head wind: A wind blowing in the opposite direction to the heading of an object in motion.

Heat units: This is same as growing degree days.

Humidity: It indicates the water vapor content of atmosphere.

Hurricane: One fierce tropical cyclone originates between 5 and 20° north and south of the equator. Tropical cyclones are called as hurricanes in the North Atlantic and Eastern North Pacific oceans.

Hydrological cycle: A water cycle which has no end and beginning. It is the triple interaction between ocean-atmosphere, ocean-land and land-atmosphere. This is the cycle that controls the Earth's water balance. It includes evaporation, transpiration, condensation, precipitation and rainfall, ground water. surface water, *etc.*, or

The water cycle describes how water evaporates from the surface of the earth and ocean, rises into the atmosphere, cools and condenses into rain or snow in clouds, and falls again to the surface as precipitation to the land and sea.

Hydrometer: An instrument to measure the density or specific gravity of liquids.

Hydrosphere: The water portion of the earth.

Inclination of the wind: At any particular location, the angle between the direction of the wind and the gradient wind.

Infrared radiation: The radiation emitted by Earth's surface, atmosphere and clouds back to space. It is also called as reradiation/terrestrial/log wave radiation. The wave length of this radiation is longer than the wave length of red colour of the visible part of spectrum. It is of electromagnetic radiation having a wavelength just greater than that of the red end of the visible light spectrum but less than that of microwaves. Infrared radiation has a wavelength from about 800 nm–1 mm, and is emitted particularly by heated objects.

Infrared thermometer: This is a non -contact thermometer to measure the temperature of the crop plants for water management and used for micrometeorology studies.

Insolation: The intensity at a specific time or the amount in a specified period of direct solar radiation incident on unit area of a horizontal surface on or above the Earth's surface.

Irradiance: The radiant flux density incident on a unit area of a real or imaginary surface. The unit used is watts/meter/meter.

Iso bar: A line of constant atmospheric pressure.

Isobath: A line drawn on a map connecting points of equal depth at the bottom of the sea.

Isohel: A line or curve formed by joining the equal points of sunshine duration over a given period of time.

Iso hyet: A line of equal quantity of rainfall amount.

Isolated system: A system in which the energy is neither lost nor gained from outside. If one considers a system in which the particles can exchange energy and interact with one another and that system can be called as isolated system.

Iso pleth: A line of equal value of given quantity.

Iso tach: A line of equal wind speed.

Iso therm: A line of equal temperature.

Jet stream: A strong and narrow stream of flowing air. Jet streams are fast flowing, narrow, meandering air currents in the atmospheres of some planets, including Earth. On Earth, the main jet streams are located near the altitude of the tropopause and are westerly winds (flowing west to east). Their paths typically have a meandering shape.

Julian day: The number of a particular day in a year starting from January first. January 1st is the first day and December 31st is the last day.

Katabatic wind: This is a valley -wind of down slope drainage type, developed mostly on clear nights when regional pressure gradients are weak. This is caused by surface level cooling of the atmosphere and the wind blows down on incline. If warm, it is a foehn; if cold it may be a fall or a gravity wind or A katabatic wind (named from the Greek word) and the meaning is "descending" wind. This is the technical name given for a drainage wind, a wind that carries high-density air from a higher elevation down a slope under the force of gravity.

Kelvin: The unit of absolute temperature. The 273.16 K = 0°C. This means that the freezing point of water is 273.16°K and the boiling point of water is 373.16°K.

Kew pattern barometer: A fixed scale mercury barometer to measure atmospheric pressure. Only one adjustment is needed before taking observation from this barometer.

Knot: The knot is one among the units to measure wind speed like Km/second/hour. One knot means one nautical mile per hour. 1 knot (kt) = 1.85200 kilometers per hour (kph).

Land breeze: A breeze from land to sea due to the difference in temperature.

Langley: One unit to measure radiant energy. This is equal to one-gram calorie/cm/cm.

Lapse rate: The rate of decrease in temperature with increase in height at a given time and location from ground level to troposphere height.

Latent heat: The quantity heat absorbed or released without any change of temperature during a change of sate of unit mass of a material. It is normally expressed in cal/g.

Latitude: The imaginary circular lines drawn horizontally connecting east-west of the globe.

Leeward: The direction towards the wind blows.

Lightening: A visible electric discharge (lightning flash) associated with thunderstorm especially it happens in cumulonimbus clouds.

Long range weather forecast: This is a forecast for decision making in agriculture and the forecast range is from one month to a season of three to four months.

Long wave radiation: Outgoing Long-wave Radiation is electromagnetic radiation of wavelengths between 3.0 and 100 μm emitted from Earth and its atmosphere out to space in the form of thermal radiation. it is also referred to as up-welling long-wave radiation and terrestrial long-wave flux. Over 99 per cent of outgoing long-wave radiation has wavelengths between 4 μm and 100 μm, in the thermal infrared part of the electromagnetic spectrum.

Longitude: An imaginary line drawn from pole to pole and this helps to compute local time. It is also called as meridian. The meridian of Greenwich is zero and UTC or GMT is based on this meridian.

Low clouds: These clouds would appear form ground level to 2 KM in the troposphere of the atmosphere.

Lucimeter: Instrument to measure global solar radiation.

Lux: The derived SI unit of illumination. In other words, it is metre candle.

Lysimeter: An instrument used to measure crop evapo -transpiration. The rate of percolation of rain water also can be measured through this instrument.

Macroclimate: Climate of a continent or over the globe.

Mamma: Under stable condition of atmosphere, the bulbous protuberances formed on cumulonimbus cloud.

Manometer: A gauge used to measure the pressure of gases, vapour and fluids.

Maritime climate: The climate of the areas or regions near to the oceans and seas.

Markov chain: The computation of the probability of wet and dry spells in rainfall data series.

Maximum temperature: The highest temperature recorded during day, week, month, season, year, etc., respectively.

Mean sea level: Mean plane at which the tide oscillates, the average height of the sea for all stages of tide. At any particular place, it is derived by averaging the hourly tide heights over a 19 years period. The mean sea level measured in meter is very important for meteorological computation.

Mediterranean climate: A climate type of which, the Mediterranean basin is the archetype and is characterized by hot dry summers and mild wet winters.

Medium range weather forecast: This is the weather forecast for next 3–10 days period from the day of issue. Being developed through NWP method.

Meridian: An imaginary great circle drawn on the earth passing through both the poles.

Meso climate: The climate of a region or location of small size like valley, forest, etc., This is intermediate between macro and micro climate

Meteorological drought: If the annual rainfall is deficit of 75 per cent from mean annual rainfall, then it is called as meteorological drought or the receipt of annual rainfall is only 25 per cent of mean annual rainfall.

Micro climate: The physical state of the atmosphere close to a very small area of Earth's surface, often in relation to living matter such as crops, insects etc.,In the case of crop microclimate, the climate that prevails from the root of the crop community to top of the canopy of the crop community.

Micron: One unit to designate the wavelength of light. This is equal to one millionth of a metre.

Millibar: This is a unit for measuring atmospheric pressure. This is equal to 100 pascals or equivalent to one thousand dynes per centimeter per centimeter.

Minimum temperature: The lowest temperature that is recorded preferably in the very early morning of the day after the escape of long wave radiation from the earth surface.

Mist: A state of atmospheric obscurity produced by suspended microscopic water droplets or wet hygroscopic particles. The term is used for synoptic purpose when visibility is equal to/or exceeds one km; the corresponding relative humidly is > 95%

Mitigation: The action taken to reduce the release of the greenhouse gases emission.

Mixing ratio: The ratio of the mass of water vapour to the mass of dry air occupying the same volume.

Monsoon: From the Arabic word of "Mausim" the monsoon word has come. The term of monsoon originally referred to the winds from Arabian sea, which blow for about six months from the northeast and for another six months from the southwest. But it is now used also for other markedly seasonal winds. In India there are two marked seasonal winds namely southwest monsoon (June-September) and northeast monsoon (October-December).

Net long wave radiation: The difference between the total incoming long wave radiation and the total out going long wave radiation.

Net pyranometer: An instrument to measure the difference of the solar radiation falling on both sides of the horizontal surface from the areas to which the exposure is directed with this instrument.

Net radiation: It is the difference between total downward flux and total upward flux and is the measure of the net energy that is available at ground surface.

Net radiometer: An instrument to measure the net radiation.

Net terrestrial radiation: The difference between the downward and upward terrestrial radiation.

Nimbus cloud: The low cloud which can produce rain, snow and sleet.

Nimbo-stratus: This is lower middle level clouds and the mean height is 2–7 km from the ground level. The cloud is greyish dark in color and thick enough to cover the sun totally. Often gives rain, snow and sleet.

Nuclei of condensation: The particles on which condensation of water vapor takes place. These are microscopic and submicroscopic in size. These have strong affinity for water. Sodium chloride and dust particles are some examples for condensation nuclei.

Numerical model: Numerical is involved in this model. The governing equations are solved step by step through numerical calculations. Up to troposphere from ground level different divisions arc made with particular height and the collected weather or computed weather data both inter and intra equations are solved to derive output on weather forecast.

Oasis effect: The vertical sensible heat energy transfer from the air to the crop canopy while flowing across the crop. The air is cooled until it reaches the equilibrium with the crop canopy. In this process, a small quantity of radiant energy may also be taken by the crop from the same air under hot dry condition.

Octa: A unit to measure cloud coverage in the sky. It needs practical experience to measure cloud coverage in the sky during meteorological observations. When the sky is open the unit octa is zero (0) and when the sky is completely covered with clouds the unit octa is eight (8). Hence the octa unit ranges from 0–8.

Orographic clouds: Cloud forms by forced uplift of moist air mass over the ground/mountain. The reduction of pressure within the rising air mass produces adiabatic cooling and if the air is sufficiently moist the condensation and precipitations start. This is common during southwest monsoon season in India.

Orographic lift: The lift caused to an air mass, when it is forced to ascend over the mountain range.

Orographic rain: The rain due to orographic clouds is called as orographic rain. Seasonal moist wind and mountain barrier are the cause for such rainfall. It is a continuous in day time and sometimes it is also nocturnal.

Ozone: The is tri -atomic form of oxygen(O^3). It is a pale blue gas with a distinctively pungent smell. It is an allotrope of oxygen that is much less stable than the diatomic allotrope O_2, breaking down in the lower atmosphere to O_2. The molecular weight is 48 g/mol. Its density is 2.14 kg/m^3. This is present in the atmosphere especially in the stratosphere (production also here) in very small amounts ranging from about 0.2–0.5 cm equivalent thickness at normal temperature and pressure. This layer does not permit high intensity rays like gamma, beta and ultra violet rays to earth, which are very lethal to living beings.

p.t.h: This is an abbreviation to indicate pressure, temperature and humidity under radio- sonde observation.

Perihlion: This is a situation, wherein the earth is little closer to sun on about Jan.3rd on each year and the distance between earth and sun would be 147 million KM.

Phenology: Study on the development stages of an organism in relation to the environment.

Physical climatology: One branches of climatology and it deals on the temporal and spatial variations in heat and moisture exchanges and also the movement of air in the atmosphere.

Physiological drought: This situation occurs when the concentration of solutes in the soil water is equal to or higher than that in the root cells so water cannot enter the plant by osmosis.

Piche evaporimeter: A filter porous paper atmometer. It is used to measure evaporation.

Pilot balloon: Small size balloon used to measure wind data.

Polar air mass: A high latitude three air masses namely, winter time continental, summer time continental and winter time maritime polar.

Polar climates: The summer - less cold climates found in the polar regions and the precipitation form is snow. The mean warmest monthly temperature is < 10°C. The natural vegetation includes lichens, masses, *etc*.

Potential evapotranspiration: The maximum amount of water loss through evaporation from soil and transpiration from crops when ground is completely covered by crops canopy under unlimited water supply.

Potometer: An instrument to measure transpiration from plants and also to measure the absorption of water by plants.

Precipitation: The falling of any condensed moisture from the clouds to the ground, depending on the temperature of the atmosphere. It may be rain, snow, hail, sleet, etc., Atmospherically precipitation includes rain, drizzle, snow, hail, fog, and frost and these are expressed in depth of water which would cover a horizontal plane, if there is no run off, infiltration, evapo-transpiration in a unit mm/day of past 24 hours.

Psychrometer: This is to measure the relative humidity of the atmosphere since it contains both dry and wet bulbs.

Pyranometer: Instrument to measure global solar radiation.

Pyrgeometer: A pyrgeometer is a device that measures near-surface infra-red radiation spectrum in the wavelength spectrum approximately from 4.5 µm–100 µm or it measures the intensity of net long wave radiation.

Pyrheliometer: Instrument to measure the intensity of direct solar radiation at normal incidence.

Pyrradiometer: It measures both incoming solar radiation and outgoing terrestrial long wave radiation.

Radar sonde: A system in which both meteorological and wind data will be obtained, as this is carried by radiosonde.

Radar: A system of detection and location of targets which are capable of reflecting high frequency radio waves (micro waves), generally in the wave length range from a fraction of a centimeter to some tens of centimeters.

Radiation balance: The computation of the amount of short wave radiation received from the sun, the amount of long wave radiation escaped to the space from earth and the net balance available at the earth. It is estimated that the earth absorbs about 124 kilo largely of solar radiation every year and radiates back 52 kilo Langley and the remaining 72 Kilo Langley is the net balance at the surface of the earth.

Radiosonde balloon: A balloon larger than pilot balloon, that carries radiosonde afloat.

Radiometer: Instrument to measure the intensity of solar radiation.

Rain: A form of precipitation that contains liquid water drops larger than 0.5 mm in diameter. The drops may fall in straight path but not in vertical direction.

Rain atmometer: An instrument used to measure evaporation from plant canopy due to sunlight.

Rainy day: A day of past 24 hours with rainfall amount 2.5 mm and more than 2.5mm. Rainfall is measured at 3GMT and under Indian condition it is 0830 hrs.

Rain day: A day of past 24 hours with rainfall amount $<$ 2.5 mm. Rainfall is measured at 3GMT and under Indian condition it is 0830 hrs.

Rainless day: It is a zero-rainfall day or dry day or sunny day.

Rain gauge: Instrument to measure the rainfall amount. Non -recording rain gauge of 100 and 200 $cm^{2,}$ in diameter, self-recording rain gauge and digital and automatic recording rain gauge are of different types are available in the market

Rain shadow area: The area where rain fall does not occur due to blocking of mountains towards leeward direction of the monsoon wind. During SWM, Tamil Nadu becomes the rain shadow area with the presence of Western Ghats.

Rainbow: One colour effect produced in the form of concentric arcs by the refraction and internal reflection of sunlight in minute droplets of rain, drizzle, fog in the atmosphere.

Rainfall: The total liquid water product from precipitation through condensation from the atmosphere as received and measured in rain gauge over 24 hours at 03 Universal Time Coordinator (UTC) or GMT.

Rainfall distribution: Spatial distribution of rainfall is highly variable and it is given as.

Isolated: When rainfall occurred over 25% or less of the total area under observation.

Scattered: When rainfall occurred over 26–50% of the total area under observation.

Fairly widespread: When rainfall occurred over 51–75% of the total area under observation.

Widespread: When rainfall occurred $>$ 75% of the total area under observation.

Rain fed farming: The practice of growing crop in rainy season entirely with rainfall received without any support from external irrigation. The field crop raised does not face any moisture stress during its growing period. This type of farming is common in humid and sub humid climate.

Resilience: The capacity to recover quickly from difficulties/impact/toughness or the capacity to adapt.

Saturated adiabat: Any statistically valid thermodynamics diagram on the graph related to adiabatic change under saturated condition of an air parcel.

Saturation vapour pressure: In micrometeorology this is used to indicate the pressure exerted by a parcel of saturated vapour and this is a function of temperature. In general meteorology, the term refers to partial pressure of water vapour in equilibrium with a plane surface of water, ice, *etc.*

Seasonal consumptive use: This means ET of the crop + water used for the crop's physiological requirement. The total amount of water consumed by the crop to meet its evaporation and transpiration during the entire growing season, expressed in depth or volume of water. Usually mentioned in mm.

Seasonal drought: A dry weather (no rainfall) with high temperature that occurs during the season of crop.

Seeding clouds: A process in which the precipitation is encouraged by injecting hygroscopic condensation nuclei like common salt, cement, silver nitrate, solid carbon di oxide, etc., in to the clouds or over the clouds. Air crafts are used to spray the seeds over the clouds or fumes developed through ground generator.

Semi-arid zone: A zone delineated by Thornthwaite and Mather moisture index and expressed as Im. This Im is negative for semi-arid zone (-66.7 to- 33.3).

Shelterbelt: A long wind break of shrub or many layers of tall plants grown, extending over an area larger than single field or farm towards leeward direction of wind, so that the micro climate of the crops grown is sustained within shelter belts for the benefit of crop productivity.

Short wave radiation: The wave length of short-wave radiation is < 4 micron.

Sleet: This is a form of precipitation falling in small particles of clear ice. These particles may be formed either due to melting of hail or snow as it descends or when the rain drops are frozen as they pass through a layer of cold air.

Smog: The situation where in the smoke and fog are mixed together and results in poor visibility. This happens in lower troposphere. Now a day's air pollution is also called as smog.

Snow: This is a form of precipitation, wherein sublimation of water vapour occurs in to snow at below freezing point temperature.

Soil moisture: The moisture content of the soil generally expressed as the percentage ratio of the mass of water to that of dry soil, but may be also expressed in terms of inches of water per given depth of soil. Soil moisture units varies from saturation- field capacity - wilting point. When water is held by tension by the soil particles in relation to pore spaces, the soil water becomes soil moisture.

Soil moisture stress: Soil moisture is available to crops when the soil is with field capacity. When moisture gets depleted with crop evapotranspiration, the soil moisture level goes towards wilting point. Soil moisture stress gets operated from 30 per cent above wilting point to wilting point.

Solar radiation: Radiation emitted by the sun. It is also referred as shortwave radiation. Solar radiation has a distinctive range of wave length as determined by the temperature of the sun.

Solstices: The maximum distance between the sun and the earth that happens twice, once each in south and north of the celestial equator during the annual path of the earth around the sun. These occur on June 21st and December 22nd.

Specific humidity: The ratio between the mass of water vapour in an air parcel of the atmosphere to the total mass of moist air.

Spectrophotometer: One instrument to measure the intensity of radiation.

Squall: Strong and sudden onset of wind and found decreases latter in minutes.

Stevenson's screen: A wooden box with wooden panels arranged in obliquely to place thermometers for recording temperature and relative humidity and this box is kept at 4-foot height from the ground level.

Stratus: This is a low-level cloud and present at 0–2 km height from the ground level. The word Stratos refers to layer. These are grey in color and has almost uniform bottom and top. This cloud may give drizzle, ice prisms and snow grains.

Sub humid climate: A zone delineated by Thornthwaite and Mather moisture index and expressed as Im. This Im is positive. There are two types of sub humid climate. One is moist sub humid where the Im is 0 to 20. The other one is dry sub humid, where th Im is -33.3–0.

Sub-tropical: The region between tropics and temperate zones.

Sublimation: This is physical process wherein the water vapour becomes ice without passing through liquid phase.

Synoptic circulation: The synoptic climatology is a valuable non-quantative approach to climatic description, emphasizing non-periodic weather episodes associated with transient weather disturbances and other circular patterns at various scales and durations. The synoptic circulation can be tracked in synoptic chart.

Synoptic meteorology: One branch of meteorology concerns with the study on atmospheric phenomenon, based on the analysis of charts on which synoptic observations are plotted for the purpose of weather analysis and forecasting.

Synoptic observations: The weather observations recorded simultaneously at required number of meteorological stations along with meteorological observations. These are useful to know the state of atmosphere at a given time and also to track the movement of cyclone and monsoon winds. Also, local convective storms could be observed.

T.H.I: Thermal humidity index. Threshold levels are fixed for both animals and human beings for their comforts. For human THI must be lesser than 15, while for animals it is < 72.

Temperature inversion: A situation in which, there is an increase in temperature with increase in height from the ground level towards troposphere. This is opposite to lapse rate.

Terrestrial radiation: The long wave radiation emitted by the Earth and atmosphere under the approximate temperature range from 200–300 K°. Earth's surface absorbs solar radiation and emits terrestrial radiation. This radiation is confined to within the wave length of about 3 and 100mm and has maximum intensity at about 10 mm.

Thermocline: The layer in a thermally stratified body of air or water within which the temperature decreases rapidly with increasing depth, usually at a rate greater than 1°C per metre of depth or distance in vertical direction.

Thermograph: One apparatus with chart paper used to give continuous record as a geographical chart of temperature with time.

Thunder: The audible noise resulting from the explosive effect of electric discharge caused by very rapid expansion and contraction of air within a cloud. This is usually associated with cumulonimbus clouds.

Thunder storm: One or more sudden electrical discharges manifested by a flash of lightning and a sharp rumbling sound(thunder). This is common in a well grown cumulonimbus cloud.

Torrential rain: Rapid and violent heavy rainfall with economic loss.

Trade winds: Winds (or tropical easterlies) that diverge from the subtropical high-pressure belts, centered at 300–400 N and S, towards the equator, from northeast in the northern hemisphere and south east in the southern hemisphere.

Trajectory: This can be referred to the curve or path in space, traced by moving air parcel.

Transpiration: The process of water vapour release to the atmosphere from the aerial parts of the plants namely leaves, stems, flowers, etc., Water is necessary for plants but only a small amount of water taken up by the roots is used for growth and metabolism. The remaining 97–99.5 per cent is lost by transpiration and guttation.

Trough: A low-pressure feature of the synoptic chart. Unlike fronts, there is not a universal symbol for a trough on a weather chart. It is characterized by a system of isobars which are concave towards a depression and have maximum curvature along the axis of the trough or trough line. The trough is said to be deep or shallow according to the maximum curvature of the isobars along the trough line being more or less, respectively; the former corresponds to the V shape referred to in the obsolete term V shaped depression or; A trough is an elongated (extended) region of relatively low atmospheric pressure, often associated with fronts. Troughs may be at the surface, or aloft, or both under various conditions. Most troughs bring clouds, showers, and a wind shift, particularly following the passage of the trough. This results from convergence or "squeezing" which forces lifting of moist air behind the trough line.

Typhoon: A hurricane in south -China sea. This is a tropical cyclone or storm with a maximum wind in excess of 32 metres per second.

Unstable: The condition of a system, which implies a particular type of imposed disturbance. The type of instability depends on the type of disturbance., convection of dynamic and baroclinic natures.

Unstable atmosphere: The condition of the atmosphere in which the decrease in temperature with increase in height from the ground is greater than dry adiabatic lapse rate.

Updraft/Up draught: The upward moving of current of air of relatively small dimensions.

Urban heat island: There are differences in temperature between an urban area and adjacent non-urban area at any given time. The urban area is warmer few degrees than its surroundings.

Veering: The clockwise change in the direction of the wind, which implies that, as the direction of the wind is changing the azimuth is increasing.

Virtual temperature: The temperature at which air would have the same density as a sample of moist air at the same pressure.

Visibility: The greatest distance at which an object of specified characteristics can be seen and identified with the naked eye (without any aid/tool). In the case of night observations, the said object can be seen and identified if the general illumination is raised to the normal daylight level. or

Visibility is a measure of the distance at which an object or light can be clearly recognized. It is reported within surface weather observations and expressed in meters or miles, depending upon the country. Visibility affects all forms of traffic namely, roads, sailing and aviation.

Visible radiation: The range from 0.4–0.7 microns of electromagnetic radiation which can be seen by human eye. This part of solar spectrum contains 41 per cent of total energy of the spectrum and is important for all life on the earth.

Vortex: A whirling mass of air, especially one in the form of a visible column or spiral as a tornado. A whirling mass of very cold air that sits over the north or south pole is called as polar vortex. Vortex is an important part of turbulent flow. Tornadoes form during thunderstorms, when warm humid air collides with colder air to form a swirling vortex that extends down from the clouds and some times reaches the ground where it can cause extensive damage.

Vulnerability: The degree to which a system is susceptible to or unable to cope with adverse effects of climate change, including climate variability and extremes. Vulnerability is a function of the character, magnitude and rate of climate variation to which the system is exposed, its sensitivity and its adaptive capacity.

Water balance: The hydrological balance between rainfall + irrigation to that of changes in soil moisture, evaporation, run off and percolation.

Water harvesting: Conservation of rainwater under non-irrigated condition by collecting runoff water from rainfall in farm ponds and other structures in order to provide supplemental irrigation at the time of soil moisture stress.

Watershed: It is a hydrological unit. Catchment basin or drainage basin are synonymous to watershed.

Weather: The day to day condition of the atmosphere at a given place and at a given instant of time in relation to temperature, humidity, wind velocity and other meteorological parameters or the status of the atmosphere with respect to weather element at a given time, may be a day or today change in the atmosphere.

Wet bulb depression: The difference in degrees between the dry bulb temperature and the wet bulb temperature of a psychrometer.

Whirl wind: One relatively small rotating air mass occasionally embedded with dust and dried straw etc.,

Water use efficiency: It is also called as water productivity. The product between the economic yield of crops and water used in mm or ETO of the crop.

Wet bulb depression: Difference between simultaneous reading of wet and dry bulb thermometers.

Wet season: A period during which the precipitation is in excess of water requirement and water accumulated in the soil and in reservoirs. As per IMD if the rainfall of all the months of a season is more than 59 % of normal rainfall of the concerned month, then it can be called as wet season.

Wind: The horizontal movement of the air related to the surface of the earth. In meteorology, the specified wind direction is that relative to true geographic North, from which the wind blows:

Northerly wind: wind from 337.6°–22.5°
North easterly wind: wind from 22.6–67.5°
Easterly wind: wind from 67.6–112.5°
South easterly wind: wind from 112.6–157.5°
Southerly wind: wind from 157.6–202.5°
South westerly wind: wind from 202.6–247.5°
Westerly wind: wind from 247.6–292.5°
North westerly wind: wind from 292.6–337.5°

Wind vane: An instrument used to indicate the directions from which the wind is blowing.

Zenith: The point on the celestial sphere directly overhead.

Zenith angle: The angle between a local line perpendicular to the surface on which the radiation falls and the position of the sun in the sky, from which the incident ray falls.

Zephyr: A pleasant, gentle and soft breeze.

For proper understanding of the text given in a book, one must know the meaning of terminologies used and hence the authors collected information for the technologies given here under from different sources including, "Guide for Agro-Meteorological Services (DST,1999) and Terminology in Agricultural Meteorology of Vasi Raju Radha Krishna Murthy,(1996) and Wikipedia".

Introduction

Agricultural meteorology and agricultural climatology are the applied sciences from the mother science of meteorology. Though both are with same structure of talking about weather elements, the science of agricultural climatology mainly discusses the application aspects of the science on agricultural productivity, while the science of agricultural meteorology talks about basic aspects of meteorology required for agriculture. Agroclimatology is a field of interdisciplinary science of agrometeorology, in which principles of climatology are applied to agricultural systems.

Greek scientists had shown much interest in meteorology first in the world. The first book on meteorology was written by the Greek author Mr. Aristotle during 350 BC. The name of the book was *"Meteorologia"*. The scientist Mr. Galileo invented thermometer by 1593 followed by the invention of mercury barometer by Mr. Evangelista Torricelli during 1643. From 19th century onwards development in meteorology was initiated especially in European countries by establishing meteorological observatory followed by measurement of weather elements.

In India, Indian Meteorological Department (IMD) was established by 1875. Through all modes namely surface observatory, balloon, satellite, ships, radiosonde, aircraft, etc., the weather data have been collected across India and used for keeping natural resources intact in addition to sustaining the welfare of the human kind. Further India gets world level meteorological data daily through global network and used for all the purposes of India's and also for other neighborhood countries development.

Now India is one among the forerunner countries in the world in terms of knowledge on pure science of meteorology and also on applied sciences like agricultural meteorology and agricultural climatology. The agricultural meteorology development was started from the work of Dr. L. A Ramdas, the Father of Agricultural Meteorology in India. Earlier to two decades from now, in agricultural meteorology, all books dealt only with atmosphere, radiation, wind, clouds, abnormal weather, crop-weather relationship partly and agricultural drought and floods. But now with the development of science, many new areas have been added both in agricultural meteorology and agricultural climatology. To quote some example, remote sensing, climate change, weather induced pest and disease initiation, crop simulation model, numerical weather prediction, integrated weather forecast (short range, medium range, extended medium range, long range, seasonal climate forecasts and nowcast), crop insurance, livestock climatology and human climatology *etc*., are some of them.

With the on-setting of climate change over the Globe and also proved through scientific results by the scientists, the science of agricultural meteorology and agricultural climatology have grown to a level of acceptance of the people, scientists and politicians. The talk of the day is climate change vs crops productivity, climate change vs animal, poultry and milk productivity, climate change vs human health, climate change vs natural resources sustenance etc.,

In this book considering students, academic staff and policy makers, both theoretical and practical aspects of crop -weather interaction and agro-met observatory, agroclimatic analysis, crop micro meteorology, remote sensing, crop simulation modelling, weather codes and their management, integrated weather forecast and agro- advisories, climate change, live stock climatology and astro- meteorology have been dealt.

When we talk about meteorological data collection, the present advancement is collection of weather data through digital mode especially through Automatic weather station. But when we go for automatic weather station for data collection, the sensors must be kept clean and dust free, otherwise, the automatic weather station becomes a white elephant. At this context, collection of weather data through manual surface observatory is very precious and serve to a high-level of acceptance. But this is somewhat costlier and needs human support around 24 hours of the day. Hence in this book, both automatic weather station and manual observatory establishment and their maintenance have been discussed. Similarly, the applied aspect of crop management has been discussed in terms of weather codes. This is a new topic promoted by Prof. Dr. M.S. Swaminathan. Under this umbrella, floods code, drought code and normal weather code have been discussed. Considering animal protein requirement, impact of climate and weather on animal productivity, livestock climatology has been also brought in this book. The role of remote sensing especially for crop monitoring, yield forecasting, drought and floods monitoring, pest and disease spread has been considered as an important area and discussed in this book. Astro-meteorology especially weather forecast gains value with agro-climatologists and hence those details have been discussed. Regarding weather forecast and conventional agro advisory, presently the technology goes with automation and reach the farmers through their mobile in time for taking weather-based crop decision on the spot. These have been brought at this book.

Farm decision making based on weather and climate seems to be utmost important in sustaining crop's productivity, considering population growth and reduction in cultivable area and this is possible only when weather and climate science becomes part of human's activity. This can be reached through this book.

The country being the prime country for agriculture, enough research and development funds is not extended to this field from Central Government Budget. Hence Government of India must provide enough R and D funds to the field of Agricultural Climatology for the benefit of higher food production, since the benevolent and malevolent weather events that occur from Southwest monsoon, Northeast monsoon season and Western disturbance, affect Indian agriculture to a greater extent. Further the present challenge to Indian agriculture is decreasing productivity per unit area from climate change related issues, reduction in cultivable area with urbanization, migration of rural population to urban area and multiplication of human population beyond the level to feed.

But, do we have enough professionals from the field of agricultural climatology/ agricultural meteorology to meet these challenges. The answer is "No". At this context, each State Agricultural University must have separate Department for this applied science of agricultural climatology/agricultural meteorology. In addition, the Agronomist also must be trained in the field of agricultural climatology/agricultural meteorology to meet the professional needs of Indian Agriculture. The Country has strong weather forecast system for agriculture along with providing agro-advisories. This is run by India Meteorological Department (IMD). This area needs modernization to reach to farmers in time under this digital world.

More research is needed on animal climatology including poultry, horticultural crops management, crops pest and disease management, response farming, integrated weather forecast and those are discussed in this book, etc., Though crop simulation modelling reduces the uncertainty to a level, still it has to be renovated to meet the present level of ground reality.

1

Crop-Weather Interaction and Agro-Met Observatory

1.1 Weather Parameters

The atmosphere is a gaseous component surrounding the earth as an integral part and its interaction with ocean (water bodies), other planets of the solar system and land (Earth) generates one physical product namely weather elements (temperature, pressure, wind, humidity, precipitation, *etc*.,) and these form the weather and climate of the Earth. The process of generation of weather and climate of the region is purely natural and cannot be modified by the human being. But they can do adaptation to the existing climate and weather for their sustainable life. The weather can be defined as day to day change in the atmosphere in terms of heat and moisture exchange, while the climate is average of long period weather covering larger area or a region.

The particular climate of a region dictates/permits the genesis of a particular bio-organism in that domain and this selected organism's growth and multiplication are triggered by the prevailing weather of that particular climate. Hence weather parameters of the atmosphere are very important to human being and other organisms of the Earth. The details on weather elements are given in question and answer section of this book (Annexure-I)

1.2 Crop-Weather Relationship

All living organism's growth and their development dependent on the type and intensity of weather elements that prevail in a domain. The usefulness of weather data is obviously known to us from our perception. Apart from regular weather forecast weather data can be used for floods and droughts monitoring, doing agricultural planning at all scales, conducting network agricultural experiments, using for crop simulation modeling, for monitoring crop pest and diseases outbreak, developing weather thumb rules at village level etc.

Crop's response to climate and weather is enormous in terms of their productivity. Hence in the history of crop production, there is always occurrence of both malevolent and benevolent weather events in the crop production system and hence the weather forecast is an important tool for sustainable crop management.

The normal weather requirement of selected crop is discussed here under (Balasubramanian and Gopalasamy, 2002):

1.2.1 Rice

Rice is an aquatic plant and it likes water impounding throughout its crop growth except the last 15 days of the harvest. Though water impounding is available to rice crop, in the absence of optimum temperature and abundant solar radiation, rice productivity would get affected drastically. From flowering to harvest, abundant availability of solar radiation is required. In other words, if the cloudy weather/environment prevails at this stage, rice yield would get reduced drastically. In Japan the rice yield is higher as compared to the country India because of prevailing optimum temperature required for rice in addition to the availability of higher solar radiation. During vegetative stage, it requires higher day temperature up to 35°C. It tolerates up to 40°C. If the temperature is below 15°C. during panicle initiation stage, flowering stage and maturity stage, there would be increase in grain infertility. The temperature requirement of rice crop for its growth stages is as follows;

Crop stage	Minimum(°C)	Maximum(°C)	Optimum(°C)
Germination	10	45	20-35
Seedling	12-13	35	25-30
Root formation	16	35	25-35
Vegetative	9-18	38	35
Tillering	9-18	33	25-31
Panicle initiation	15-30	38	33
Flowering	22	35	30-32
Grain development and maturity	12-18	30	20-25

The ET of rice crop is 1500 mm under submerged condition. If it is grown under alternative wetting and drying condition, the water requirement is 800–1000 mm. Under rainfed, rice requires uniform distribution of rainfall from sowing to milking stage and the water requirement is from 600–800 mm. The optimum RH is between 60–65 per cent. High wind speed of >40 km/hour along with heavy rainfall if it prevails during maturity stage, the crop would get lodged. Considering the weather, optimum sowing window must be selected for getting higher yield for each region under climate change scenario.

1.2.2 Wheat

Wheat is a cool weather loving crop. The minimum, maximum and optimum temperature requirement of the crop is 3–4; 25 and 30–32°C. respectively, but stage wise, it differs greatly. High temperature combined with low relative humidity would reduce its vegetative growth and finally the yield level also. Sunny weather in the day with low day temperature both in the day and night is the pre requisite weather required during grain maturity stage for getting higher yield. In other words, it requires low temperature from germination to flowering and thereafter it requires warm temperature up to harvest for getting higher yield. In an average for the 90 days of the crop growth from germination, a mean temperature level of 15°C is required. In the tropical situation, the yield will be reduced beyond the expectation and this crop is not suitable. However, it can be cultivated during winter season with reduced yield. Wheat can be cultivated both under irrigated and rainfed eco systems. The water need is 450–650 mm based on the variation in the weather that would prevail during cropping season.

1.2.3 Maize

The crop maize comes up very well in the intermediate climate between tropical and temperate climate. The average temperature requirement is around 25°C and when the temperature falls below 15°C, the crop gets affected. Temperature above 30°C during night also affects the growth and the productivity of this crop.

The requirement of temperature for different stages is as follows:

Seed germination	Minimum requirement 7°C
Emergence of tassel	< 1 to 3°C affects emergence
Tasseling stage	< 3°C would affect the pollen development And temperature >35°C also causes pollen sterility

The crop requires well distributed rainfall of 450–600 mm under rainfed condition. The ET of the crop is 500–800 mm. Under irrigation, germination stage, tasseling stage, silking stage, and milking stage are critical for water requirement. Maize crop is sensitive to both moisture stress and excessive moisture. Regulate optimum moisture availability during the most critical phase (45–65 days after sowing); otherwise yield will be reduced to a greater.

1.2.4 Millets including minor millets

The millet crops whether small or bold, belong to tropical crops category. Under adverse and unfavorable weather condition also, these crops would provide some yield. Seed germination requires a temperature level of 8 to 10°C. During growing period, these crops require 25–30°C. Lower temperature below 20°C and high temperature above 40°C would affect crops grain productivity. These crops are highly suitable for both rainfed and dry land conditions. They are drought tolerant. A minimum of 300 mm of well distributed rainfall is required under dry land to harvest 60–70 per cent of the normal yield. The ET of the crop is 400–500 mm. Minor millet crops like tenai (fox tail millet), samai (little millet), varagu (kodo millet), panai–varagu (proso millet) and kudraivalli (barn yard millet) are highly suitable for climate change scenario.

1.2.5 Cotton

Cotton crop belongs to tropical climate category. During its growing period of 4–5 months, the crop requires 28–45°C temperature. Along with high temperature, it requires bright sun shine hours and high relative humidity. Cotton crop is sensitive to prevailing weather condition. The required weather condition is as follows:

Vegetative growth	21–29°C is optimum, Fog and mist would affect the growth
Flowering to boll maturity stage	27–32°C + 8 to 9 hours of bright sunshine hours +70 per cent RH are the requirements
Boll bursting stage	Open sky, temperature between 30–32°C and bright day are the requirements
Vegetative stage, flowering and bolling and boll picking stage	Continuous high rain fall for two to three days would affect productivity.
Peak vegetative stage, flowering and bolling stages	High RH would invite insects and diseases

The ET of the crop is 700–800 mm. This crop can be raised under irrigated, rainfed and dry land conditions.

1.2.6 Sugarcane

This crop belongs to tropical climate. However, this crop is being cultivated under intermediate climate also, that is between temperate and tropical climate successfully.

The temperature requirement of this crop is as follows:

Crop stage	Temperature requirement (°C)
Germination of setts	29–30
Crop growth or vegetative stage	32–35
Crop growth or vegetative stage	If the temperature is below 20, the crop growth will be affected
First 4 to 5 months	Average temperature between 30 and 35°C is the requirement
Last 2 months before harvest	Mild and cool temperature with long bright sun shine hours.

The crop is cultivated under irrigation in majority of the countries. The ET of the crop is 1500–2500 mm. Like rice, when the ET requirement is more, the productivity is also more with this crop.

1.2.7 Groundnut

This crop also belongs to tropics. However, this is cultivated in countries in between 45°N–30°S latitude. Temperature requirement varies with different crop stages. However high temperature above 35°C along with mist affect crop's productivity. For germination of seeds, an average temperature between 14 and 16°C is required. For higher productivity, a mean temperature of 25 °C is required. At the time of pod development, very low soil temperature and air temperature of below 20 °C bring empty pods. This crop is cultivated both under rainfed and dry land conditions. Irrigated condition also could be seen wherever assured irrigation is available. Under irrigation the productivity is more as compared to dry land situation. Under rain fed and dry land situation, a minimum of 400–500 mm of well distributed rainfall is required. The ET of the crop is 500–700 mm. Optimum soil moisture is required from germination to pegging stage.

1.2.8 Sunflower

The crop sunflower also is one among the tropical climate crops. Though this crop belongs to tropical climate, this is being cultivated in temperate countries successfully. Regions with high diurnal variation in temperature may not be suitable for cultivation of this crop. At a temperature level of 40°C, seed numbers and seed weight in the flower disc get reduced. The crop comes up very well under a day temperature level of 30–40°C and night temperature level of 18–20°C and under RH level of 60–70 per cent. Optimum soil temperature favors excellent crop growth. During seed maturity period, low temperature and long sunshine hours are required. If the air temperature is above 35°C during flowering, the effective fertilization gets reduced greatly and the number of sterile seeds get increased. It can be raised both in red and black soils under rain fed and dry land situations. The productivity is higher under black soils as compared to red soils under dry land situation. Both surplus and deficit soil moisture affect crop's yield. The crop ET is 350–500 mm. Even though, it comes up very well when well distributed rainfall of 300 mm is received under rain fed and dry land situations.

1.2.9 Soybean

This crop is of from temperate climate. However, this crop is being cultivated in India, where sub-tropical to tropical climate exist. This crop seems to be weather sensitive. For seed germination a minimum of 5°C temperature with a maximum of 40°C is required. If the night temperature is above the normal, vegetative growth gets increased. Enhanced flowering occurs with a temperature level of 20–30°C. During maturity of seeds that is from milking to seed maturity period, optimum temperature other than low temperature is required. During cropping season, RH must be between 60–70 per cent rather than 50 per cent. This crop can be cultivated under rain fed situation; however well distributed rainfall of 400 mm is required. The ET of the crop is 450–700 mm.

1.2.10 Summary

The information on weather requirement given for selected nine crops revealed that, crop specific weather requirement is the need for reaping higher productivity from the concerned crop. Both direct and indirect effect from weather affect crop production and this may be positive or negative as given hereunder.

Crop- weather interaction

Mode	Example
Direct + ive	Increased crop yield
Direct–ive	Decreased crop yield
Indirect +ive	Triggering crop physiology activity
Indirect - ive	Pest and disease increased load

1.3 Agricultural Meteorological Observatory

Measurement of weather elements is essentially required for carrying out all human activities on day to day basis including crop production. Hence meteorological observatories are of paramount important. At meteorological observatory, all weather elements are measured at particular international time [(Greenwich Meridian Time (GMT)] or [Universal Time Coordinator (UTC)] both for automatic and manual instruments. As per the World Meteorological Organization's (WMO) guidance, weather elements are being recorded at different countries of the world and being exchanged between countries to understand weather system movement within the country and between countries (*e.g.*, cyclone/hurricane/typhoon) and also to develop weather forecast, may be now casting, short range, medium range, long range *etc.* Measurement of weather elements is done at surface observatory, ship, aero plane, balloon and also through other radio sound instruments.

There are different types of observatory based on the objectives set in, though similar types of meteorological instrument are used and those are as follows (IMD,1987; Balasubramanian *et al.*, 2009)

a. Climatological observatory
b. Hydrological observatory
c. Meteorological observatory
d. Agro-meteorological observatory

a. Climatological observatory

This is to understand climatology of a new region. This type of observatory is jointly set by different countries for a common mandate on climate summaries of different seasons in a new domain like Arctic/Antarctic poles. Highly advanced and automatic precious instruments will be installed and climate summaries will be reviewed very carefully and validated over seasons.

b. Hydrological observatory

This is established at catchment area of a dam to record rainfall, run off etc., to regulate inflow and out flow of water from a dam.

c. Meteorological observatory

This is established and maintained by concerned Country Meteorological Department. In India it is maintained by India Meteorological Department (IMD). In addition to meteorological instruments, earth quake measuring instrument also being installed. They use weather data for multiple uses of the human requirement including weather forecast.

d. Agricultural meteorological observatory

The purpose of the observatory is to measure weather elements for the benefit of crop production and also to understand crop weather relationship/response. In this type, based on size of the observatory and also based on the type of instruments used namely manual or automatic, there are three categories of agro meteorological observatory.

i. Principal Agro-met observatory **(A type)**
ii. Ordinary Agro-met observatory **(B type)**
iii. Auxiliary Agro-met observatory or Village level agro-met observatory **(C type)**

i. Principal Agro-met observatory (A type)

This type of agro met observatory can be established at State Agricultural University campus and districts headquarter of the of the State. This observatory needs lot of land space for laying out and also it is very costlier to establish and also to maintain it. The size of the observatory is 55 × 36 m (1980 m^2). The shape of the observatory is rectangular and the length of 55 m must be parallel to north-south direction of the observatory. The hand drawn (not to scale) lay out plan is given in the Figure 1.1. In the lay out plan leaving the outs given, there are 15 boxes of each 9 × 8.66 m. In the box number 4 either wind vane or anemometer can be installed. In the centre of each box, pit may be dug out for the installation of instruments as discussed in the latter pages. The box given under "open" can be used for the installation of Automatic Weather Station (AWS) and other needy instruments of that region. The entrance gate may be kept in the centre of the eastern side. Fence must be given around the observatory for the protection of the instruments kept against the anticipated damage from dogs and other animals.

a. Essential instruments required for this observatory

- Single Stevenson's screen
- Maximum and minimum thermometers
- Wet and dry bulb thermometers
- Soil thermometers (5, 10, 15, 30, 60 cm depth)
- Grass minimum thermometer
- Ordinary and self-recording rain gauges
- Wind vane and anemometer
- Open pan evaporimeter
- Sunshine recorder
- Assman psychrometer
- Dew gauge
- Double Stevenson's screen
- Thermograph, hygrograph
- Barometer both manual and automatic
- Soil moisture measuring instruments
- Solar radiation instruments like pyranometer

b. Optional instruments

- Lysimeters
- Thermopile sensing elements for short and long wave radiation.

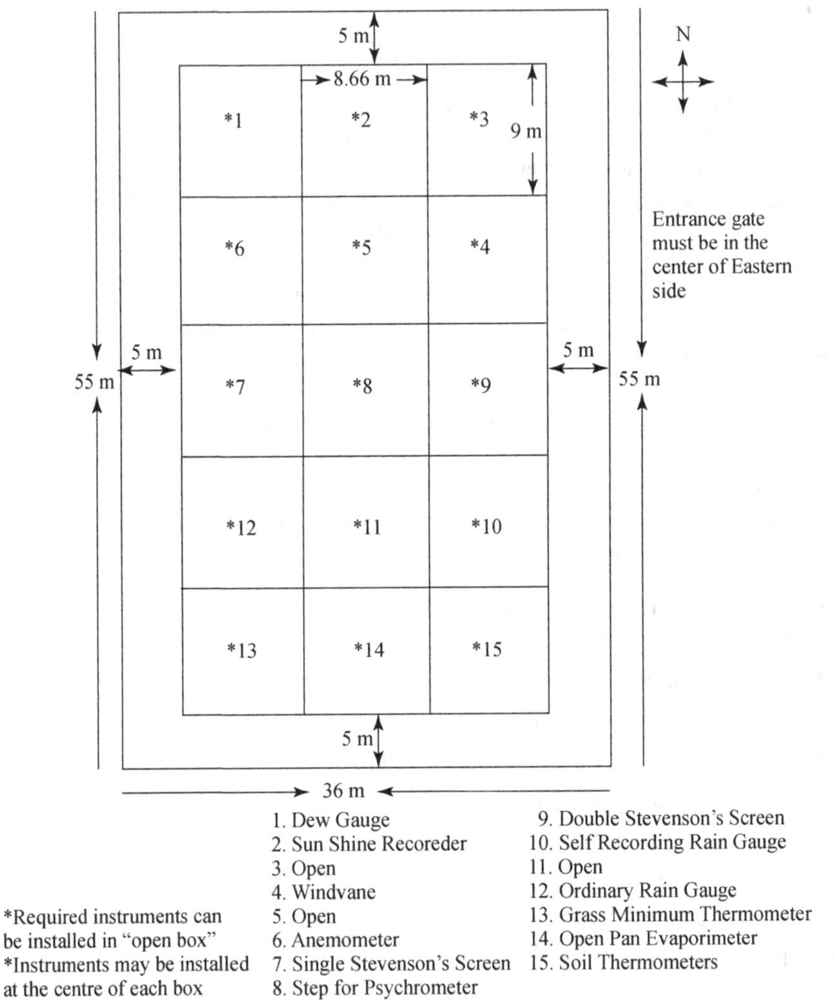

Fig. 1.1 Layout plan for Principal Agro-met observatory (A type)

ii. Ordinary Agro-met observatory (B type)

This type of Agro met observatory can be established at all research stations of State Agricultural university and also at taluk level/block level of the State. This observatory requires less space as compared to A type observatory as given above. The size of the observatory is 31 × 15 m (465 m^2). The shape of the observatory is rectangular. Similar to A type and the length of 31m must be parallel to north-south direction of the observatory. The hand drawn (not to scale) lay out plan is given in the Figure 1.2. In the lay out plan, leaving the outs given, there are 15 boxes of each 5 × 4.

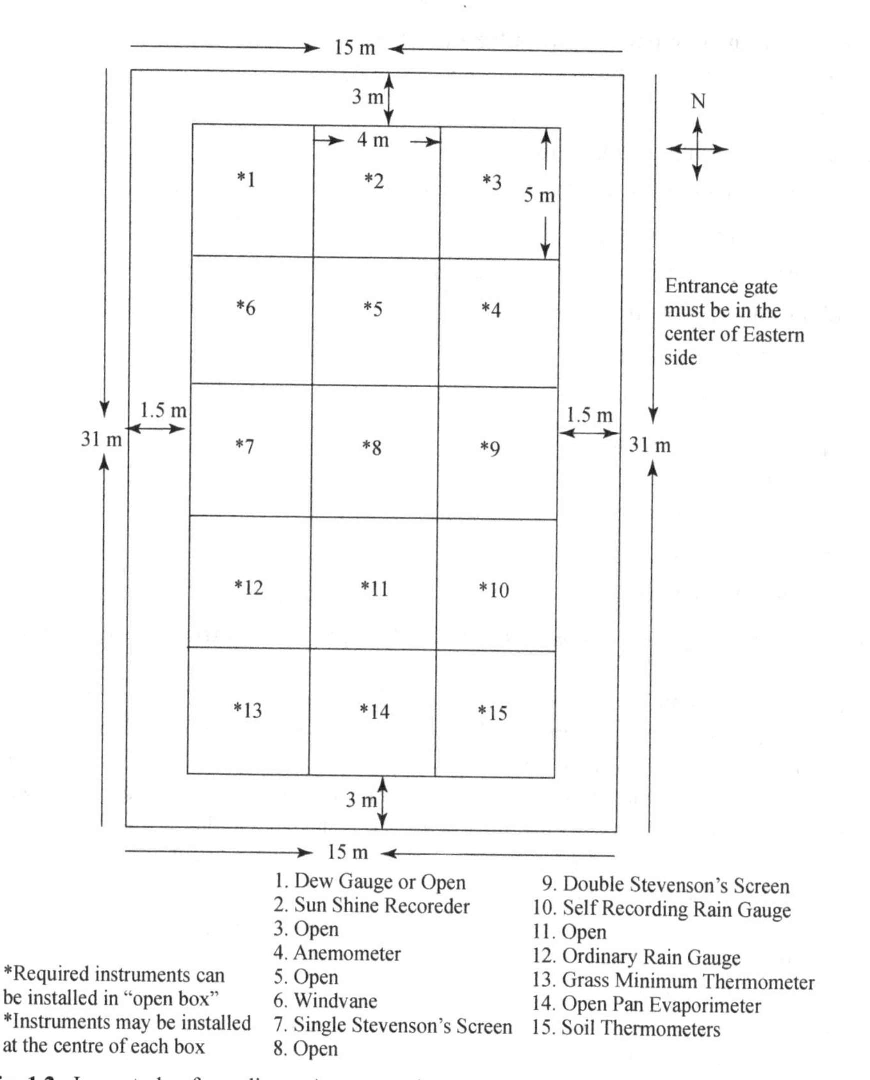

1. Dew Gauge or Open
2. Sun Shine Recoreder
3. Open
4. Anemometer
5. Open
6. Windvane
7. Single Stevenson's Screen
8. Open
9. Double Stevenson's Screen
10. Self Recording Rain Gauge
11. Open
12. Ordinary Rain Gauge
13. Grass Minimum Thermometer
14. Open Pan Evaporimeter
15. Soil Thermometers

*Required instruments can be installed in "open box"
*Instruments may be installed at the centre of each box

Fig. 1.2. Layout plan for ordinary Agro-met observatory (B type)

In the box number 4, either wind vane or anemometer can be installed. In the centre of each box, pit may be dug out for the installation of instruments as discussed in the latter pages. The box given under "open" can be used for the installation of Automatic Weather Station (AWS) and other needy instruments of that region. The entrance gate may be kept in the centre of eastern side. Fence must be given around the observatory for the protection of the instruments kept against anticipated damage from dogs and other animals.

a. Essential instruments required for this observatory

- Single Stevenson's screen
- Maximum and minimum thermometers
- Wet and dry bulb thermometers
- Soil thermometers (5, 10, 15, 30, 60 cm depth)
- Grass minimum thermometer
- Ordinary and self-recording rain gauges
- Wind vane and anemometer
- Open pan evaporimeter
- Double Stevenson's screen

b. Optional instruments

- Dew gauge
- Barometer (manual)
- Sunshine recorder
- Thermo -hygrograph

iii. Auxiliary Agro-met observatory or Village level Agro-met observatory (C type)

This type of Agro met observatory can be established at village level This observatory is very simple and easy to operate by the villagers with enough training. The size of the observatory is 8 × 3 m (24 m^2). The shape of the observatory is lean rectangular and the length of 8 m must be parallel to north-south direction of the observatory. The hand drawn (not to scale) lay out plan is given in the Figure 1.3. In the lay out plan, in the center of the lay out namely at 1.5 m, instruments are to be installed at 2 m interval in the north-south direction. Here there is no provision for the installation of Automatic Weather Station (AWS), since it is not farmers friendly now. The entrance gate may be kept in the center of eastern side. Fence must be given around the observatory for the protection of the instruments kept against the anticipated damage from dogs and other animals.

Among the three types of observatories presented namely A, B and C, the auxiliary type or otherwise called as village level mini agro- met observatory is farmers friendly. This observatory consists of simple, limited and durable and in- expensive weather measuring instruments, which can be easily operated by the villagers.

The data from this observatory would provide opportunities to the farming community to develop their weather-based thump rules for crop and animals' management by integrating collected and real time weather data with field crop observations done over seasons and years. Thus, the concept of seeing is believing will be known to the village community. There is also chance for skill development in this process. Over time the villagers would become climate managers of that region.

a. Essential instruments required for this observatory

- Single Stevenson's screen
- Maximum and minimum thermometers

- Wet and dry bulb thermometers
- Ordinary rain gauge

b. Optional instrument
- Anemometer
- Wind vane

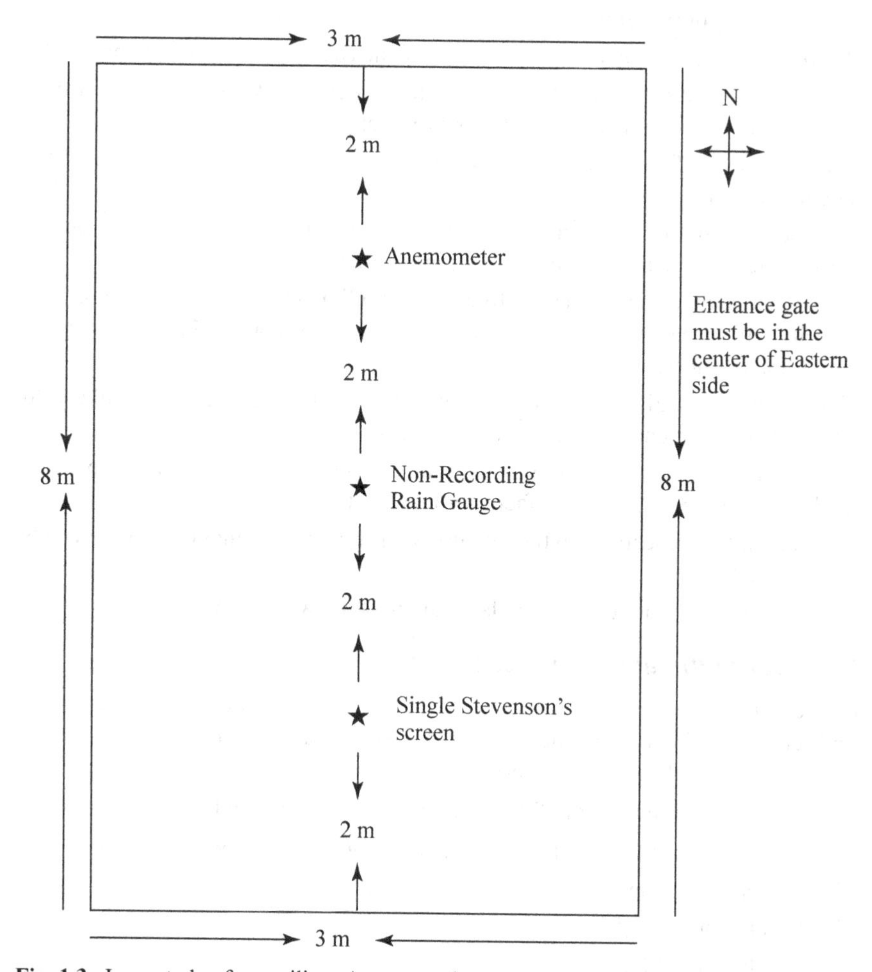

Fig. 1.3. Layout plan for auxiliary Agro-met observatory (C type)

1.3.1 Selection of site for meteorological observatory

Establishment of meteorological observatory should be done with rules and regulations as given by IMD. If the observatory is not established with standard measurements, the readings/data taken or collected from the observatory may not be useful for any practical applications and also not useful for any comparison with any standard observatory meteorological data collected with in India and elsewhere in the world.

Hence meteorological observatory site must be selected with enough care. They are as follows:

i. The site must be in an open field without any physical and biological obstructions like tall buildings, trees, industries etc., At least the site to be selected should be 10 times away from the height of the obstruction in that area.
ii. The site must be in the ridge point of the area so as to have free drainage against water logging during rainy period.
iii. There should not be any hills or mountain on the east-west direction in the nearby
iv. The selected site must be easily accessible for observer to take reading in time and also for easy maintenance of the instruments
v. It must be away from human and animal trespassing and also not vulnerable for any instrument's theft
vi. It must be away from main roads against dust deposit on the instruments. Dust is an allergic to the sensitivity of the instrument to be fitted or used
vii. Locations should be representative of the climate of the region and not heavily influenced by unique local factors. This primary criterion is spatial representativeness.
viii. Consideration is given to whether the area surrounding the site is likely to experience major change within 50–100 years
ix. The risk of human encroachment over time and the chance the site will close due to the sale of the land or other factors are evaluated.
x. Flood plains and locations in the vicinity of orographically induced severe winds are avoided.
xi. The site must be from the leveled land and must be well exposed.

1.3.2 Layout of the agro-met observatory

After selecting the site, layout has to be done. In the case of lay out demonstration, let us take B class or ordinary agro-met observatory as an example. This type of layout is common to all types of agro-met observatory.

For laying out observatory, the following materials are required.

1. Wooden cross staff (available at village office or any Engineering college)
2. Coir ropes/Nylon ropes (300 m)
3. White chalk powder (5 kg)
4. Wooden hammer (2)
5. Wooden bamboo sticks/iron rods of 3' ht. (70 no's)
6. Metallic tape for measurement-50 m

Procedure for layout

By using magnetic campus, the direction of magnetic north has to be identified first from any one point of selected site and by taking 3° approximately to east from the magnetic north, the true north can be fixed in general. Then select a true north point at the southeast corner of the proposed site and by placing the cross staff over this point, identify another point of approximately 35 m in length at the northeast corner direct to southeast corner where the cross staff is placed. Peg these two points of south east and northeast with bamboo rods/stakes and tie the rope and see that the line is straight to true north that we have identified already. In the length of 35 m rope from the southeast point, measure 31 m length as required for the observatory at the north and peg the point with bamboo stake. Please see that the rope is tight. From this northeast corner of 31 m by using cross staff see another point at northwest approximately 20 m in length and peg it. Tie a rope between the bamboo stake of northwest and northeast and the rope must be tight and straight. Normally, since we use cross staff, the intersection point between line of southeast-northeast and northeast-northwest would be at 90°. It shows that the lines are straight. If you have any doubt, use some engineering tools to verify 90° or use Pythagorean theorem rules ($a^2 + b^2 = c^2$: 3, 4, 5). Mark 15 m with tape in the rope line from northeast and northwest and peg it with bamboo stake. From this point, by using cross staff, rope and bamboo stakes, complete the line from north west to southwest for 31 m length and from southwest to southeast for 15 m length. All the corners of the rectangular of 31 × 15 m must be at 90°. That is, one line is perpendicular to another line. In this case of this angle accuracy, the measured distance of 31 × 15 m of this rectangular would be correct by measurement. The object of the work is to get perpendicular rectangular of 31 × 15 m. You can use any engineering survey instrument for this purpose.

'Outs' of 3 m at the N and as well as at south as shown in the lay out must be given. Similarly, in the east and west 1.5 m 'outs' may be given. In the plot size of 12 × 25 m, 15 boxes of 4 × 5 m can be marked for installing instruments by using white chalk powder.

The center of each box also may be marked for digging pit of standard size as required for each instrument as shown in the Table 1.1.

1.3.3 Installation of meteorological instruments

In the case of installation of instrument, let us take B class or ordinary agro met observatory as an example for learning installation of instruments. This type of installation is common to all types agro met observatory.

The required pit size and pit base concrete for various instruments to be installed are presented in the Table 1.1. The measurement is given in feet and inches as per the original document of IMD. It can be converted in to centimeter @ one inch is equal to 2.5 cm.

For the installation of instruments, the following building materials are approximately required;

Bricks : 3000 numbers

Cement : 35 bags of normal cement (industry packing)

Baby jelly : 2.0 units

Sand : 3.5 units

Table 1.1. Pit size and base pit concrete height (in feet) requirement for meteorological instruments of B class observatory

S. No.	Name of the instrument	Required pit size at the centre of box (feet) - L* × B* × D*	Initial pit base concrete height (feet)	Remarks for pit shape
1.	Dew Gauge (DG)	2 × 2 × 3	2	Length must be in east west direction
2.	Sun Shine Recorder (SSR)	4 × 3 × 4	1.5	–do–
3.	Anemometer (ANE)	2 × 2 × 3	1	–do–
4.	Wind Vane (WV)	2 × 2 × 3	1	–do–
5.	Single Stevenson's Screen (SSS)	3 × 2 × 2.5	1.5	–do–
6.	Double Stevenson's Screen (DSS)	5 × 2 × 2.5	1.5	–do–
7.	Ordinary Rain Gauge (ORG)	3 × 3 × 3	1	–do–
8.	Self-recording Rain Gauge (SRG)	3 × 3 × 3	1	–do–
9.	Grass Minimum Thermometer (GMT)	0.5 × 0.5 × 0.5	–	–do–
10.	Soil Thermometers (ST)	6 × 2 × 2	–	–do–
11.	Open Pan Evaporimeter (OPE)	5 × 5 × 9	–	–do–

*L; Length, B; Breath, D; Depth in feet

For preparing plastering, cement and sand may be mixed at 1:4–1:5

For preparing concrete mixture, cement, sand and jelly may be mixed at 1:2:4

Cautions

1. Both the boxes of SSS and DSS must be placed so as to open the door of the box facing True north.
2. Soil thermometers must be placed facing south.
3. After 1.5 feet concrete filling in the pit, the pillar for SSR must be constructed with bricks and cement up to 10 feet height with a size of 3 × 3 feet square pillar. Top of the pillar must be leveled like table top for fixing SSR precisely. The base of the instrument must be adjusted to local latitude and the instrument must be placed to face true North by adjusting through the hole of the instrument which is kept at the base of the instrument.
4. The wind vane must be installed so that the pointer N must face True North
5. The maximum depth of the pit for open pan evaporimeter must be 9 feet in the case of loose soils, but if any murram structure (hard structure) is noticed in the soil profile at any depth from 1 to 9 feet, the depth may be restricted to murram level itself.
6. The wind vane and anemometer must be installed at 10 feet height from the ground level. GI pipe of 2.5 inch with C class in nature must be purchased.

Considering the instrument height as supplied by the Indian company, 10 feet length GI pipe may be purchased for wind vane, while for anemometer, it is 10.5 feet height. Required reducer may be purchased to couple the instrument with GI pipe. This length of pipe includes pipes to be inserted below ground level also. Support may be obtained from local engineer if you have any doubt.

1.3.4 Installation of instruments

In the observatory, the instrument to be fitted first is wind vane, since it shows the direction of true north, which is useful to install other instruments very easily. Hence let us start from the installation of wind vane

i. Wind vane

The wind vane is important in the field of agriculture to take decision for spraying chemicals effectively on crops, otherwise wastage of spray solution would occur. Irrigation water applied through rain gun and sprinkler can be minimized based on wind direction.

It can be installed either in the 4th or 6th box of the lay out, preferably it can be installed in 4th box. In the centre of the box, dig a pit of 2 × 2 × 3 feet and concrete may be given up to one foot from the base depth of the pit and wait for its curing. After getting cured, top of the concrete may be leveled for placing the base of the wind vane pipe. The 10 feet height GI pipe of 2.5 inch coupled with holed flange at the top of the pipe to fit the base of wind vane with bolt and nut may be brought to the observatory and the bottom of the pipe up to 2 feet depth below ground level may be erected on the leveled centre of the pit concrete, holding straight and again concrete may be given up to 1 feet 10 inch and press the concrete for well settlement. Please see that the pipe with its top of flange must be perpendicular to the ground level by using rope with plummet. Please leave it for curing one or two days by scaffolding with cashew poles. By using bricks and cement, construct a pillar of 2 × 2 feet from the 2 inches below ground level to one foot above the ground level and plastered smoothly and leave it for curing for two days. Then fix the wind vane instrument in the pipe of 8 feet height in the flange by bolts and nuts. Since the instrument is near 2 feet in height, the wind vane would be at 10 feet height now from the ground level. First by verifying true north by scientific means (separately discussed in 1.3.10) the wind vane indicator N may be placed in the instrument and thereupon other directions indicator may be fitted followed by fitting the front arrow(head) and tail of the instruments. Now the wind vane is ready for use. Lubrication

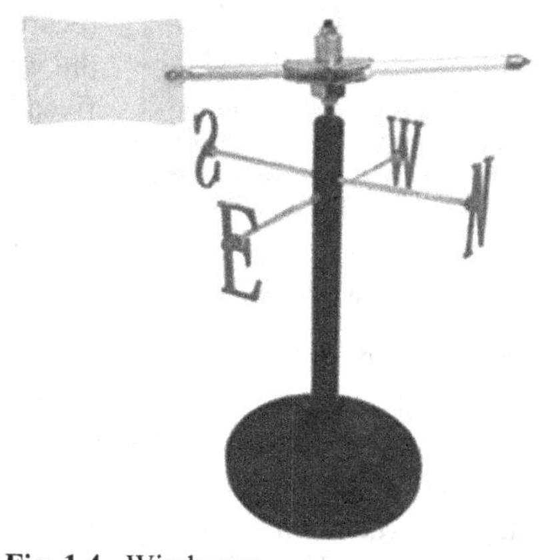

Fig. 1.4. Wind vane
Courtesy: https:/https/www.school speciality.com/gsc-internationalwind-vane-1559718? gdid

must be given once in 15 days by using standard lubricants. Normally the oil used for tailoring machine can be used. The arrow of the instrument indicates the wind ward direction, while the tail indicates the lee ward direction of the wind.

Let us start from the box no. 1

ii. Dew gauge (optional instrument to this observatory)
Dew is very important for crop production especially for *rabi* crops in India. The maximum dew recorded quantity in India is around 2 mm. The temperature of air and earth surface decreases to such an extent that water condenses over the surface and this is called as dew. It occurs in the early morning hours. It generally forms at night due to radiational cooling (clear, cool nights) which causes the temperature of the air to fall below dew point (dew point is the temperature at which saturation of the air occurs).

Dig a pit of 2 × 2 × 3 feet in the centre of th box No. 1 of the lay out given. Apply two feet concrete in the pit, press it and level at the top of the concrete and leave it for curing one day. For the stem of the dew gauge, buy 5 feet height one-inch hollow GI pipe with cork closure at the top and erect the pipe bottom at the centre of the concrete already given for two feet and apply concrete for another 3/4 feet over the base where two feet concrete was given already. Pl see that the pipe is kept at straight in the centre of the pit and perpendicular to the ground level. Pl check with plummet for the pipe's perpendicular to ground level. With enough support to the pipe against slanting, leave it for one day for curing. In the next day, fill the 1/4 feet of the pit with native soil and fix four hands and coated dew wooden plate on the hand from the ground level at 0.5, 1, 2 and 4 feet heights, so that dew deposit can be monitored at different heights. The deposited dew drops will be referred through an album and amount will be recorded. On every evening of a day after sunset, the wooden plate should be placed on the hands fixed in the stem of the dew gauge for recording dew amount and this will be read before sunrise of the next day and the wooden plate will be wiped with dry muslin cloth and kept safely inside the room for subsequent day uses.

Fig. 1.5. Dew Gauge
Courtesy: https://www.britannica.com

iii. Sun shine recorder
The sunshine recorder is necessary to record bright sunshine hours of the day. As per the available literature, only during bright sunshine hours, the plant does photosynthesis or the Hill reaction happens in the plant. From this bright sunshine hours, it is possible to compute solar radiation received in a particular location. And also, the cloud amount in terms of octa could be also computed preciously.

In box No.2 in the centre, pit may dig out at 4 × 3 × 4 feet and concrete may be given up to 1.5 feet from the base and left for curing for one day. In the next day, pillar of 3 × 3 feet construction may be started with bricks and cement from the top of the 1.5 feet concrete given in the pit. The height of the pillar must be continued up to 10 feet from the ground level and at the top of the pillar, fine leveling may be done to place sun shine recorder. The four sides of the pillar must face the four-cardinal direction of E, W, N, S as per the indicators of wind vane. Necessary steps may be provided from the southern side of the pillar up to 7 feet height of the pillar height of 10 feet from the ground level to change sunshine card daily. At 4 feet from the ground level in the northern side, one bookshelf of 1.5 feet height and one feet width and necessary inward depth of 3/4 feet with wooden door may be created to place both used and unused cards and needy other small tools like rain measuring cylinders, dew gauge wooden plate, thermo-hygrograph charts etc., safely.

After plastering we can install the sunshine recorder at the top of the pillar. The base of the instrument must be adjusted to local latitude and the instrument must be placed to face true north by adjusting through the hole of the instrument, which is kept at the base of the instrument. The concerned card may be placed and tested for burning. This type of test must be done for 10 days continuously and once satisfaction occurs, the instrument can be fixed permanently against any theft and damage. Three types of card are being used according to the season. The respective cards should be inserted very carefully in the given slot in the instrument before 30 minutes of sun rise of the day and be removed after 30 minutes after sunset of the that day. The position, season and types of card used are as follows.

Fig. 1.6. Sunshine recorder
Courtesy: https://www.ebay.com/itm292183658406

Table 1.2. Sun card details.

Position of the slot in the instrument	Season	Type of card	Dates
Bottom slot	Summer	Long curve card	12th April-31st August
Top slot	Winter	Short curved card	13th Oct- to the end of February
Middle slot	Equinoxes	Straight card	1st March 11th April: 1st Sep-12th October

iv. Anemometer

Anemometer is to measure wind speed in km per hour or some other units. Wind speed is one among the weather parameters to trigger evapotranspiration of the crops or otherwise crop water requirement. Wind speed is also important to dilute and spread CO_2 for aiding photosynthesis by the plants. Wind speed is important to take decision for plant protection spraying also. The anemometer is kept at 10 feet height for agricultural purposes from the ground level.

The installation of anemometer is similar to the installation of wind vane except the pipe length. It can be installed either in the 4th or 6th box of the lay out. In the centre of the box, dig a pit of 2 × 2 × 3 feet and concrete may be given up to one foot from the base depth of the pit and wait for its curing. After getting cured, top of the concrete may be leveled for placing the base of the anemometer pipe. The 10.5 feet height GI pipe of 2.5 inch coupled with reducer at the top of the pipe to fit the base of the anemometer may be brought to the observatory and the bottom of the pipe up to 2 feet may be placed on the leveled centre of the pit concrete, holding straight and again concrete may be given up to 1 feet 10 inch and press the concrete for well settlement. Please see that the pipe with its top with thread reducer must be perpendicular to the ground level by using rope with plummet. Please leave it for curing one or two days by scaffolding with cashew poles. By using bricks and cement, construct a pillar of 2 × 2 feet from the 2 inches below ground level to one foot above the ground level and plastered smoothly and leave it for curing for two days. Then construct small steps near the pillar constructed already up to 4 feet, so that the observer can take reading very easily during daily observation. Please see that the proposed steps to be constructed should not touch the pipe of the wind vane. There must be a gap of 0.5 feet between the post and the brim of the steps. Then fix the anemometer instrument in the pipe of 8 feet height with the thread of the reducer of the pipe. Since the instrument is near 1.5 feet in height, the anemometer would be at 10 feet height from the ground level. Now the anemometer is ready for use. Lubrication must be given once in 15 days by using standard lubricants. Normally the oil used for tailoring machine can be used.

Fig. 1.7. Anemometer
Courtesy: https://en.Wikipedia.org/wiki/anemometer

v. Single Stevenson's Screen (SSS)

The Single Stevenson screen is made up of teak wood and wooden vents are placed in all directions of the box. This box is important to provide ventilated shading to keep maximum, minimum, dry and wet bulb thermometers which are kept inside the box.

This instrument is to be installed in box No.7. In the centre of the box, a pit with a size of 3 × 2 × 2.5 feet (L × B × D) may be dug out. The length of 3 feet must be in the east-west direction, or, parallel to east west direction. A base concrete up to 1.5 feet is to be laid in the pit first and leveled and let it be cured for one day. Place the four wooden poles/posts or iron poles/posts with all its attachment of cross bars etc., over this concrete preciously to hold the four corner legs of the SSS correctly, otherwise after concrete settlement, the box of four legs cannot be fitted correctly over the slots of the four posts. After placing the four legs of the SSS, add concrete up to another ten-inch height in the pit and scaffold the structure with cashew poles against any sliding of the poles of the SSS from its original position placed before concreting. Let it be cured for one day. After curing, fill the balance five inch of the pit with native soil and place the box over the four poles correctly. Please see that the opening of the box must be towards north direction. Instruments like thermometer to be placed inside SSS will be discussed at latter. The following points may be looked in to:

- Door must face North
- Top of the post or pole must be 1.25 m (50 inch) from the ground level *i.e.,* out of 5 feet height of post, four feet of the post must be above ground level and one foot below ground level.

Fig. 1.8. Single Stevenson's screen
Courtesy: https://en.wikipedia.org/wiki/stevenson-screen

vi. Double Stevenson's Screen (DSS)

This box is longer than the single Stevenson's screen box otherwise it is similar to SSS. In this DSS, auto rotated thermograph, hygrograph, barograph and other self-recording instruments will be kept.

This instrument has to be installed in pit No. 9. The installation of DSS is similar to the installation of SSS except the length of the pit and box.

Fig. 1.9. Inside view of Double Stevenson's screen
Courtesy: https://www.tradeindia.com/fb2608476/double-stevenson-screen-html

vii. Self-recording rain gauge (200 CM^2)-SRG

Rain gauge is to measure the amount of precipitation that received in a time for a particular location. Since rainfall has spatial variation greatly, dense network of rain gauge stations is required. Rainfall measurement is very important in terms of computing water balance/ water budgeting of a location/region/catchment. It is recorded across the countries in a standard time of 3.0 GMT/UTC as per WMO for international comparison of values and weather system movement across the globe. Both manual and self-recording instruments are available to measure rainfall. Unless self-recording instruments are maintained well, it may not be useful for any practical purpose.

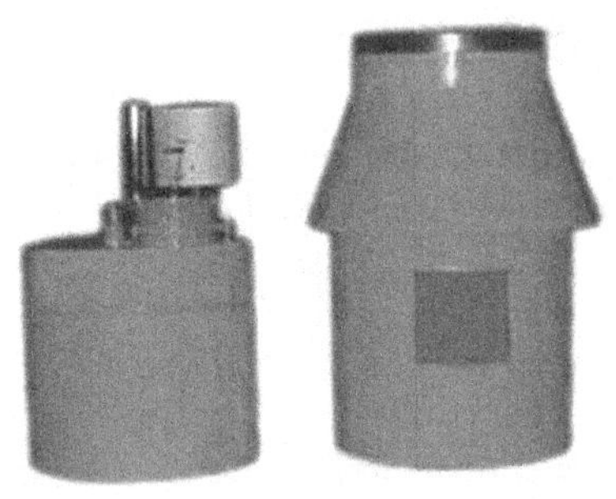

Fig. 1.10. Self-recording rain gauge
Courtesy: Rajinstruments.com/products/ self-recording rain gauge

The Self-recording rain gauge is used to measure a continuous record of daily or weekly rainfall from non-corrosive fiber glass reinforced plastic container. It consists of a funnel-shaped collector at the top of the gauge as one part of instrument and a float siphon chamber and recording mechanism just below it as second part. The rain water collected by the collector is led into the float chamber. The float is having a central stud on which a pen assembly is fixed. The pen is moving over a daily chart or weekly chart wound over a clock drum. After every 10 mm of rainfall, siphoning occurs and recording of rain starts.

The self-recording rain gauge is to be fixed in 10th box of the observatory lay out. In the centre of box, a pit size of 3 × 3 × 3 feet may be dig out and base concrete of one feet height may be laid out, cured and leveled. In the next day two to four layers of brick construction of a pillar with the size of 2 × 2 feet may be started with bricks and cement from the top of the concrete laid out already. Place the SRG instrument over the brick layer of the construction and measure the height of the brim of the SRG from the ground level. It must be 2.5 feet above the ground level. Accordingly, the brick layer construction may be modified through removal or additional construction of brick layer. The standard height of 2.5 feet height from the ground level to brim of the SRG has to be maintained at the end of construction. The float, clock and floating chamber would be above ground level and to be fitted with the basal part of the instrument. One hole with tube may be connected at the base of the instrument to drain the rain water as a result of float -siphon mechanisms of SRG. Fit the instrument and test for its functioning by pouring fresh water through the rain receiving broad open funnel at the top of the instrument.

Key type of SRG is better than battery operated one. In the case battery operated one, enough care may be taken against leakage of battery cell. In the case of key type when the daily chart is changed, key can be given with standard twist of 6 to 8 times daily. This will maintain the health of the cloak very well. Daily, the chart has to be changed by 3.0 GMT/UTC in India in all stations.

viii. Ordinary rain gauge (200 CM²)-ORG

Rain Gauge is made from non-corrosive fiber glass reinforced plastic container. The essential parts of an ordinary rain gauge are a funnel through which the rain water is collected in a receiver and a measuring glass (rain measure) with which the rain water collected is measured.

The installation of this rain gauge has to be done in the 12th box of the lay out. In the center of the box, a pit size of 3 × 3 × 3 has to be dug out and concrete may be laid up to one foot, cured and leveled. In the next day two to four layers of brick construction of a pillar with the size of 2 × 2 feet may be started with bricks and cement from the top of the concrete laid out already. Place the ORG instrument over the brick layer of the construction and measure the height of the brim of the ORG from the ground level. It must be one foot above the ground level. Accordingly, the brick layer construction may be modified through removal or additional construction of brick layer. The standard height of one feet height from the ground level to brim of the ORG has to be maintained at the end of construction. Above the ground level, brick and cement work continued up to the lower portion of the key locking system given in the instrument and plaster the construction well. Cure it for one or two days and thereafter rain measurement can be started.

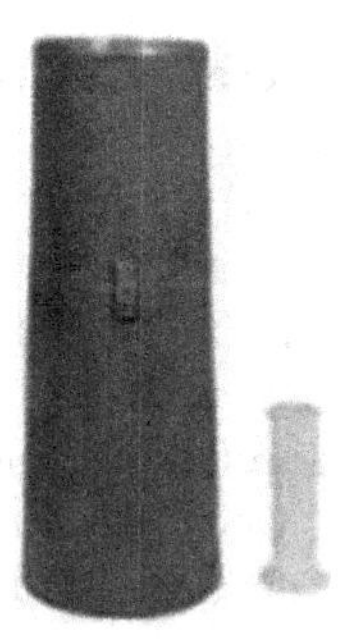

Fig. 1.11. Ordinary rain gauge with rain measure cylinder
Courtesy: https://the constructor. org>water resources>types-of-rain-gauges

ix. Grass minimum thermometer

The grass minimum temperature is the temperature being recorded in open air at ground level over short grass turf and the bulb of the thermometer is in contact with the grass blade tips. Normally the thermometer is placed in horizontal position over the grass vegetation at a height of 2 inch above ground level. The temperature would indicate the progress of frost to occur.

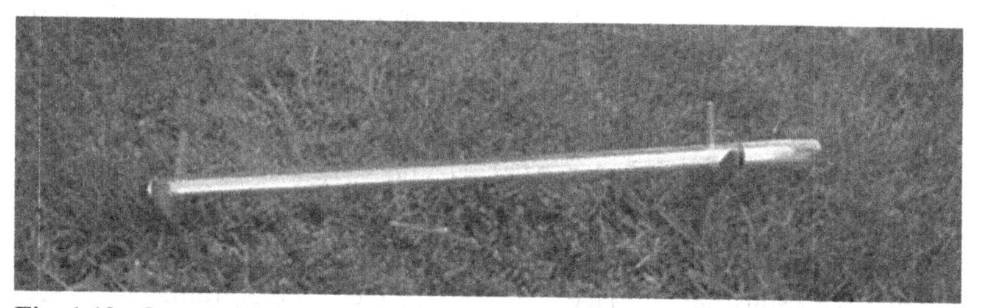

Fig. 1.12. Grass minimum thermometer
Courtesy: https://www.researchgate.net/figure/grass-minimum-thermometer_fig3-3225

In the box No.13, installing of grass minimum thermometer can be done. In the centre of the box, fix two 12-inch length of bi-fork (" quarter inch" thickness) iron thin rods in to the soil up to six inches leaving the balance six-inch rod with the bi-fork top above the ground level. The two iron thin rods can be inserted in the soil with a spacing of

12 inches in the east west direction. On every evening after sun set, the grass minimum thermometer has to be placed horizontally over the bi fork stand fixed in the soil.In the next day morning before sunrise, take the reading and keep the thermometer inside safely. If it is kept outside over day by mistake, the line column will be broken and the thermometer cannot be used further.

x. **Class A pan evaporimeter (OPE)**

US. WB, Class A pan evaporimeter measures the water loss from free surface by evaporation based on the atmospheric demand. This pan is made up of either with galvanized iron or with copper sheet of 20 gauge in thickness. The pan has a diameter of 122 cm with 22.5 cm depth and covered with mess cover of hexagonal in shape. One still well is supplied, which is to be kept at 30 cm away from northern side of the pan. This gives reference point to add water to the pan based on the evaporation or to remove water, whenever rain occurs. One water temperature measuring thermometer is also being supplied. The pan rests on a carefully leveled, wooden base. Normally the reading will be taken by 3 GMT/UTC. The evaporation recorded in mm on every day has to be multiplied by pan coefficient of 0.8 and recorded in the register. It is now accepted that the pan evaporation reading can be taken as Potential Evapo-Transpiration (PET of the area). The PET recorded on that day is equal to the total atmospheric demand of the day of the region. For computing water balance these collected data would be very useful.

Evaporation is measured daily as the depth of water (in mm) evaporates from the pan. The measurement day begins with the pan filled to exactly two inches (5 cm) below, from the pan top. At the end of 24 hours, the amount of water to refill the pan to exactly two inches below from its top is measured.

This evaporimeter has to be installed in the box No 14 of the lay out. In the centre of the box, 5 × 5 × 9 feet pit has to be dug out. The maximum depth of the pit for OPE must be 9 feet in the case of loose soils, but if any murram soil structure is noticed at any depth from 1 to 9 feet, the depth may be reduced to murram level. For our example we consider that 9 feet depth has to be taken. Fill the pit with broken bricks up to 4 feet from the pit depth of 9 feet. Over that, full bricks are arranged in layer up to height of another three feet. In the top of two feet, loose native soil may be filled up and leveled. This is done to avoid transfer of heat from the earth to the pan bottom. If murram structure is noticed, accordingly the layer of broken bricks, full bricks and loose soil application can be proportionally reduced. After leveling the ground level with native soil, the horizontal nature of the ground level has to be checked with spirit level.

Place the wooden frame over the leveled area and check the horizontal nature of the wooden frame with spirit level. Place the pan over the wooden frame and check the horizontal nature of the pan again with spirit level. The brim of the pan should be 36.5 cm from the leveled ground level. If it is ok, add water up to the reference point of the still well. Please check for leakage of water and go on reading the evaporation in mm if leak does not happen.

Fig. 1.13. Open pan evaporimeter:
Courtesy: https://www.india,art.com/proddetail/open-pan-eavaporimeter-1

There are four steps to measure the evaporation for rainy day and non-rainy day and those are as follows: Daily evaporation is to be taken based on the four rules in mm (no rainfall situation, drizzling, heavy rainfall under no overflow and too heavy rainfall under over flow).

i. **No rainfall situation**

   Evaporation (mm) = water added through cane*;

ii. **Drizzling situation (under water added situation to the ref. point in addition to drizzling rain amount)**

   Evaporation (mm) = water added through cane + drizzled rain amount;

iii. **Heavy rainfall under no overflow situation from the pan**

   Evaporation (mm) = Recorded rainfall amount (mm) - water removed from the pan through cane to make the water level to tip of reference tip point

iv. **Too heavy rainfall that made the water in the pan to get overflow**

   Recorded as over flow.

   Remove the water from the pan through cane to make the water level to the tip of reference point for next observation

   *Cane capacity (20 cm height) = 2 mm of water added or removed per cane

   After taking reading from the pan it has to be corrected to avoid error from material used for making drum of the pan. Normally a pan coefficient of 0.8 is considered.

   Example: Measured evaporation from the tank; 6 mm/day

   To be corrected as $6 \times 0.8 = 4.8$ mm.

   Hence the free water surface evaporation is 4.8 mm/day

Implications.

The value is in depth. It cannot be used as it is.

Hence the data collected from the pan can be used to quantify water loss from a tank of one hectare, where water is kept under free surface;

4.8 mm/1000 mm = 0.0048 m (mm is converted in to meter)

Per hectare = 10,000 m^2

Then water loss from one hectare = 0.0048 × 10,000 = 48 m^3

1 m^3 = 1000 litres

and hence from the one-hectare tank the water loss on the particular day = 48 × 1000 litres

= 48000 litre/day.

For computing ET, this value has to be multiplied by Kc value of the crop

Example.

Crop; Sorghum

Stage; Flowering

Kc value collected from FAO book; 0.9

ET (mm) from sorghum = 4.8 × 0.9 = 4.32 mm of the day.

xi. Soil thermometers

About 40 per cent of solar radiation is being absorbed by the earth surface. The soil thermometer is used to measure soil temperature up to 60 cm for agricultural meteorological studies. Different layers of soil get heated differently due to low conducting power of the soil and hence there is diurnal variation in temperature. The surface of the soil gets heated up quickly during day time and loses heat quickly during night hours. Normally surface soil temperature and soil temperature at 5, 10, 15, 20, 30 and 60 cm depth would be studied for agricultural study purposes. From the past studies there is dynamic change in soil temperature up to 30 cm on daily and weekly basis but beyond 30 cm the change in soil temperature occurs only with change in season.

Soil thermometer has to be installed in the box 15th of the layout. In the centre of the enclosure, in the east west direction 2m length and 45 cm width with a depth of 40 cm pit may be dug out and cured for one day. In the next day five iron stands as shown in the figure may be made from local shop enough to support thermometers height of 5, 10, 15, 20 and 30 cm depth and fitted at 45 cm interval in the pit permanently along the east west. The bulb of each thermometer may be placed at the appropriate depth in the pit and tie with the strand already fixed in the soil. Carefully fill up the pit up to the ground level with the native soil. The soil should be so nice without any stones and pebbles. Allow one or two days to settle the soil and fill it up again with the soil still we get stabilization up to the ground level and start to take observation. The soil thermometers should face south direction. Normally observation may be taken from half to one foot away from the bulb of the thermometer buried in the soil.

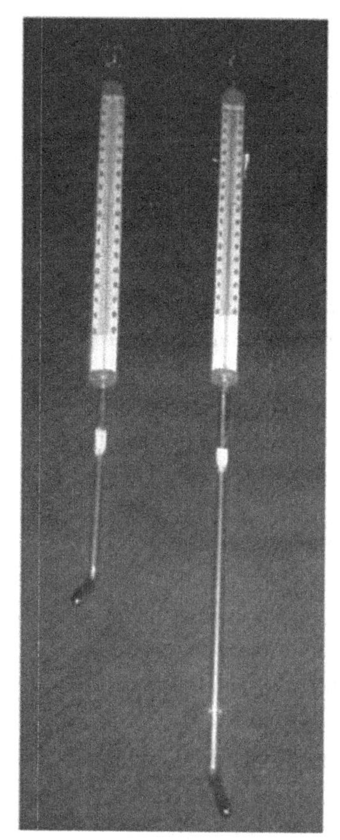

Fig. 1.14. Soil thermometer
Courtesy: https://dir.indiamart.com/impact/soil-thermometer.htm

Fig. 1.15. Iron stand to hold soil thermometer

xii. Automatic weather station, its installation and maintenance
An Automatic Weather Station (AWS) is defined as a "meteorological station, from which observations are made and transmitted automatically" (WMO, 1992). Normally for agricultural purposes, in addition to manual instruments, AWS is an optional one for Principal Agro-met observatories. However, the AWS helps in observing

accurate weather measurements. Its usage for agriculture becoming more, owing to the development in the science of Information and Communication Technology (ICT). Now many of the States in India are trying to establish network of AWS mainly for agriculture purposes. Under National Agricultural Development Programme, Tamil Nadu, a network of 385 AWS was established in each block of Tamil Nadu by Tamil Nadu Agricultural University (Tamilnadu Agricultural Weather Network) and this is a typical example of such net work which is being followed in many of the Indian States.

Purpose of AWS

1. It helps in getting more accurate measurements in real time as there are incidence of erroneous manual measurements or some time observations are recorded without actual measurements.
2. Any observing interval in a day can be recorded which is not possible in manual observatories. For example, it is easy to record wind gusts and rainfall intensity in AWS. In other words, it is possible to record the extreme values at the exact moment of occurrence.
3. Helps in reducing observational costs as there is no requirement for observers.
4. It will ensure the uniformity in the measurements of observations
5. New kind of measurements as required is possible like measurement of leaf wetness that can be related to development of fungal diseases in crops
6. The observed data can easily be ingested in to simulation models and get alerts in advance for taking necessary steps.
7. This will also help in Automated Agro Advisory development which can be passed on to farmers in real time.

Though there are many advantages of installing AWS, it is often noticed that many networks have failed owing to the poor maintenance. Hence, periodic maintenance and calibration of the sensor should be given top most priority in the AWS network. Another important aspect is the quality of sensors which should not be compromised for their price and one should take care to install reliable, standard and approved sensors. One has to remember that visual or subjective observations like cloud cover etc., cannot be made fully automatic and it is yet to be achieved.

Installation requirements

Like any other agro-meteorological observatory, installation care has to be taken in installing the AWS. The following points may be considered while installing the AWS.

1. The location selected should satisfy the exposure conditions set out for a normal observatory *i.e.,* away from shaded areas and obstructions (tall trees and buildings etc.)
2. It is better to install AWS in the centre of a farm, away from water sources so that it gets exposed to the actual field conditions.
3. The height at which measurements required for agriculture, the sensors should be placed accordingly.

4. Avoid shading of different sensors as well obstruction between them to prevent vitiation of observation.
5. Select sensors which is durable, standard and approved by India Meteorological Department (IMD). It is better to use sensors traceable to the United States National Institute of Standards and Technology (NIST) which is being followed by IMD.
6. Ensure un-interrupted power supply to satisfy the energy requirement of the sensors and loggers.
7. For AWS networks, it is better to keep spare sensors, which can help in immediate replacement due to sensor problems, so that continuous measurement is maintained.
8. Periodical maintenance of AWS to be done by way of clearing weeds and removing dusts in solar panel as well as on the sensors.
9. Accurate metadata should be maintained for each AWS installation
10. WMO (2008) recommends site not less than 25 m × 25 m area for installing AWS for general meteorological observation where the wind observation is recorded at 10 m height which makes it necessary to have a bigger site. However, for agricultural purpose the height of the measurements is restricted to 3m accordingly smaller site area is sufficient. The Tamil Nadu Agricultural University for its automated weather network utilized only 6mx5m area for all its AWS sites.

Types of AWS

Automatic weather stations are sometimes used as an optional one in agro-meteorological observatory to help the observer at manned stations or to complete replacement of observers at fully automatic stations. Either the AWS can be installed to record and communicate in real time or it can be deployed to just record the data automatically so that it can be retrieved whenever required. Hence, WMO (2008) classified the AWS as Real-time AWS and Off-line AWS

Real-time AWS: A station providing data to users of meteorological observations in real time, typically at programmed intervals say hourly or three hourly synoptic intervals. If warranted on emergency requirements or upon external request even AWS can be configured to record less than a minute. During Solar Eclipse on 15.01.2010 two of the AWS of Tamilnadu Automated Weather Network was configured to record for each minute to understand the influence of eclipse on surface weather.

Off-line AWS: A station recording data on site on internal or external data storage devices possibly combined with a display of actual data in same location. The stored data can be retrieved whenever required or at the completion of a cropping season so that it can be analysed in conjunction with crop observations.

Components of AWS

The major components of AWS consist of sensors for weather observation, data logger with digital and analog channels for receiving respective signals from sensors, a microprocessor to process the logged signals and provide appropriate instruction to the sensors, a communication unit for data transmission, a power unit consisting of battery and solar panel. The sensors for observing agricultural weather variables may be digital or analog has to be installed at appropriate height as required. Depending up on the number of sensors, the number of digital or analog channels for the data logger has to be designed for receiving appropriate signals. The microprocessor not only process the data logged by the logger and converts to the appropriate format. It also manages the recording interval of sensors and pass it on to the communication unit as and when required. Based on the communication protocol there are different communication devices available and is discussed in detail separately. The voltage of battery in power unit is normally 12 Volts which is being recommended by WMO (2008) and however, the battery's Ampere Hour (AH) depends on the power utilization by all the components of AWS which is to be calculated before setting up the AWS. Appropriate solar panel has to be installed to supply the required power to the battery and it again depends on the climate condition of the AWS location as there are differences in solar radiation of individual location. The AWS site should also be protected with suitable fence.

Data transmission

The logged data in the AWS is being communicated to the Central Unit of the network or to the clouds or to the satellite for onward transmission to the base station in the earth. The mode of communication in the technological era being made as cheaper as possible. The following techniques are currently used for transmitting data from the AWS to the required locations.

a. Telecommunication
b. Satellite communication
c. Global System for Mobile communication (GSM)

The telecommunication requires cable connection for individual AWS and calls need to be made to fetch the data every time. Hence, the stations which are in remote areas cannot be accessed. Moreover, only one-way communication is possible which restrict its usage. The satellite technology which has wireless connection is the most suitable for remote areas but again the communication is one-way. The Time Division Multiple Access (TDMA) is the technique used for satellite communication which implies that each AWS can transmit the data in that particular time division and failure to do so there are possibilities for data loss. This kind of data loss is common in uplink satellite transmission and we cannot recover such data unless otherwise it was retrieved from that AWS by actually visiting it. Transmitting to satellite is costly affair because it requires special antennae to be installed in the AWS location. Another major issue is the charges levied by the Department of Telecommunication in India for data transmission is huge compared to other communication modes.

The third technology is Global System for Mobile (GSM) communication which also uses TDMA technology. Some GSM providers use the Code Division Multiple Access (CDMA) technology for mobile communication. The CDMA technology becoming obsolete as it prevents the use of different devices unlike that of GSM-TDMA in which any mobile phones can be used by switching the Subscriber Identification Module (SIM). The General Packet Radio Services (GPRS) is a packet-based wireless communication embedded on GSM paved way for the cheapest data transmission. Compared to higher rate of data transmission capabilities of GSM-GPRS, the requirement for data transmission from AWS is very less. Another important advantage is the possibilities for two-way communication and hence, it is easy to configure the AWS over the air. The Tamil Nadu Agricultural University used GSM-GPRS technology for data transmission from 385 AWS installed across entire Tamil Nadu for providing real-time data to the farmers. As this mode of communication is cheaper and possess two-way communication, the same is being followed in other States and also by India Meteorological Department, which hither to utilized the satellite transmission.

Data sampling requirement

The WMO (2008) recommends 1 to 10-minute averages for atmospheric pressure, air temperature, air humidity etc., while 2-min or 10-min averages are recommended for wind observation. Though it is averaged values, it has to be considered as the "instantaneous" values at the top of the hour in which the recording is done. However, the wind gusts, maximum and minimum temperature are the values that are occurred in a specified period and normally for the 24 hours of the day.

Calibration and maintenance

Like manual observing surface instruments, We need to calibrate the sensors periodically as specified by their manufacturer. Travelling standards having similar filtering characteristics to the AWS measuring chain and with a digital read-out are to be preferred for calibration. Instruments at the end of their calibration interval, instruments showing an accuracy deviation beyond allowed limits during a field inspection and instruments repaired by the maintenance service should be returned to a calibration laboratory prior to their re-use as per the guidance of World Meteorological Organisation. Depending on the requirement, a schedule should be drawn for periodic calibration. Similar to the calibration, periodic maintenance of the AWS location like removing weeds from the site, cleaning dusts on the sensors and solar panel is important. It is better if the maintenance visit is made every fortnightly or at least once in a month for better performance of AWS. Annual painting of masts of sensors and fence is another essential task in the maintenance schedule

xiii. Fencing the observatory

The observatory has to be fenced appropriately with barbed wires or with mesh wires around the lay out by fixing poles of five feet height at six feet interval with concrete pegging. The fence must be at four feet height from the ground level. It must be painted with white. The gate may be given at the center of eastern side of the lay out.

One sign board with details as given below may be erected on the left side of the eastern gate. The size of th sign board is 4 × 3 feet.

i. Name and type of the observatory with the name of organization
ii. Latitude, longitude and MSL
iii. Instrument fitted
iv. Times of observation

1.3.5 Fixing Thermometers in SSS

In the inside of the SSS four thermometers have to be fixed and those are maximum thermometer, minimum thermometer, dry bulb thermometer and wet bulb thermometer. These thermometers should not be exposed to the sun directly.

i. Maximum thermometer

The maximum thermometer mounted on a wooden frame along with a "U" shaped tin handle at the chamber side of the thermometer will be supplied by the firm in India. This maximum thermometer is used to measure the maximum temperature that prevails in a particular day. Normally in India the maximum temperature will occur between 2 and 2.30 pm IST (8.30 and 9.00 GMT). In the thermometer there is a small constriction near the neck of the bulb, where the capillary tube hole is extremely fine. When the temperature of the air rises, the mercury in the thermometer bulb expands and forces its way in to the stem past this constriction, but when the temperature goes down, the mercury above the constriction does not have sufficient pressure to flow back in to the bulb. hence the length of the mercury column remains just the same as it was when the bulb was warmest. The end of the mercury column farthest from the bulb thus indicates the maximum temperature that reached on a day. Two horizontal frames running east -west is nailed with two vertical frames at the extreme end of horizontal frames and fitted inside of the SSS as shown in the following figure. In the top of the horizontal frame, this thermometer along with its mounted frame is to be fixed on a hook freely and horizontally. The chamber end would be quarter of an inch higher than the bulb end.

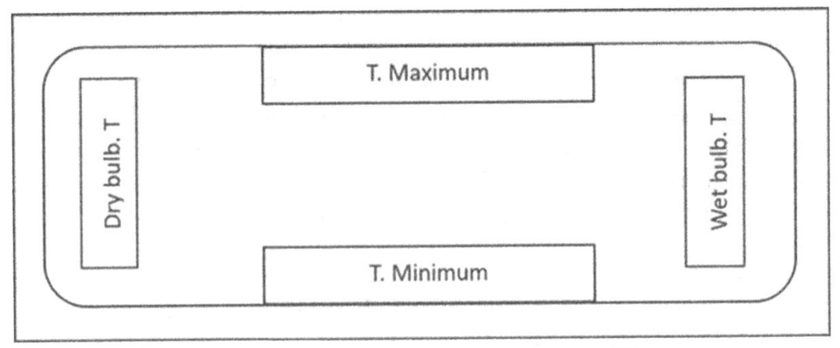

Fig. 1.16. Inside Frame to mount four thermometers in the SSS

ii. Minimum thermometer

The minimum thermometer mounted on a wooden frame along with a "U" shaped tin handle at the chamber side of the thermometer will be supplied by the firm in India. This minimum thermometer is used to measure the minimum temperature that prevails in a particular day. Normally in India the minimum temperature will occur between 5 and 5.30 am IST (11.30 and 12.00 GMT). The thermometer is filled with spirit. A dumb bell-shaped index is immersed in the spirit. When the temperature falls, the spirit drags the index along with it towards the bulb end. But when the temperature rises, the spirit expands and runs past the index without any disturbance to the index position. Thus, the end of the index farthest from the bulb gives the lowest temperature recorded in the day. This thermometer along with its mounted frame is to be fixed freely in a hook at the lower horizontal frame of the SSS.

iii. Dry bulb and wet bulb thermometers

Both dry and wet bulb thermometers are same except for wet bulb thermometers, wherein, the bulb is made to cool to depress the temperature. The readings from these two thermometers will be used to get relative humidity of the day. In both the thermometers mercury is used to measure the temperature. The dry bulb thermometer is fixed vertically on the left side of the frame as shown in the figure, while the other wet bulb thermometer is fixed vertically in the right side of the frame. The bulb of the wet bulb thermometer is always kept cool/damp with the help of a moist muslin cloth. One small bottle of distilled water kept aside of the bulb of wet bulb thermometer is connected to wrapped muslin cloth of wet blub through starch free hygroscopic cotton thread of four strands. This creates coolness around the bulb region due to evaporation of water from the muslin cloth. This leads to drop in temperature in the wet bulb thermometer and therefore it indicates a lower temperature than that of dry bulb thermometer, where in the air temperature can be read. The difference between the readings of dry and wet bulb thermometers is known as wet bulb depression. The wet bulb depression is directly proportional to the dryness of the air. The readings of the wet and dry bulb thermometers will be used to refer hygrometric table of concerned location mean sea level and find out relative humidity, vapor pressure and dew point temperature

1.3.6 Setting and Maintenance of Thermometers

The maximum thermometer must be read in the morning in the local time fixed for the observatory and that is given in the following paragraph. The reading now taken indicates the past 24 hours maximum temperature recorded by the thermometer. No evening resetting is required. After taking reading from the thermometer, it has to be reset for next day morning reading. It has to be done as follows; To reset the thermometer, remove it from the hanging position outside SSS and hold the upper end of the wooden mount (chamber end), keeping the bulb end downwards. Take care not to exert any pressure on the thermometer stem against breaking. Then stretch out your arm and swing it briskly from the shoulder, beginning the swing with the arm raised more than 45° above the horizontal and ending about 30° beyond the vertically downward position. The swing must be vigorous but smooth. While swinging the thermometer, stand in a clear space and take care to see that the thermometer does not get damaged by striking against any hard bodies, including your body side and shirt. Repeat the swinging till the mercury gets in to bulb of the thermometer. Then place the thermometer on the hook in horizontal position. Verify that the thermometer reads the same as dry bulb thermometer reading with a variation between 0.2 to 0.3 decimals of °C.

For resetting the minimum thermometer, it must not be removed from the SSS and resetting must be done to dry bulb reading. Resetting must be done both after morning and evening readings by lifting gently the bulb end to adjust the moving index within the alcohol column at chamber end. On any account the minimum thermometer should not be taken out from the screen and swung it. This may lead to break in alcohol column.

a. Important points to be taken to maintain the thermometers

i. At the local time of observation, in the morning, take reading from the maximum thermometer and reset it to dry bulb temperature reading. Avoid parallax error

ii. The minimum thermometer must be read in the morning and reset to dry bulb reading. In the afternoon also the thermometer has to be re set it again.

iii. In the event of alcohol column break in minimum thermometer, boil the water for 100°C and keep it to cool to 80 °C. Insert the bulb of the thermometer in the hot water, simultaneously, keeping the chamber(the other end of the thermometer) cool by keeping wet cloth covering the safety chamber OR keep the thermometer in vertical-tilted position with bulb downward over a lighted candle. The distance between the bulb and candle must be around four inches. No direct heating of the bulb over the candle is done. This will lead to bursting of the bulb. Heat the bulb gently till the column joins in safety chamber, provided at the top of thermometer that is the safety chamber must be kept under cool gently by keeping with wet cloth.

iv. If there are some bubbles in the alcohol column of the minimum thermometer, then hold the thermometer horizontally and warm the bulb by grasping it with the palm until the bubbles rises in the tube. Now shake out the bubble and finally keep the thermometer in a cold bath for some time

v. Sometimes, the index of the minimum thermometer may go into the bulb or get stuck in the bend above the bulb. To rectify this defect, simply raise the bulb until the thermometer is vertical and gently tap the instrument. The index will slide down to the tube.

vi. The index of the minimum thermometer sometimes is thrown out of the sprit and may stick in the upper part of the thermometer stem (chamber end). In this case, hold the instrument vertically in the right hand with the bulb end at the lowest and gently tap the lower end of the thermometer mount against the palm of the left hand. If repeated tapings do not succeed in displacing the index, turn the thermometer upside down so as to transfer the greater portion of the spirit column to the end farthest from the bulb. Then reverse the instrument and allow the index to fall to the lower end of the transferred column of spirit. Restore the spirit column by repeating this action and finally keep the thermometer immersed in cold water for about an hour.

vii. The bulb of the wet bulb thermometer should be covered with only single layer of thin and soft muslin cloth. The cloth should be washed with boiling water first to remove starch and then washed with distilled water four to six times before wrapping over the bulb of the wet bulb thermometer. The muslin cloth around the wet bulb is kept moistened by wicks of cotton thread let in to bottle of water. Four strands of darning cotton are looped round the neck of the bulb over the muslin in the form of a noose and let in to the bottle of water. The bottle must have a small neck so that the air inside SSS is not moistened by evaporation of water from the water kept in the bottle. Alternatively, it should have cover with a small hole through the cotton strands pass. One empty ink bottle can be used effectively for this purpose after enough cleaning. The bottle must be kept a little away (1 to 2 cm) from the bulb of the thermometer. On every day the bottle may be filled with distilled water after observation taken from the SSS. The cotton threads and muslin cloth must be changed once in 15 to 30 days during fine weather and once in a week during dusty weather.

viii. The graduations in the thermometer often become invisible through long use. To restore graduations visibly, rub gently the stem of the thermometer with black lead pencil followed by wiping off the superfluous black graphite with a piece of soft cloth.

b. Important points to be taken to maintain other meteorological instruments

i. The dew wooden plates and grass minimum thermometer must be placed after sunset for recording and remove them before sunrise of next day and keep it safely in the room.

ii. The concerned sun seasonal cards must be removed after sunset of that day and insert next day card simultaneously.

iii. The wind vane and anemometer must be lubricated with tailoring machine oil once in 15 days

iv. The SSS door must be closed immediately after reading

v. The receiver and the body of the ordinary rain gauge must be checked on every day for any leakage. Only the specific measuring cylinder must be used for measuring rain fall. Ordinary lab measuring jars must not be used. In the market both 100 cm^2 and 200 cm^2 rain gauges are available along with their respective measuring cylinders. It is recommended to use 200 cm^2 rain gauge for our use.

vi. The charts of SRG, whether it is weekly or daily, it must be changed on the scheduled time. Similarly, if the SRG is with key clock, daily give key for 6 to 8 rounds and not more than that. If it is battery attached, check for battery leakage and change it accordingly. Regularly daily check water siphoning mechanism *i.e.*, the float and test the pen nip rises to 10mm for every full up. The pen may be cleaned on every day and inked

vii. Check the wooden frame of open pan evaporimeter for termite attack. Paint it with anti-termite paints

viii. Annual painting may be given with white to all instruments and fence

1.3.7 Instruments to be Placed in DSS

Hygrograph, thermograph, barograph and other automatic machines may be kept at this DSS.

The following attentions are required;

i. On every Monday, the chart has to be changed in the specified time for weekly chart machines. For daily chart machine, change it daily.
ii. Fix the chart with the help of the clip given
iii. Clean the pen and ink it daily or weekly as per the schedule

1.3.8 Registers to be Maintained

a. In the observatory

i. Pocket register
ii. Basic register

b. In the Office

i. Daily register
ii. Weekly register
iii. Monthly register
iv. Seasonal register
v. Annual register
vi. Computer programme for data storing and retrieval

1.3.9 Fixing Time for Taking Observation in a Proposed Observatory

The time of taking observations from an observatory is very important in terms of adding value to the collected data. The data are of national and international value if it is collected at Indian Standard time. Indian Standard Time (IST) is fixed on the meridian line passes through India *i.e.,* at Allahabad, which is called as reference longitude of India and this longitude point is 82.5°. Based on the local time as computed from IST, the data have to be collected. For an observatory at Coimbatore the local time is computed for an example.

The longitude of Coimbatore: 77 °E

The reference longitude: 82.5 °E

Difference between these two locations: 5.5°

Delay in sunrise for each 1°: 4 minutes

Delay in sunrise in Coimbatore; 5.5 × 4 = 22 minutes

Time of observation in the morning at this observatory; IST + delay in sun rise = 0700 + 22 = 0722 hours

Time of observation in the afternoon at this observatory; IST + delay in sun rise = 1400 + 22 = 1422 hours

Measurement of rainfall and evaporation at standard UTC time of 3.0 GMT = 3 + 5.30 = 0830 hrs.

Hence for Coimbatore the observation times are as follows:

0722; 0830; and 1422 hours.

1.3.10 Order of observation

Before sunrise

i. Grass minimum thermometer reading and taking to room for safety
ii. Dew deposit reading and taking to room for safety
iii. Setting sunshine card for recording bright sun shine hours
iv. Solar radiation reading and setting the instrument

0722 and 1422 hours

i. Maximum temperature from maximum thermometer and re - setting to dry bulb reading after morning reading only
ii. Minimum temperature from minimum thermometer and re-setting to dry bulb reading both morning and afternoon reading
iii. Wet and dry bulb reading for computing relative humidity
iv. Soil temperature from soil thermometer
v. Wind direction and wind velocity
vi. Cloud cover

0830 hours

i. Rain fall
ii. Evaporation

After sun set

i. Placing dew plate and grass minimum thermometer
ii. Removing burned sunshine card

1.3.11 Methodology for fixing True North

Different methods are available and three methods are given below.

i. Punjab method

In this method, one straight ranging rod (without any bent) may be taken and fix it in the proposed observatory in the previous day evening firmly at any point in the eastern side of the laid-out plot vertically. By early morning of the next day, when sunrises, mark the shadow of the ranging rod in the soil in the western side and mark it with white chalk powder. In the evening also mark the shadow of the ranging rod again in the soil in the eastern side. Draw a perpendicular line to the shadow line drawn in the direction of east west. The northern point of this perpendicular line indicates true north.

ii. IMD method

This method is suggested for installing wind vane by the India Meteorological Department (IMD). In India the magnetic North lies to the east or west of the true North at an angle of the order of 3°. This angle varies from place to place and should be accurately determined from a map showing lines of equal magnetic declination.

To determine magnetic North, fix a pin vertically very close to the place where wind vane is to be installed. Standing at a distance of about a metre from the place, approximately to the South of it, sight the pin through the prismatic campus. Move slightly till the 0° mark on the floating dial of the campus coincides with the pin. Remaining in the same position and still sighting the pin through campus, ask an assistant to fix another pin behind or in front of the first pin at a distance of about 30 cm. So that both lie in the same line of sight. Draw a straight line indicating the direction of magnetic North. To obtain the direction of true North, read off the declination of the place from the map showing lines of equal magnetic declination and mark the true North to the east or west of the magnetic North as the case may be. Now draw a true North-South line indicating its direction towards North by an arrow head. This can be done through plane table survey instrument.

We can also use ranging rods for this purpose against pin, but enough care is required.

iii. Equation of time method

This method also being recommended by IMD. Equation of time is available in online for longitudinal value of a place for 12 months and days of each month. We have to select a day and do the exercise to identify true North.

This can be calculated as detailed below;

Local mean time ± Equation of time + solar noon time *i.e.,* 12.00 hours.

The equation of time may be ± based on the location.

Example for Coimbatore;

The longitudinal value for Allahabad where from IST is computed, where the longitude is 82.5°E

The longitude of Coimbatore is 77°E

The difference is (82.5-77) = 5.5°

For one degree it is 4 minutes (1440 minutes of a day/360° = 4)

For 5.5° it is 22 minutes (5.5 × 4 = 22) and this local mean time (LMT)

Time for identifying shadow for fixing true North (July, 18) = 22 minutes (LMT)

Equation of time for July, 18 from table (-) 5.88 minutes

Solar noon = 12.00 hrs.

= 12.00 + 0.22 =12.22 minutes

= 12.22 minutes– 5.88 minutes (6 minutes, 28 seconds)

= 12 h, 15 minutes and 32 seconds.

Keep a ranging rod in the observatory site and mark the shadow by 12 h, 15 minutes and 32 seconds on July, 18 and this shadow mark will indicate the true North. If the day is cloudy repeat on the next day and fix the true North.

1.3.12 Inspection Report

The senior author as a professional man visited many observatories for their lay out, establishment, etc., and copy of selected report is given below for reference:

I. Establishment of Agro-met Observatory at Grapes Research Station at Annaimalayanpatti, Theni District, Tamil Nadu

The following details have been sent for action already.

Needed instruments and costs:

1. List of firms dealing Met instruments
2. IMD Specification Number of Met instruments
3. Material required to lay out observatory on 21.1.2015

Observatory type: A

a. Size of the observatory; 55 × 36 m (49.5 cents)
b. Shape of the observatory; Rectangular, length of 55m will run parallel to North-South.

1. Needed instruments for A class or Principal observatory and costs (varying with time)

S. No.	Name of the Instruments	Approximate maximum costs (INR) per piece	No of units required	Total cost (INR)	Remarks
1.	Sunshine recorder	32,000	1	32,000	To be purchased form firm
2.	Sunshine cards (3 types) for three years	1,500/year	3	4,500	To be purchased form firm
3.	Anemometer	7,000	1	7,000	To be purchased form firm
4.	Pole for anemometer	2,000	1	2,000	To be Purchased locally
5.	Wind vane	7,000	1	7,000	To be purchased form firm
6.	Pole for wind vane	2,000	1	2,000	To be Purchased locally
7.	Single Stevenson's Screen with teak legs	12,000	1	12,000	To be purchased form firm
8.	T max thermometer with one spare	2,000	2	4,000	To be purchased form firm

(Contd.)

S. No.	Name of the Instruments	Approximate maximum costs (INR) per piece	No of units required	Total cost (INR)	Remarks
9.	T min thermometer with one spare	2,000	2	4,000	To be purchased form firm
10.	Dry bulb thermometer with one spare	2,000	2	4,000	To be purchased form firm
11.	Wet bulb thermometer with one spare	2,000	2	4,000	To be purchased form firm
12.	Container for wet bulb, muslin cloth and threads etc.,	500	1	500	To be purchased locally
13.	Double Stevenson's Screen with teak legs	22,000	1	22,000	To be purchased form firm
14.	Weekly Thermo graph	22,000	1	22,000	To be purchased form firm
15.	Weekly charts for Thermo graph for three years	1,500/year	3	4,500	To be purchased form firm
16.	Hygrograph	22,000	1	22,000	To be purchased form firm
17.	Weekly chart for hygrograph	1,500/year	3	4,500	To be purchased from the firm
18.	Barograph	32,000	1	32,000	To be purchased form firm
19.	Weekly chart for barograph	1,500/year	3	4,500	To be purchased form firm
20.	FRP Ordinary rain gauge(200CM^2)	8,000	1	8,000	To be purchased form firm
21	Measuring cylinder for 200CM^2 with one spare	2,000	2	4,000	To be purchased form firm
22.	Self-recording rain gauge with key	25,000	1	25,000	To be purchased form firm
23.	Daily chart for self-recording rain gauge	1,500/year	3	4,500	To be purchased form firm
24.	Soil thermometers surface,5cm,10cm, 15cm,.20cm,30cm and 60cm with one each spare	2,000	$7 \times 2 = 14$	28,000	To be purchased form firm
25.	Stand for soil thermometers 2sets	1,000	1	1,000	To be purchased locally

S. No.	Name of the Instruments	Approximate maximum costs (INR) per piece	No of units required	Total cost (INR)	Remarks
26.	Pyranometer	1,50,000	1	1,50,000	To be purchased form firm
27.	Stand for pyranometer 5-6' ht.	1,000	1	1,000	To be purchased locally
28.	Grass minimum thermometer with one spare	2,000	2	4,000	To be purchased form firm
29.	Stand for grass minimum thermometer	300	1	300	To be purchased locally
30.	Class A open pan evaporimeter with all accessories copper bottom	25,000	1	25,000	To be purchased form firm
31.	Dew gauge (stand, four blocks and one album)	8,000	1	8,000	To be purchased form firm
32.	Assman psychrometer	5,000	1	5,000	To be purchased form firm
33.	Whirling psychrometer with spare	2,000	2	4,000	To be purchased form firm
34.	Cost of building materials (cement, sand, baby jelly, bricks) + masons cost + other assistance	—	—	1,00,000	—
35.	Fencing the observatory with post at 1m distance (5'height) with gate at eastern side = single eye wire mesh of 4'ht (Running m = 182M @₹1,200/m	—	—	2,18,000	—
36.	Optional instruments Lysimeters Thermopile sensing elements for short and long wave radiation	 1,00,000 1,50,000	 3 1	 3,00,000 1,50,000	

The cost given is ± 30 per cent of the cost of 2013-'14

2. List of firms dealing met instruments [Guidance letter of DDG (AM)]

Office of the Deputy Director General of Agricultural Meteorology
Meteorological office, Shivajinagar, Pune - 411 005

Instruction for Procurement of Meteorological Instruments

1. While placing orders for meteorological instruments or accessories, concerned firms may be asked to supply the instruments with ISI mark or IMD test certificate. A copy of the IMD test certificate may be forwarded to this office.
2. All thermometers, Stevenson screens, wind vane, rain measure glasses and evaporimeters must be routed through this office for test and standardization. After testing and standardization at this office, if found satisfactory, these instruments/equipments will be dispatched to you by the firms concerned. Free testing facility is provided only for those Agri -met observatories that are established in consultation with this office. Other parties may approach Deputy Director General of Meteorology (Surface instruments), Meteorological Office, Shivajinagar, Pune - 411 005 for testing and standardization of their instruments on payment basis.
3. Anemometer and autographic instruments (Self-recording rain gauge, thermograph, hair hygrograph) must be routed through Deputy Director General of Meteorology (Surface instruments), Meteorological office, Shivajinagar, Pune - 411 005. Amount charged by Deputy Director General of Meteorology (Surface instruments) for testing of instruments has to be borne by the concerned party/Agri-met observatory. However, repairing of these instruments will not be undertaken by the above-mentioned office, the same may be got done by the concerned firms.
4. Instruments/equipment which are not manufactured as per ISI/IMD specifications, will not be tested/standardized/installed by this office.
5. Copies of the correspondence made with the firms may be enclosed to this office. Regarding fragile instruments such as thermometers etc., it may be clarified whether the party would like to receive them duly insured or otherwise, they will be dispatched accordingly after testing.

Firms dealing with recording ink and charts for self - recording instruments.

1. Kedar, Enterprises, Krupashree, 31, Shukrwar Peth, Bhau Maharaj.Lane, Pune - 411 002 Ph. No. 020-24485797
2. Bells Graphic Recording Charts Ltd., S-297, Track Road, Kalina, Mumbai-400 029.
3. Chenna Corporation, A-6, Narayana Industrial Area, Phase II, New Delhi-110 028
4. Technograph Corporation, Sara Ramdwara, Paharganj, New Delhi
5. Larson & Toubro Ltd., L. T. House, Dougall Road, Ballard Estate, Mumbai - 400001.
6. Bharat Carbon and Ribbon Manufacturing Co. Ltd., Tollyganj, Kolkata
7. UTILE Equipments, 13, Jal Tarang, Prabhat Road Pune- 411 004. 020-22348652
8. Mahalakshmi Enterprises, 700/B, Sarnaik Colony, Old Washi, Naka Road, Shivajipeth, Kolhapur - 416012.

Note: While placing order for sunshine cards/autographic charts one sample may be sent to the concerned firm.

List of Firms Manufacturing Meteorological Instruments (as per IMD)

1.	M/s National Instruments Ltd 1/1 Raja Subodh Chanda Mullick Road Jadhavpur, Kolkatta-700032	Thermometers Sunshine Recorder Radiation Instruments
2.	Kinson Instruments Co. Railway Road, Jind Haryana - 126102.	Soil Thermometers All other Thermometers
3.	M/s New Technolab Instruments Poornam, Centre point, Opp. Gaikwad Classes kanherewadi, Near CBS Nashik-422 001 MS Ph/FAX 0253-2501437 Mobile: 9823153987 e-mail pkarandel@rediffmail.com www.newtechnolabinstruments.com	All Meteorological Instruments
4.	M/s Hindustan Clock Works 246/2. Shaniwar Peth, Pune–411030 020-24455462,020-24451223	Self-Recording Rain gauge, Thermograph Hair Hygrograph & Clock Drums for Recording Instruments
5.	Premier Instruments & Controls Ltd. PB No. 4209, Perianaiekenpalyam Coimbatore - 641020	Cup Counter Anemometer
6.	National Industrial Designers 2, Union Co-op Industrial Estate Rakhial, Ahmedabad - 380023 Ph. 079-2743521	Self-Recording Rain-gauge, Sunshine Recorder, Cup Counter Anemometer etc.
7.	Borosil Glass Works Ltd, Khanna Construction House, 44, Montana Abdul Gallar Road, Mumbai 400018.	Measuring Glass for use with Non Recording Rain gauge, Snow gauge.
8.	M/s Adept Recording Instruments Pvt. Ltd. Plot No 4, Survey No 17/1B Kothrud Industrial Estate Kothrud, Pune 411029.	Recorders
9.	Record Well D 290/B, Street No 11, Laxmi Nagar Vikas Marg, Delhi 110092.	Charts for Recording Instruments
10.	M/s Kedar Enterprises, Krupashree 31, Shukrwar Peth, Bhau Maharaj Lane Pune-411002. 020-24485797	Thermometers, Charts for Recording Instruments & Recording Ink, Sunshine cards
11.	M/s Chenna Corporation A- 6, Naraina Industrial Area, Phase II New Delhi.	Charts for Recording Instruments.

(Contd.)

12.	Technocrats Equipments, Sales & Service No 12, BFW Layout, Ganapathi Nagar Beenva III Phase Bangalore - 560058.	Self-Recording Rain gauge Non Recording Rain gauge
13.	M/s Technograph Corporation Rara Ramdwara, Paharganj, NewDelhi	Charts for Recording Instruments & Recording Ink
14.	M/s Computech India 403, Brookfield, Lokhandwala Complex Off Four, Bungalows, Andheri (W) Mumbai -400058	Charts for Recording Instruments.
15.	M/s Hydromet Instruments 1, Gujarati House, Opposite Victoria Garden, Bhadra Ahmedabad-380001 Ph. 079-25716157, 079-25716213	FRP Non recording Rain gauge Mechanical Wind vane, Sunshine Recorder, Stevenson Screens and Open Pan Evaporimeter
16.	M/s Ramkala Shop No.1, Prashant Apartment Narhe Road, Dhayary Phata Near Deshpande, Garden Society Pune–411041 Email: sanjay.r.karandikar@gmail.com Ph: 020-24336010 Mob. 09822422102	FRP Non recording Rain gauge, Rain measure Glass, Anemometer, Mechanical Wind vane, Sunshine Recorder, Stevenson Screens, Dew gauge set, Stands for Soil Thermometers, All Thermometers, Open Pan Evaporimeter, Muslin Cloth, Crochet Thread and Soil moisture Equipment
17.	M/s Utile Equipments 13, Jal Tarang, Prabhat Road Lane No 1, Pune - 411004 Ph. 020-22348652	All Thermometers, Sunshine Recorder and Radiation Instruments (Agent for M/s National Instruments Ltd. Kolkata)
18.	M/s Lawrence & Mayo (I) Ltd 3, Dr. Ambedkar Road Opp. Nehru Memorial Hall Pune - 411001	FRP Non recording Rain gauge, Self-Recording Rain gauge, Electrical Anemometer/Mechanical Wind vane, Sunshine Recorder, Whirling Psychrometer and Open Pan Evaporimeter,
19.	M/s Dynalab G-2 Bldg. C-3, Bramha Memories Bhosale Nagar, Pune - 411007 Ph. 020-25537109	FRP Non recording Rain gauge, Self-Recording Rain gauge, Electrical Anemometer/Mechanical Wind vane, Sunshine Recorder, Whirling Psychrometer and Open Pan Evaporimeter,
20.	Deputy Director General of Meteorology (Surface instruments), Shivaji Nagar Pune-5 (Please See web site imd.gov.in)	All instruments
21.	A.S. Industries, 206, Station Road Opp. Buckau Wolf, Pimpri Pune - 4110 I8 020-27120965,020-27127775	FRP Non Recording Rain gauge.

22.	M/s. Pharma Pack Corporation 23/171, Vardan, Behind Hindustan Bakery Opp. Elpro International Chinchwad, Pune 411033 020-27120965	FRP Non Recording Rain gauge
23.	A.Paul Instruments Co. Aravali Shopping Complex First Floor, Alaknanda New Delhi - 110019.	Soil Thermometers All other Thermometers and All Instruments
24.	Haryana Scientific Instruments 507/114, Bhavani Road, Jind (Haryana) 126102	Soil Thermometers All other Thermometers
25.	Universal Traders 20-Pachratan Bldg., 5th Floor Almeida Park Road, Bandra Mumbai-400050. PhoneNo, 26427686/9820510231 Email: vdusija2003@yahoo.com	Assamann Psychrometer
26.	Elron Instruments Co. Pvt. Ltd. B-2, Satkar Building, 79-80 Nehru Place, New Delhi - 110019	Soil Moisture Equipments
27.	M/s Toshbro Ltd. Hyderabad Groondale Building First Floor, Next To Green Park Hotel Ameer Peth, Hyderabad 50016	Soil Moisture Equipments
28.	M/s Mahakakshmi Enterprises 700/B, Sarnaik Colony, Old Washi Naka Road, Shivajipeth, Kolhapur-416012 09372479079/09850084686.	All types of recording graphs, Sunshine cards & Meteorological instruments
29.	J.R. Mulilick & Co. Pvt. Ltd. 72-A, Virvwani Industrial Estate Goregaon (E), Mumbai-400063 Ph. No. 09821070255 02228740385, FAX 022 28745781 e-mail maheshmullick@jrm.in	Screen Thermometers

Out of these firms with the introduction of AWS, many companies are not interested to offer quotation and as well as for supply. Hence request may be given to all firms listed.

3. IMD Specification Number

For calling quotation and as well as for supply order the following IMD specific number must be quoted along with meteorological Instruments

All instruments must be obtained through Deputy Director General Meteorology (Surface Instruments) Shivaji Nagar, Pune-411005 from the firm by the buyer to check for quality. This may be quoted as one among the conditions.

S. No.	Name of the instruments	Specification number
1.	Thermometers (WB, DB, Tmax, Tmin)	IS9681/1970
2.	All soil thermometers	IS6692/1972
3.	Single Stevenson's screen and Double Stevenson's screen	IS5948/1970
4.	Anemometer	IS5912/1970
5.	Wind vane	IS5799/1970
6.	FRP rain gauge/ordinary	IS5225/1969
7.	Self-recording rain gauge with key	IS5235/1969
8.	Rain measuring cylinder (200 cm^2)	IS4849/1968
9.	Class A open pan evaporimeter zinc/copper bottom	IS5973/1970
10.	Sun shine recorder	IS7243/1974
11.	Whirling psychrometer	IS 5946/1970
12.	Assman psychrometer	IS 6805/1973
13.	Thermograph (bi metallic)	IS 5901/1970
14.	Hair Hygrometer	IS 5900/1970
15.	Other instruments	Indicate as per IMD approval

Note: The number and year given under specification varies since every year IMD gives specification. Hence if the firm supply with IMD approval number of different, we can agree for purchase. But without number, purchase is prohibited.

4. Materials required on 21.1.2015 at the time of layout

1. Wooden cross staff (can be available with VAO's office)
2. Coir/Nylon rope (300 m)
3. Chalk powder white/grinded white stone powder (5 kg)
4. Wooden hammers (2)
5. Wooden/iron rods of 3 'ht (70 no's)
6. Site area of three to four locations within the farm (each 50 cents)
7. Metallic tape for measurement-50m

On 21.1.2015 visited the Grape Research Station, Annaimalayanpatti, Uthamapalayam, Theni district along with Professor and Head, ACRC, TNAU, Coimbatore-3. Started by 5.30 am in the taxi and reached the farm by 12 noon and started selection of site for A class agro met observatory at the farm of Grape Research Station. Considering the hills on the east and building on the north and proposed buildings and area to be brought under vine farming, one site at the south east side of the farm was selected. The soil is deep red and though the slope is more than 2 per cent one land with uniform plain was selected and lay out was started.

Dr. Parthiban, the Prof and Head, GRS and Dr. Subbiah, AP and their staff Mr. Mani AO participated in this process. As per the layout, 55 m along NS and 36 m along EW was measured in the selected site with due caution to maintain 90° at the intersect points. Fifteen boxes of 9 × 8.66 m were laid out. Over all, border of 5m on both side's perpendicular to EW and 5 m border perpendicular to NS were given as per the rules and regulations.

II. Met Instrument Establishment report of JKKM College of Agricultural Sciences and Guidance Tips for the Maintenance of Agro met instrument

1. General

Prof. Dr. T. N. Balasubramanian and Mr. Veerabadiran of ACRC did jointly install meteorological instruments at the B class observatory of JKKM college of Agricultural Sciences, T. N. Palayam from 8.10.15 to 10.10.15 for the educational purpose of under graduate students and also for interpreting the results for future research.

2. Instrument Status Report

Before installation, all the instruments were verified for their suitability for installation and the comments are as follows.

(a) The thermograph and hygrograph were purchased individually with daily chart arrangement against the standard availability of thermo- hygrograph (both together) with weekly chart arrangement. In the case of daily chart instrument, we have to change the chart daily and this involves hard task and also, we require more charts per year. Further these two instruments have been purchased on battery support against key pattern. The battery has to be monitored on daily basis for its leakage against damage to the instrument. With the availability of these two thermograph and hygrograph instruments, they have been assembled and positioned in the Double Stevenson's Screen. Daily, the charts have to be changed at 7.20 am and check for ink availability in the pen. If charts are not changed in time, the nib may rub with central metal pin and get damaged.

(b) The Wind vane tail plate was found missing at the time of installation (it was told that it was missed with frequent transfer from one place to another place), though, all parts are in position. One aluminum tail of 30cm length with 25 cm width was purchased from TN. Palayam and fitted. This is fitted on temporary basis. The company may be contacted for fresh piece and original may be fitted.

(c) The anemometer head pin was missing again due to frequent shifting from one place to other place and this problem was eliminated by fitting with alternate nuts. Mr. Mohan Kumar will take action to buy new one from the available source and get fitted.

(d) With frequent transfer from one floor to another floor, the open pan evaporimeter drum had leakage when it was tested with water. However, this problem was minimized by using M. seal. Mr. Sarath Kumar and Mr. Mohan Kumar have been requested to monitor further leakage and if leakage is found subsequently, they have been requested to use M.seal.

(e) All the parts of the Sun Shine Recorder were good and temporarily positioned in the 10' pillar after assembling and the card burning was watched on 10.10.15 and further, the burning has to be watched for right direction and if it is ok, we can plaster the marble of the recorder with the base of the top of the pillar (Mr. Sarath Kumar and Mr. Mohan Kumar) with cement.

(f) Both Single Stevenson's Screen (SSS) and Double Stevenson's Screen (DSS) were ok and installed. The maximum, minimum, dry and wet bulb thermometers were positioned in the SSS. Among the four thermometers, the minimum thermometer's alcohol column is found to be weak and it may break at any time. But this can be brought back by dipping the bulb of the thermometer in hot water of 80°C, simultaneously by keeping wet cloth over the other end of the thermometer (chamber end).

(g) There was no problem with non-recording rain gauge and self-recording rain gauge and both were installed. The charts for self-recording rain gauge have to be changed daily by 8.30 am.

The role of Mr. Mohan Kumar, Mr. Sarath Kumar and Mr. Gopal is commendable to give 100 per cent back support to establish the met-observatory. They deserve all appreciation from our team.

3. Guidance

The observatory and met-instruments in the observatory have to be maintained well for collecting precious weather data and for that guidance tips are given below:

- Based on the longitudinal position of TN. Palayam, the observations are to be taken as follows
- 7.20 am; Maximum, minimum, dry and wet bulb (For RH) temperature, wind speed, wind direction
- 2.20 pm; Maximum, minimum, dry and wet bulb (For RH) temperature, wind speed, wind direction
- 8.30 am; rainfall and evaporation
- Concerned day sunshine card has to be changed daily either in the previous day of 6.30 pm or 5.30 am of the current day.
- Re-setting of maximum thermometer has to be done by 7.20 am after recording the value (demonstrations given already). The reading has to be set to the value of dry bulb reading. But minimum thermometer has to be set twice (7.20 am and 2.20 pm) by slightly raising the chamber end (Demonstrations given).
- Daily evaporation is to be taken based on the four rules (no rainfall situation, drizzling, heavy rainfall under no overflow and heavy rainfall under over flow).
- One (+12) student may be trained for taking observations daily on the scheduled time and documenting. He must be responsible for the maintenance of met instrument and keeping met records. He must be under the meticulous control of either Dr. Sarath Kumar or Mr. Mohan Kumar as nominated by the Dean. The selected observer can be trained at ACRC, TNAU after obtaining permission from Prof and Head, ACRC, TNAU, Coimbatore.
- One 1000 mb RH table book may be obtained from DDG (Agrl. Meteorology), Shivaji Nagar, Pune.5 or DG, IMD Mausam Bhavan, Lodi Road, New-Delhi on cost. Till

then, one xerox copy may be taken from the RH table book of ACRC, Coimbatore after obtaining permission from Prof. and Head.

- All daily recorded data may be documented in the register and based on the accumulation of data over years, separate register may be opened for weekly, monthly, seasonal and yearly data set.
- All charts of thermograph, hygrograph and self-recording for daily use will come for 10 months only and hence new charts for three years may be purchased from standard company. The quality of the charts must be in way that its quality of the paper used for chart should not imbibe ink when it is fitted in the instrument. Similarly, high quality sun shine card must be purchased for three years.
- Spare thermometers one each or two in maximum, minimum, dry, wet bulb and rain measuring cylinder for 200cm^2 may be purchased from Standard company (May be from Ramkala) and kept ready for emergency use.
- Annually all the instruments, fence and posts must be painted with white paint.
- One 4*3' board may be kept at the outside of the observatory on the left side with the following information. Name of the observatory, Latitude, longitude, mean sea level, date of lay out. date of installation of met instrument, date of regular operation, daily information with date, rainfall received in mm, maximum temperature, minimum temperature in °C, wind speed (kmph), wind direction (in degrees), evaporation (mm) etc.
- Link may be established with the ACRC, TNAU, DDG (ag-met), Pune and DDG, RMC Nungambakkam, Chennai for National level action.
- For wet bulb, muslin cloth and threads must be changed very carefully once in 20 to 30 days. The distilled water must be used in the ink bottle kept below the bulb of the wet bulb thermometer.
- The anemometer and wind vane must be lubricated with Singer oil (3in one oil) once in 20–30 days.
- Open pan evaporimeter may be cleaned once in 15–20 days against dust and mass accumulation.
- Individual label may be given with small board for each instrument
- As indicated in the site selection report, trees pointed out near the observatory may be removed forth with to get precious BSSH data.

Coimbatore T. N. Balasubramanian
11.10.15

III. Met Instrument Establishment report of JSA College of Agriculture and Technology, Thittakudi and Guidance Tips for the Maintenance of Agro met instrument

1. General

Prof. Dr. T. N. Balasubramanian assisted by Mr. C. Veerabadiran, Observer (Rtd.) of ACRC did visit JSA COA&T from 15.9.16 to 17.9.16 and installed meteorological instruments at the B class observatory of JSA COA&T, Thittakudi for educational purpose of under graduate students and also for interpreting the results for future research to be done.

2. Instruments Report

Before installation, all the instruments were verified for their suitability for installation during last visit on 19.7.16 and the comments are as follows.

(a) The thermograph and hygrograph were purchased individually with weekly chart. They work very well. With the availability these thermograph and hygrograph instruments, they have been assembled and positioned in the Double Stevenson's Screen. Weekly on every Monday, the charts have to be changed at 7.14am and check for ink availability in the pen. If charts are not changed in time, the nib may rub with central metal pin and get damaged and the instrument will go out of order over weeks.

Both Single Stevenson's Screen (SSS) and Double Stevenson's Screens (DSS) are ok and installed. The maximum, minimum, dry and wet bulb thermometers have been positioned in the SSS. In the DSS, thermograph, hygrograph, disc RH meter and aneroid barometer have been fitted.

(b) The Wind vane is ok and fitted.

(c) The anemometer is ok and fitted.

(d) The open pan evaporimeter is ok and installed.

(e) All the parts of the Sun Shine Recorder are good and hence positioned at 11°5″ N in the 10′ height pillar after assembling and the card burning was watched on 15.9.16 to 17.9.16 and the burning is in order and hence pasted with cement at the base of the recorder.

(f) There is no problem with non-recording rain gauge and self-recording rain gauge and both have been installed. The charts for self-recording rain gauge have to be changed daily by 8.30 am.

(g) Since all the four soil thermometers are ok, they have been installed.

Though all the instruments have been installed, urgent action may be taken to put fence around the observatory, otherwise animals by trespass, may damage the installed instruments. After fencing, quality white paints may be given to the pillar of sunshine recorder, posts of both wind vane and anemometer (not instruments), stand of soil thermometers (take care not to damage the fragile glass thermometers), outside of open pan evaporimeter, single Stevenson's screen and double Stevenson's screen.

The role of Mr. Bojan and Mr., Venkatesh, AP (Agro) is very important to maintain instruments installed at the observatory in future.

I appreciate the Dean, Civil engineers, Mr. Bojan and Mr., Venkatesh, AP (Agron) and other teaching staff especially Dr. Mohan, AP (Animal science) and II-year students for their dedicated involvement to install the instruments in time during the processes of installation. I also appreciate the first-year students for their NSS work done at the observatory.

3. Guidance

The observatory and met- instruments in the observatory have to be maintained well for collecting precious weather data and for that guidance tips are given below:

- Based on the longitudinal position of College (79°E) the observations are to be taken as follows
- 7.14 am; Maximum, minimum, dry and wet bulb (For RH) temperature, wind speed, wind direction, soil temperature
- 2.14 pm; Maximum, minimum, dry and wet bulb (For RH) temperature, wind speed, wind direction, soil temperature
- 8.30 am; Rainfall and evaporation
- Concerned day sunshine card has to be changed daily either in the previous day of 6.30 pm or 5.30 am of the current day.
- Re setting of maximum thermometer has to be done by 7.14 am after recording the value (demonstrations given already). The reading has to be set to the value of dry bulb reading. But minimum thermometer has to be set twice (7.14 am and 2.14 pm) by slightly raising the chamber end (Demonstrations given).
- Daily evaporation is to be taken based on the four rules (no rainfall situation, drizzling, heavy rainfall under no overflow and heavy rainfall under over flow).
- One +12 student/any other staff may be trained for taking observations daily on the scheduled time and documenting. He must be responsible for the maintenance of met instrument and keeping met records also. He must be under the meticulous control of either Mr. Bojan and Mr., Venkatesh as nominated by the Dean. The selected observer can be trained at ACRC, TNAU after obtaining permission from Prof and Head, ACRC, TNAU, Coimbatore,
- One 1000 mb RH table book may be obtained from DDG (Agrl.Meteorology), Shivaji Nagar, Pune-5 or DG, IMD Mausam Bhavan, Lodi Road, New Delhi on cost. Till then, one xerox copy may be taken from the RH table book of ACRC, Coimbatore after obtaining permission from Prof. and Head.
- All daily recorded data may be documented in the register and based on the accumulation of data over years, separate register may be opened for weekly, monthly, seasonal and yearly data set.
- All charts of thermograph, hygrograph, sun shine recorder and self-recording will come for one year only and hence new charts for three years may be purchased from standard company. The quality of the charts must be in way that its quality of the paper used for chart should not imbibe ink when it is fitted in the instrument (company address given).
- Spare thermometers one each or two in maximum, minimum, dry, wet bulb and rain measuring cylinder for 200 cm^2 may be purchased from standard company and kept ready for emergency use.
- Annually all the instruments, fence and posts must be painted with white paint.

- One 4*3′ board may be kept at the outside of the observatory on the left side with the following information. Name of the observatory, Latitude, longitude, mean sea level, date of lay out, date of installation of met instrument, daily information with date, rainfall received in mm, maximum temperature, minimum temperature in °C, wind speed (kmph), wind direction (in degrees), evaporation (mm) etc.,
- Link may be established with the ACRC, TNAU, DDG (ag-met), Pune and DDG, RMC, Nungambakkam, Chennai for National level action by providing data.
- For wet bulb, muslin cloth and threads must be changed very carefully once in 20–30 days. The distilled water must be used in the ink bottle kept below the bulb of the wet bulb thermometer.
- The anemometer and wind vane must be lubricated with Singer oil (3 in 1 oil) once in 20 to 30 days.
- Open pan evaporimeter may be cleaned once in 15–20 days against dust and mass accumulation. The wooden platform may be protected from termites by spraying ant-termite pesticide.
- Individual label may be given with small board for each instrument.
- One permanent room with hollow cement block may be constructed on the NW direction of the observatory, may be 200 m away from the observatory to keep records and aids. In future, building and trees must not be constructed or planted around the observatory.
- One dew gauge and grass minimum thermometer may be purchased and installed (company address given).

Coimbatore
18.09.17

T. N. Balasubramanian

2

Agro-Climatic Analysis

The climate, the past weather event that existed over larger region and covering long years of more than 30 years is very important for the emergence and successful sustenance of an organism including human kind. Normally scientists talk about environment and this includes not only the soils but also the climate of the particular region. Based on this, during 1989, the Planning Commission of India divided the entire India in to 15 agro-climatic zones based on physiography and climate. This division is only for planning purposes. Subsequently under NARP, the Indian Council of Agricultural Research (ICAR), New Delhi in consultation with the Planning Commission, requested all State Agricultural Universities (SAU) to divide their State in to different sub climatic zones considering soils, temperature and other weather parameters preciously with in the major agroclimatic zone of India in which the concerned State falls. Accordingly, each SAU has divided their State into different agro-climatic sub-zones and this resulted in 129 agro-climatic sub zones in India and limited to 127 and these all fall within 15 agro climatic zones of India as done by the Planning Commission of India.

The National Bureau of Soil Survey and Land Use Planning (NBSS&LUP), Nagpur not satisfied with this division, considering soils, temperature, rainfall, vegetation and all factors responsible for ecosystem (bio-climate, length of growing period, moisture index) divided the whole India in to 20 agro-eco regions by 1992 and revised preciously again by 2015 (Mandal *et al.,* 2016) and the revised approach is being considered as scientific tool and taken for micro-level planning in India now and those are discussed at this chapter.

2.1 Agro-climatic Zones and Agro-ecological Regions of India

The Planning Commission of India, as a result of mid-term appraisal of the planning targets of VII Plan (1985–1990) divided the entire country in to 15 broad agro-climatic zones based on physiography and climate. The emphasis was on the development of resources and their optimum utilization in a sustainable manner within the frame work of resource constraints and potentials of each region. This is a departure from the previous practice of planning with the focus on specific crops and fertilisers, treating the State as a unit of planning (Sehgal *et al.,* 1992).

This approach involves integration of homogenous climate locations in to one region for multipurpose activities including development. The benefits from this approach are somewhat interesting and those important ones are; opportunity to increase agricultural productivity, to understand the problems related to agriculture, opportunity to introduce new crops and variety etc. The 15 agro climatic zones are

S. No.	Name of the agro-climatic zones	States and parts included	Special features
1.	Western Himalayan region	Jammu and Kashmir, Himachal Pradesh and hills of UP	Lower productivity
2.	Eastern Himalayan region	Sikkim and Darjeeling hills, Arunachala Pradesh, Meghalaya, Nagaland, Manipur, Tripura, Mizoram, Assam, Jalpaiguri and Coochbehar districts of West Bengal	High rain- fall and high forest covered area
3.	Lower Gangetic plains	This is in West Bengal consists of four sub-regions namely, Basind plains, central alluvial plains, alluvial coastal plains and Rark plains	Floods prone area
4.	Middle Gangetic plains	12 districts of UP and 27 districts of Bihar plains	Cropping intensity is around 140 %. Flood prone area
5.	Upper Gangetic plains	There are 32 districts divided in to three sub zones	Irrigation intensity is 131%, cropping intensity is 144%, Problem soils exists
6.	Trans Gangetic Plains	Punjab, Haryana, Union territories Delhi, Chandigarh and Sriganganagar in Rajasthan	Potential area for agriculture
7.	Eastern plateau and hills	Madhya Pradesh eastern hills and Orissa islands, Northern Orissa and Madhya Pradesh eastern hills and plateau, Chotanagpur and eastern hills, Chotanagpur south and West Bengal hills, Chhattisgarh and south western Orissa hills	Topography is undulating
8.	Central plateau and hills	46 districts from Madhya Pradesh, Uttar Pradesh and Rajasthan	1/3rd of the land is not available for cultivation
9.	Western plateau and hills	Major parts of Maharashtra, parts of Madhya Pradesh, and parts of Rajasthan	Fifty per cent of National sorghum production comes from this zone
10.	Southern plateau and hills	35 districts from Andhra Pradesh Karnataka and Tamil Nadu	Major comes under rain fed/dry land

S. No.	Name of the agro-climatic zones	States and parts included	Special features
11.	East coast plains and Ghats	Orissa coast, Andhra north coast, North coastal of Tamil Nadu, Thanjavur of Tamil Nadu, south coastal Tamil Nadu	20 per cent National rice production comes from this zone
12.	West coast plains and Ghats	Running along the west coast covering Tamil Nadu, Kerala, Karnataka Maharashtra and Goa	Suitable for rain water management, minor irrigation developments, crop diversification and fishing development
13.	Gujarat plains and hills	19 districts from Gujarat	Rain fed farming
14.	Western dry region	Nine districts from Rajasthan	Hot sandy desert
15.	Islands region	Andaman and Nicobar Islands, Lakshadweep	Lie in the equatorial line

As per ICAR direction, these 15 agro-climatic zones have been divided in to initially 129 number of sub- agro-climatic zones and limited to 127 latter and those are as follows:

State	No of sub agro-climatic zones
Andhra Pradesh (undivided)	7
Assam	6
Bihar	6
Gujarat	8
Haryana	2
Himachal Pradesh	4
Jammu and Kashmir	4
Karnataka	10
Kerala	8
Madhya Pradesh	12
Maharashtra	9
North east	6
Orissa	10
Punjab	5
Rajasthan	9
Tamil Nadu	7
UP	10
WB	6

Climatologically the IMD has established it's Agro-Met Field Units in 127 agro-climatic sub zones of India to disseminate its medium range weather forecast along with agro advisories to the farmers. The Agro-Met Field Units have been attached with either ICAR, SAU, Private and KVK's. Initially this work was established by 1991.

Not satisfied with this division, the NBSS&LUP, Nagpur has divided this country India in to 20 agro - ecological zones/regions, which are relevant for any use. By combining 19 physiographic and 16 broad soil types, 24 widely distributed soil-physiographic regions have been selected. Superimposing dominant 9 bio-climatic cum length of growing periods over these dominant 24 regions, finally they got 20 agro-ecological regions. An agro-ecological region (AER) is the land unit on earth's surface, carved out of agroclimatic region by superimposing climate on landforms and soils, which are the modifiers of climate and length of growing period. The AER is designed to address the issues related to agricultural production through agro-technologies. The agro-ecological region information of 1992 (Table 2.1 and 2.2) is being widely used by scientists and planners for regional level agricultural land use planning. Nevertheless, it has been found limiting due to the use of a relatively small-scale soil map (1:7 million) and climatic data on a coarser resolution. Therefore, AER map of the country has been re-delineated using soil resource data acquired from 1:1 million scale soil map and climatic resource database from 600 weather stations across the country and also laying greater emphasis on soil quality parameters. Thus, the soil quality-based AER map developed during 2015 opens a new vista of research to link the potential of natural resources and crop performance for better and realistic regional agricultural land use plan.

Table 2.1. AER developed during 1992 by NBSS and LUP, Nagpur

AER (1992)	Description
1	Western Himalayas. cold arid eco region, with shallow skeletal soils & length of Growing Period (GP) <90 days
2	Western Plain. Kachchh and part of Kathiawar Peninsula. hot arid eco region, with desert & saline soils & GP <90 days
3	Deccan Plateau. hot arid eco region, with red & black soils & GP <90 days
4	Northern Plain and Central Highlands including Aravalli's, hot semi-arid eco region, with alluvium-derived soils & GP 90-150 days
5	Central (Malwa) Highlands. Gujarat Plains & Kathiawar Peninsula. hot semi-arid eco region, with medium & deep black soils. & GP 90-150 days
6	Deccan Plateau. hot semi-arid eco region, with shallow and medium (with inclusion of deep) black soils. & GP 90-150 days
7	Deccan (Telangana) Plateau and Eastern Ghats. hot semi-arid ecoregion, with red & black soils & GP 90-150 days
8	Eastern Ghats. TN uplands and Deccan (Karnataka) Plateau. hot semi-arid eco region with red loamy soils & GP 90-150 days

AER (1992)	Description
9	Northern Plain. hot sub humid (dry) eco region, with alluvium -derived soils & GP 150–180 days
10	Central Highlands (Malva. Bundelkhand & Eastern Satpura), hot sub humid eco region, with black and red soils. & GP 150–180 (to 210) days
11	Eastern Plateau (Chhattisgarh), hot sub humid eco region, with red & yellow soils, & GP 150–180 days
12	Eastern (Chotanagpur) Plateau and Eastern Ghats. Hot sub humid eco region with red & lateritic soils. & GP 150–180 (to 210) days
13	Eastern Plain. hot sub humid (moist) eco region, with alluvium-derived soils & GP 180–210 days
14	Western Himalayas. Warm sub humid (to humid with "inclusion of per humid) eco region with brown forest and podzolic soils. & GP 180–210 + days
15	Bengal and Assam Plain. hot sub humid (moist) to humid (inclusion of per humid) eco region, with alluvium-derived soils & GP 210 + days
16	Eastern Himalayas. warm per humid eco region, with brown and red hill soils. & GP 210 + days
17	North-eastern Hills (Purvachal), warm per humid eco region, with red and lateritic soils & GP 210 + days
18	Eastern Coastal Plain. hot sub humid to semi-arid eco region, with coastal alluvium -derived soils & GP 90-210 + days
19	Western Ghats & Coastal Plain. hot humid-per humid eco region, with red. lateritic and alluvium-derived soils. & GP 210 + days
20	Islands of Andaman-Nicobar and Lakshadweep hot humid to per humid island ecoregion, with red loamy and sandy soils. & GP 210 + days

Table 2.2. Revised AER developed during 2015 by NBSS and LUP, Nagpur

Ecosystem	AER. regions (2015)	Description	Area in m. ha (%)
Arid	1 A13E1	Western Himalayas, cold arid eco-region with shallow skeletal soils and LGP <90 days	18.5 (5.6)
	2 M12E1(2)	Western plains and Kutch Peninsula, hot arid ecoregion with desert saline soils and LGP <90 days (inclusion of 90-120 days)	24.7 (7.5)
	3 K5E2	Deccan plateau, hot arid eco-region with mixed red and black soils and LGP 90-120 days	1.8 (0.6)

(Contd.)

Ecosystem	AER. regions (2015)	Description	Area in m. ha (%)
Semi-arid	4 N10D(C)23(4)	Northern Plain (Upper Gangetic) semiarid to sub humid eco-region with coarse loamy alluvial soils (sandy loam to sandy clay loam) and LGP 90–150 and 150+ days	7.0 (2.1)
	5 N9D3(2)	Northern Plain (Rajasthan Upland and Gujarat plains) hot semiarid eco-region with old alluvial soils and LGP 120–150 days (inclusion of 90–120 days	22.9 (7.0)
	6 N8D(C)4(3)	Northern Plain (Middle Gangetic Plain) hot semiarid to sub humid eco-region with alluvial and Tarai soils and LGP 150–180 days (inclusion of 120–150 days)	21.7 (6.6)
	7 K6D3(4-2)	Deccan Plateau (Malwa Plateau, Gujarat plains and Kathiawar peninsula) hot, semiarid eco-region with moderately deep black soils (inclusion of shallow soils) and LGP 120–150 days (inclusion of 150–180 days and 90–120 days)	49.0 (14.9)
	8 K5D4-3	Deccan Plateau, hot semiarid eco-region with mixed red and black soils and LGP 150–180 and 120–150 days	25.9 (7.9)
	9 KID4(5)	Deccan Plateau, hot semiarid eco-region with red loamy soils and LGP 150–180 (inclusion of 180–210 days)	18.7 (5.7)
Sub humid eco system	10 J6C4(3)	Eastern Plateau (Satpura range and Mahanadi Basin) hot sub- humid eco-region with moderately deep black soils (inclusion red soils) and LGP 150–180 days (inclusion of 120–150 days)	15.2 (4.6)
Sub humid eco system	11 J3C4(3)	Eastern Plateau (Bundelkhand Upland) hot sub-humid eco-region with red and yellow soils and LGP 150–180 days (inclusion of 120–150 days)	21.7 (6.6)

Ecosystem	AER. regions (2015)	Description	Area in m. ha (%)
Sub humid eco system	12 J2C5(4)	Eastern Plateau, hot sub-humid eco-region with red and lateritic soils and LGP 180–210+ days (inclusion of 150–180 days)	26.6 (8.1)
	13 N8C5(4)	Northern Plains (Lower Gangetic) hot, sub-humid ecoregion with alluvial soils (calcareous) and LGP 180–210 days (inclusion 150–180 days)	5.3 (1.6)
	14 A13C(B)3(6)	Western Himalayas, warm to hot sub-humid to humid eco-region sub montane shallow and skeletal hill soils and LGP 120–150 and >210 days	15.6 (4.7)
	15 O8C5(6)	Bengal basin, hot, sub-humid eco-region with loamy to clayey alluvial soils and LGP 180–210 days (inclusion of >210 days)	5.2 (1.6)
Humid/per humid eco system	16 Q8B(A)6	Assam and North Bengal Plain, warm humid to per humid eco-region with alluvial soils and LGP >210 days	9.7 (3.0)
	17 C13-1A6	Eastern Himalayas, warm per humid eco-region with shallow and skeletal red soils and LGP >210 days	8.7 (2.6)
	18 D3A6	North Eastern hills (Purvanchal), warm per humid eco-region with red and yellow soils and LGP >210 days	10.0 (3.0)
Coastal eco system	19 ST7C(B/A)4-5	Eastern Coastal Plains and Island of Andaman and Nicobar, hot sub-humid (with humid to per-humid inclusion) transitional zone with coastal and deltaic alluvial soils and LGP 150—210 + days	11.3 (3.5)
	20 E2B-A(C)6(5)	Western Ghats (Coastal Plains and Western Hills) hot humid to Per humid (inclusion of sub-humid eco-region) with red and lateritic and alluvium derived soils and LGP >210 days (inclusion of 180–210 days)	9.5 (2.9)

The division of ecosystem is drawn from the threshold level of Moisture Index (Im) as reported by Thornthwaite and Mather (1955*). As per them, the Im can be computed from the following formula;

$$Im = 100(P - PE/PE) \text{ or } 100(P/PE - 1)$$

Where, P = Precipitation in mm

PE = Potential evapotranspiration in mm

Based on the output Thornthwaite and Mather (1955*) made a wonderful climate classification across the world and still it is valid.

Climate symbol	Climate type	Values of Im (moisture index)
A	Per-humid	100 and >100
B_4	Humid	80–100
B_3	Humid	60–80
B_2	Humid	40–60
B_1	Humid	20–40
C_2	Moist sub-humid	0–20
C_1	Dry sub-humid	(–) 33.3 to 0
D	Semi-arid	(–) 66.7 to (–)33.3
E	Arid	(–) 100 to (–) 66.7

2.2 Agricultural Climatological Characterization

This area needs higher attention, if India wants to increase food productivity per unit area, considering the decrease in cultivable areas over years. In India based on 2019 enumeration, there are 28 States and 9 territories totaling to 37. There are 732 districts and out of which, 5500 blocks are available. Now a question arises whether State level climate characterization is useful than district level than block level. Micro level planning is better than macro level planning. It is very difficult to have village level climate characterization presently though it is very precious. Since the medium range weather forecast will be given at the block level in the coming years, let us consider to have block level climate characterization. For doing this at block level, three products seem to be very important and they are block level climate characterization, establishing block level night climate school and nomination of block level climate manager/block.

2.2.1 Block level characterization

Climate characterization at block level can be done by collecting climate data for longer period of 30 years if possible and mean arrived and go for the analysis of the following climate characterization agenda (Table 2.3). The formulae given under Equation and Formulae in Annexure-III also may be used.

Table 2.3 Processed climate information required for block level

S. No.	Name of the climate characterization	Remarks
1.	Weekly rainfall data and graph	
2.	Weekly T. Max temperature (°C) data and graph	
3.	Weekly T. Min temperature (°C) data and graph	
4.	Weekly mean temperature (°C) data and graph	
5.	Relative temperature disparity	
6.	Growing degree days for important 3–4 crops	
7.	Conditional probability and initial probability for seasonal rainfall	
8.	Markov chain analysis	
9.	Moisture index, aridity index and humidity index	
10.	Length of growing periods for upland crops	
11.	Weekly morning RH (%) and graph	
12.	Weekly evening RH (%) and graph	
13.	Weekly mean RH (%) and graph	
14.	Weekly mean wind speed data and graph	
15.	Weekly mean wind direction data in degrees	
16.	Return period for extreme weather events	
20.	Weekly mean thermal humidity index data	
21.	Weather forecast information on weekly basis along with agro- advisories	
22.	Traditional knowledge on climate and crop management	
23.	Developed climate thumb rules by the villagers	
24.	Monsoon rainfall onset and withdrawal dates	

These documents must be open to all at any time. These may be kept in common place and updated once in three years.

2.2.2 Establishing Night School on Climate Literacy at Block Level

One-night climate school/block can be established in Government school during early night time for an hour. This is an optional school and not compulsory. Villagers may discuss their weather anticipation and all problems related to climate and weather. Once in a month a climate specialist from nearby village may be invited for one to one interaction. Joint venture to solve their climate problems is the moto of the establishing block level night school on climate. In the night school village level or block level climate pre and post seasonal workshop can be conducted. Earlier, village level night school was established in Indonesia country. The present position is not known.

2.2.3 Nomination of Block Level Climate Manager

A man with above 60 years with wisdom and visionary mode may be nominated as block level climate manager. He is the coordinator for all climate activities at block level. He must be a social worker and he must be away from politics and caste-based organization. His duties are exhaustive but some are given;

- Organizing pre and post seasonal workshop
- Organizing meeting to prepare contingency plan for floods and drought
- Weather based agro-advisory preparation and discussion
- Pest and disease surveillance
- Crop planning
- Disaster management preparations at block level
- He is the bridge between officials and villagers
- He is responsible for livestock and crop management activities based on weather forecasts information received (long range, medium range, short range etc.).
- Taking lead for discussion at block level night school on climate

3

Crop Micrometeorology

The subject on micrometeorology deals with small scale meteorology of near ground surface with its related elements of both physical and biological processes. When this micrometeorology is integrated for crop production purposes, then it becomes crop micrometeorology, which is very important in the science of agricultural climatology. This crop micrometeorology gets varied with the architecture of the crops grown, available soil moisture, agronomical practices followed, soil texture, topography of the area, the nature of climate and weather that prevails etc. The soil surface heat, exerts a profound frictional effect on wind movements and hence it is different from the air layers at higher levels. These special features constitute the microclimate, which is modified compared to a flat bare surface by factors like topography, nature of soil surface, presence of vegetation and type of vegetative cover. The presence of crop canopy leads to distinct differences in the thermal and humidity regimes of the soil, the air layers immediately above it and the distribution of radiation and wind profiles inside the canopy. The study on the physical exchanges of the transfer of heat, momentum and moisture in the microclimatic region is called as micrometeorology (Venkataraman and Krishnan, 1992).

Radha Krishna Murthy (2002) reported that the micrometeorology is the micro scale of both the physical and meteorological factors that are taking place within the ecosphere. The ecosphere means the area between just above the crop canopy and just below the root zone of a crop. Further he was on the opinion that the microclimate of valley is important as compared to the entire climate of the mountain for the bio-life of valley. For the crops purpose, the microclimate is the climate of the ground surface to top of the crop canopy or otherwise it can be called as crop's climate, a modified one from the Stevenson's Screen climate/weather observation.

Lenka (1998) opined that in micrometeorology, the crop scientists are concerned with surface layer of the ground and air layers immediately above, up to two metres. Further he stated that the ground surface is an active surface layers which gets heated during day time and gets cooled during night time and also get slight warming up during night time due to upward movement of heat from lower soil layers. Micrometeorology of the soil layer gets changed greatly with crop's presence.

The micrometeorology also ironically can be defined as; the study of atmosphere at the ground surface level. It is a small horizontal and short temporal scales of motions in meteorology and specifically less than one kilometer and one hour.

All the definitions clearly indicate the importance of one modified climate that exist near ground surface and that is micrometeorology.

3.1 Importance of Micrometeorology in Crop Production

In the presence of crops, the climate of the area between the root of the crop and top of the crop canopy is modified and it is entirely different from the climate of the crop above its canopy. This may be good for some time and bad for some time. For example, with irrigation and thick crop canopy, the microclimate becomes highly humid and invites diseases in cotton at its 60 days of growth. If we analyse the crop's height, it can be divided in to four categories and those are:

(i) Very short up to10 cm(chick pea) from ground surface;

(ii) short up to 45 cm from ground surface(pulses);

(iii) Medium height up to 120 cm from ground level (rice, cotton) and

(iv) Crops > 120 cm (banana, sugarcane crops). The microclimate of these crops get differed greatly and accordingly the productivity also. This is mainly due to many reasons, but the important ones are:

- The architecture of the plant including, plant height, leaf angle, leaf types, number of nodes and internode length
- Leaf area index
- Dry matter production
- Spacing adopted
- Number of irrigations given
- Fertiliser applied
- Greenness of the crop

There is also good relationship between the plant architecture and crop evapotranspiration. Like banana, sugar cane and rice, the crop ET falls from 1200 mm to 2500 mm and hence the yield is also more. For very short and short height crops the crop evapotranspiration is lower up to 250 mm and hence it meets with lower productivity. This conveys the message that the microclimate of the crops based on the architecture of the plants plays vital role in enhancing its productivity or otherwise, how the crop canopy with due support from soil surface modify the microclimate that prevails above the crop canopy for their life duration.

3.2 Important Microclimatology Instruments

There are many agrometeorological instruments available in the market. But it varies with different countries need. Some important instruments name is given hereunder.

1. Infrared thermometer to measure leaf temperature
2. Photosynthetic analyser to understand photosynthesis
3. Ventilated psychrometer to measure temperature and humidity in crops
4. Soil heat flux plates to measure energy balance or Bowen ratio flux system
5. Steady state porometer to measure leaf temperature, conductance, pressure etc.,
6. Quantum sensor to measure radiation that drives photosynthesis
7. Sun fleck Ceptometer to measure PAR

8. Portable leaf area meter
9. Portable photosynthesis system (Bowen ratio system)
10. MINCER to measure micrometeorology of rice canopy
11. Gas analyser
12. Sap flow meter

3.3 Recent Research Output on Microclimatology of Crops

3.3.1 Maize-Wheat Cropping System

A Bowen ratio tower to measure temperature, humidity, wind speed in five heights (0.5 m, 1.0 m, 2 m, 4 m and 8 m) with net radiometer and PAR sensor at 2 m height were installed inside the crop field. Soil heat flux plates were also installed at 5 cm and 15 cm soil depth. The grain yield of maize was 5.84 t/ha, while the wheat yield was 4.25 t/ha. At physiological maturity stage of maize, the maximum R_n (total net radiation at the canopy layer) was 245.48 W m^2 at 1330 hrs. and its partitioning into LE(amount of energy consumed for ET or latent heat term), H(sensible heat term) and G (storage of heat in the soil) was 54, 35 and 15 per cent, respectively. In wheat crop, at CRI stage, the highest value of R_n recorded was 518.56 Wm^2 and it gets partitioned into LE, H and G and the values were 85.4, 8.8 and 4.67 per cent, respectively. Moist soil did contribute more to LE as compared to H and G. The contribution to G was low due to higher LAI. The initial stage of wheat crop had the lowest ETc of 0.48 mm/day followed by development stage (1.56 mm/day) and at the terminal stage, it was 1.9 mm/day (Joy Deep Mukherjee *et al.,* 2020).

3.3.2 Microclimate Profiles of Pigeon Pea

Rajesh Kumar *et al.,* (2020) studied the microclimate profiles and PAR interception in pigeon pea under different sowing dates and cultivars at Hissar during *kharif* season. The study revealed that maximum reflection of 7. 6 per cent and maximum transmission of 14.6 per cent were recorded with crops sown during second fortnight of June followed by first fortnight of June and first fortnight of May, whereas maximum absorption of 82.5 per cent was noted for the crops sown during first fortnight of May followed by first fortnight of June and second fortnight of June. The maximum reflection of 6.7 per cent and maximum transmission of 14.5 per cent was seen with the cultivar Manak followed by Paras and UPAS- 120, whereas, maximum absorption of 87.6 per cent was seen with the cultivar UPAS-120 followed by Paras and Manak.

3.3.3 Spatial and Temporal Variation in Microclimate in Capsicum

The experiment was conducted both under greenhouse and open condition at Anand, Gujarat from October 2017 to March, 2018. The microclimatic condition was observed at 4.75 m interval ground area of 4 × 3 regular grids. Air temperature, relative humidity and soil temperature were recorded at 0930 hrs. to 1000 hrs. and PAR was measured at 1240 hrs. to 1250 hrs inside the green house and open condition. The day time observation was recorded at 0930, 1240 and 1530 hrs. at 15 days interval. Assaman psychrometer (to measure temperature and relative humidity) and line quantum sensor (PAR) were used. Soil temperature at 5 cm depth also was measured. The result revealed that the PAR inside the greenhouse at the top of the crop canopy was about 17–26 per cent of open field condition irrespective of growth stages because of the polythene roof of 200 micron and shade net of 40 mesh. The values of PAR inside the greenhouse at the top of crop canopy was found decreased during the month of December and recorded low at northern and central part of greenhouse. The vertical profile of PAR showed extinction pattern with depth within canopy. The relative humidity was more under greenhouse condition as compared to ambient air values. The soil temperature was lower under greenhouse condition. Finally, they concluded that, the spatial variation observed in microclimatic variables did not favour or suppress capsicum growth differently at different area inside (Chauhan Kalyan Singh Kiransingh and M. M. Lunagaria,2020).

3.3.4 Micrometeorology study in pearl millet

Anil Kumar *et al.,* (2020) studied the microclimate of pearl millet under different environment with different cultivars at Hissar during *kharif* 2017 and 2018. Point quantum sensor was used to measure PAR (watt/m^2). Observations on incoming, transmitted and reflected components were made at hourly interval on clear weather condition during noon time at different growth stages of the crop. Canopy temperature was measured through infrared thermometer. Wind speed(m/s) was studied through digital portable experimental anemometer and RH by digital psychrometer. The soil thermometer was measured at 5 and 10 cm depth. The temporal profile indicated decreasing trend in soil temperature, transmitted PAR, soil heat flux, thermal conductivity pattern from advanced vegetative stage to grain filling stage, which is ascribed to complete cover of ground and this resulted in higher absorption of PAR. Early sowing did show higher yield under favourable microclimate.

3.3.5 Variation in Energy Fluxes Over Wheat Ecosystem

The investigation of energy exchange over irrigated wheat ecosystem was done using eddy covariance technique from Jan. 2015 to May, 2015 in Saharanpur region. The average Bowen ratio for the growing season was 0.11. It was small (0.02) in the active growing period, while it was large enough during January (0.26) representing initial growth phase. The current results indicated the portions of net energy converted to other energy flux components (LE, G, H) and was affected by crop stages and soil moisture availability (Shweta Pokhariyal and Patel, 2020).

3.3.6 Radiation use efficiency in potato under different microclimate (Raktim Jyoti Saikia et al., 2020)

At Assam Agricultural University farm, to study the RUE in potato, four planting dates and three mulching treatments were screened during *rabi*, 2018-19. The incident, reflected and transmitted PAR were measured periodically over the crop canopy with line quantum sensor and daily incident radiation was calculated from measured PAR. The incident PAR varied considerably irrespective of plating dates and mulches studied. The reflected PAR was found to be lowest under early planting (10th November) owing to its better canopy coverage and among mulching treatments the reflected PAR was lowest under black polythene (65 μ mol $s^{-1}m^{-2}$) and highest under non-mulched treatment (77 μ mol $s^{-1}m^{-2}$). The transmission of PAR through the canopy was lowest under the water hyacinth mulch treatment (296 μ mol $s^{-1}m^{-2}$) due to good coverage of canopy as compared to black polythene mulch (326 μ mol $s^{-1}m^{-2}$) and non-mulched plot (594 μ mol $s^{-1}m^{-2}$). The interception PAR was found varied considerably among different treatments studied, however it was highest at 55 DAP (74.8%) in all planting dates and mulching treatments. The RUE was highest for 10th November planting (2.44 g MJ^{-1}) and water hyacinth mulched plot (2.35 MJ^{-1}).

3.4 Modifications of Micrometeorology for Enhancing Crop's Productivity

The existing unfavorable micrometeorology can be modified into certain extent to partly favourable microclimate by introducing some agronomy techniques and others and those are discussed hereunder

3.4.1 Shelter belt/Wind break

The wind break (shelter belt) is a planting usually consists of one or multiple rows of trees or shrubs under closed spacing or under wider spacing as the case may be, planted in such a way as to provide shelter from the wind and to protect soil from erosion. Very light or moderate wind is enough to optimise transpiration, CO_2 supply to crops and other photosynthetic activities if an area is vulnerable to tornado or high wind speed that triggers soil erosion, the best way is to go for shelter belt or wind break. The wind break is a structure that reduces wind speed. In the case of shelter belt, multiple rows of tall trees like cashew are planted parallel to east west direction. Reduced wind speed, high maximum temperature, low minimum temperature, increase in vapour pressure, increase in relative humidity may occur during night. During summer the relative humidity is lesser in shelter belt area.

The barrier of the wind breaks must be moderately denser so as to provide cushioning effect on the leeward side for better CO_2 flux movement.

3.4.2 Intercropping and Paired Row Cropping

These agro-techniques would reduce wind speed in moderate wind blowing areas and alters the microclimate favorable to crops cultivated.

3.4.3 Mulching

The straw mulching or black polythene over cultivable area alters soil temperature required for crop management in addition to controlling weeds also.

The optical properties of mulch material is important as it may likely to favour or otherwise crop growth. For example, transparent polythene mulch may increase soil temperature than coloured ones. With respect to season of study, at TNAU, Coimbatore-3, mulching did increase 30 per cent yield in groundnut during winter, while the result was negative during summer. In another research, newspaper mulch in cotton did reduce sucking pests as a result of reflection of light in the lower canopy.

3.4.4 Irrigation

The weather-based crop irrigation would always provide favorable microclimate to the crops cultivated.

3.4.5 Other Practices

The other practices like plant's spacing, soil tilth, land configuration, alley cropping etc., would make the microclimate suitable for the crops to sustain their productivity.

Different land configuration for crop cultivation like planting in ridges and furrows, beds and channels, *etc*., play an important role in modifying the micro climate. For example, in north-south running ridges, there is difference in heating in east-west soil slope up to 3 °C and the west facing slope gets heated more than east facing slope.

3.4.6 Fruit Maturity

The canopy temperature, soil temperature and air temperature was found to have good relationship with one another. For example, in tomato, during near harvest time, if air temperature increases by 1–2 °C suddenly and this will trigger the maturity of fruits as GDD increases. This may lead to economical loss. This can be avoided by giving irrigation when the air temperature found increasing.

The crop micro meteorology study was lacking in the earlier decades because of non-availability of appropriate and sensitive instruments. But presently micro meteorological instruments are available in the market. The only problem is lack of finance and lack of technical knowledge to operate and maintain sensitive instruments with the Research Institutions in India. In India there must be a network for micro meteorological study in selected food crops to enhance the productivity to feed the ever-increasing population.

Under climate change scenario, many crops commercial cultivation especially for flower, fruits and vegetables have been done under greenhouse, shade net house etc., to get favourable micro meteorology against the outside harsh climate. Many research studies also being done in growth chamber to understand crops productivity against global warming.

4

Remote Sensing

4.1. Introduction

Many countries have set up a network of agrometeorological observatories to regularly record important weather parameters such as clouds, rainfall, surface temperature, humidity, wind and wind direction, solar radiation, soil moisture, temperature and humidity profile of the atmosphere, *etc*. The observation process is performed either manually or through Automatic Weather Stations (AWS).

This recorded information is used for weather forecasting, natural resources management and for tracking agricultural production. However, these networks, which handle data of high spatial climate variability, are typically less robust than they should be. Moreover, the agrometeorological data collected from different ground stations is not transmitted in real time to a central collection point which limits the use of information.

Satellite remote sensing technology is an alternative system that is gaining popularity in the recent times which has allowed us to obtain frequent and precise measurements of vital agrometeorological parameters such as rainfall, solar radiation, temperature, evapotranspiration, albedo etc. Satellite data has many advantages over traditional ground stations, such as high spatial resolution (50 m–5 km), wide spectral bandwidth range (0.4–3.75 um) and high temporal frequency (even half an hour) and the prospect of collecting data from remote areas. In addition to routine repeated observations, satellites often provide synoptic views of the earth's surface from space on a wide scale and have archive of information to create baseline data.

The preparation, planning and use of agricultural technologies for agriculture, needs application of agricultural meteorology. The role of agrometeorology in the last three decades has increased in terms of increasing food and nutritional security for the burgeoning population and natural resource management particularly in the context of global climate change. Agrometeorology can help to minimize inputs, thus clarifying the contribution of ecosystems and agriculture to the carbon budget in the event of global change. Agrometeorological information needs to be disseminated to users in time and space quite accurately, as agricultural activities are linked to the very local situation where the outputs of modern tools such as local area models, climate models, and dynamic crop simulation models are reduced to finer grid resolution and combined using GIS to extract practical suggestions. This chapter describes the recent applications of remote sensing in Agrometeorology.

4.2. Remote Sensing

Remote sensing provides spatial information from the earth's surface and the surrounding atmosphere by observing the reflected and transmitted electromagnetic radiation through a wide spectrum of wavebands either from the space using satellites or from the air using aircrafts. The amount of radiation from an object is influenced by both the properties of the object and the radiation hitting the object (irradiance). Each waveband emitted as a result of reflected and emitted radiation from the atmosphere and land surface that could be detected by satellites provides a specific information about surface temperature, clouds, solar radiation, processes of photosynthesis and evaporation. Various tools are used to make electromagnetic radiation including visible range (400–700 nm) and outside this range *viz.*, near infrared, middle-infrared, thermal-infrared and microwaves to interpret the colours. The scale of observation can vary from a few square meters to continental scale using a scanner mounted on a vehicle to a meteorological satellite.

By measuring the energy that is reflected by targets on earth's surface over a variety of different wavelengths, spectral signature (ratio of reflected energy to incident energy as a function of wavelength) for that object is built and by comparing the response pattern the objects are distinguished. The reflectance characteristics of the earth's surface features are expressed by spectral signature. Typical Spectral Reflectance curves for vegetation, soil and water are presented in Figure 4.1.

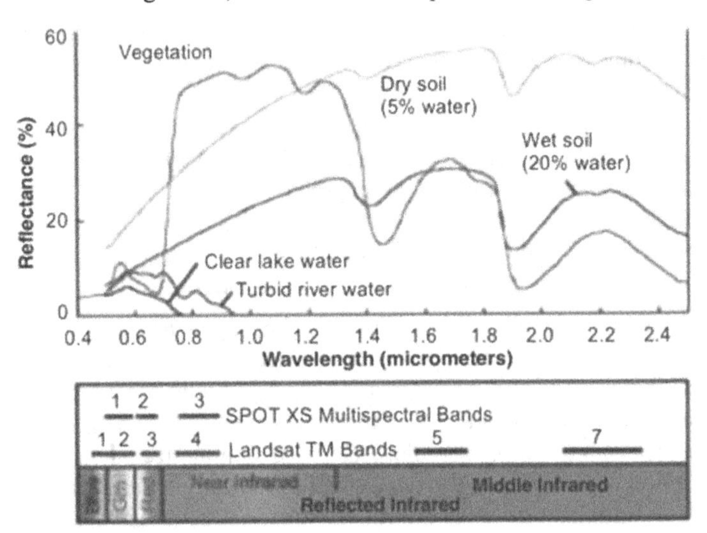

Fig. 4.1. Typical Spectral Reflectance curves for vegetation, soil and water (Aggarwal, 2004) (*Source*: Jiang *et al.*, 2013)

In case of vegetation, chlorophyll in the leaves strongly absorbs red and blue wavelengths and reflects green wavelength. The internal structure of healthy leaves acts as diffuse reflector of near infrared wavelengths and by measuring and monitoring the near infrared reflectance healthiness of the vegetation could be assessed. As far as water is concerned, majority of the radiation incident is either absorbed or transmitted and not reflected. Longer visible wavelengths and near infrared radiation is absorbed more by water than by the visible wavelengths. Thus, water looks blue or blue green due to stronger reflectance at these shorter wavelengths and darker if viewed at red or near infrared wavelengths. The factors that affect the variability in reflectance of a water body are depth of water, materials within water and surface roughness of water. In case of soil, majority of radiation incident on the surface is either reflected or absorbed and little is transmitted. The characteristics of soil that determine its reflectance properties are its moisture content, organic matter content, texture, structure and iron oxide content. The soil curve shows less peak and valley variations. The presence of moisture in soil decreases its reflectance.

4.3. Satellites for Agrometeorology Purpose

Satellite image data can display wider areas than data from an aerial survey and, because a satellite constantly flies over the same area collecting new data each time, like land use/land cover changes can be tracked periodically. There are two kinds of satellite orbits viz., the geo-stationary orbit and the polar orbit. The geo-stationary satellites are positioned at an altitude of 36,000 km in an equatorial plane rotating from west to east, keeping pace with the Earth's rotation focusing between 70°N to 70°S latitudes and covering the same location to provide 24 hours continuous near-hemispheric coverage of the same area. Geo-stationary satellites such as Meteosat, MSG and GOES are mainly used for communication and meteorological applications and they take 30 minutes a picture of the weather conditions over the same locations (Figure 4.2a). Satellites in a polar orbit, with an inclination of approximately 99 degrees with the equator to maintain a sun synchronous overpass over an altitude of approximately 650 to 900 km above the MSL, cycle the earth from North Pole to South Pole. The satellite passes over all places on earth and gives much finer details images over a particular area (Figure 4.2b).

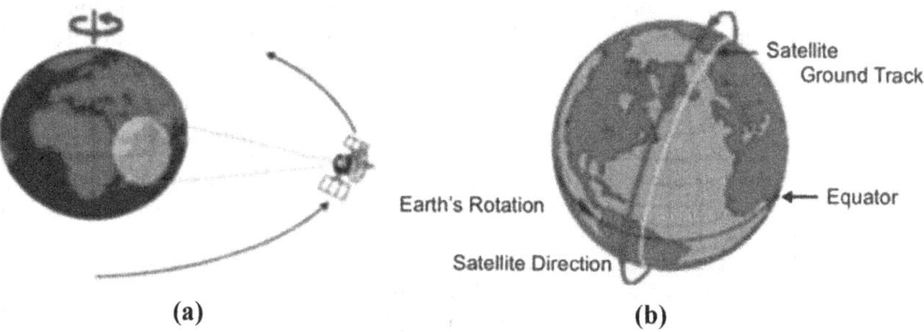

Fig. 4.2. Satellite Orbits **(a)** Geo-stationar Orbit **(b)** Polar Orbit
(*Courtesy:* http://sar.kangwon.ac.kr/etc/fundam/chapter2/chapter2_2_e.html)

Some of the examples of satellites that get information for the agrometeorology purpose are listed in Table 4.1.

Table 4.1. Some of the important satellites used in agricultural meteorology, the products derived from the satellites and their applications

Satellite	Field	Parameters/products	Applications
HySIS (PSLV C43) (India)	Agriculture and allied areas	To provide hyperspectral imaging combining the power of digital imaging as well as spectroscopy.	Agriculture, forestry, geology, soil survey, coastal zones, environmental studies, inland water studies and detection of pollution from industries
RISAT-2B (India)	Agriculture Insurance	Area affected by flood and drought	Can be useful for civilian purposes like agriculture and disaster relief management
Resourcesat-2 (PSLV-C16) Resourcesat-2A (PSLV-C36) India	Disaster Management Support.	Land & Water Resources development and to provide multispectral images	For resource inventory and management of natural resources, crop production forecast, wasteland inventory,
Cartosat-1 (PSLV-C6) India	Land elevation details	Provide high resolution images for Cartographic mapping, Stereo data for Topographic Mapping & DEM	DEM Applications– Contour, Drainage network, etc.
RISAT-1 (PSLV-C19)	Natural resource management	Provide all weather imaging capability useful for agriculture	Particularly paddy and jute monitoring in *kharif* season and management of natural disasters
Kalpana-1 (PSLV-C4) India	Weather services	Provide meteorological data.	to enable weather forecasting services
INSAT-3D (Procured launch) INSAT-3DR (GSLV-F05) India	Meteorological observations	Vertical profile of the atmosphere in terms of temperature and humidity	Designed for enhanced meteorological observations, and for improved weather forecasting and disaster warning

Satellite	Field	Parameters/products	Applications
SPOT (France)	Lands	Vegetation index maps; Land cover and land use classification maps; Digital Elevation Model; Ortho photo map	High resolution land use analysis; biomass and vegetation hydrological conditions; Agricultural and natural resources planning and managing; Cartography
LANDSAT TM (LandSat 1 to 5) USA	Atmosphere	Cloud cover maps Albedo maps Aerosol maps	Meteorological forecasts, numerical models; Climatology, floods forecasting; Energetic assessment, general climatology
	Inland and coastal waters	Water quality monitoring; Sea surface temperature maps (SST)	Pollution monitoring; environmental impact assessment; Coastal zones monitoring; coastal streams, Dynamics; sediments mapping
	Lands	Land surface temperature maps (LST); Normalized difference vegetation index maps (NDVI); Land cover and land use classification maps	Temperature distribution maps for urban areas; high resolution land use analysis; biomass and vegetation, hydrological conditions; fire risk maps and damage assessment; Agricultural and natural resources planning and managing
AVHRR-NOAA (USA)	Inland and coastal waters	Oil slick maps and monitoring; Sea surface temperature maps (SST)	Pollution monitoring, environmental impact assessment; Coastal zones monitoring; coastal streams dynamics; sediments mapping

(Contd.)

Satellite	Field	Parameters/products	Applications
	Lands	Land surface temperature maps (LST); Normalized difference vegetation index maps (NDVI); Land cover classification maps	Temperature distribution maps for urban areas; high resolution land use analysis; biomass and vegetation hydrological conditions; fire risk maps and damages assessment; Monitoring of the temporal variations of the coverage
ERS-1 SAR	Inland and coastal waters	Oil slick maps and monitoring Ship tracking	Monitoring and evaluation of accidental oil slicks area assessment; Temporal evolution forecasting of oil slicks and identification of coastal areas at risk in case of accidental oil slicks; Coordination support for antipollution actions; Ship identification in case of illegal polluting slicks; Maritime traffic control also in fishing areas
	Lands	Flooded areas maps	Emergency actions support in flooded areas; Area and damage assessment for flooded areas; Strong temporal soil moisture variation evaluation
GOES-1 to 7 (USA)	Atmosphere	Weather	Weather satellite for precise weather information
Soil Moisture and Ocean Salinity Satellite (SMOS)	Irrigation, Ocean salinity	Soil moisture data Ocean salinity data Ocean circulation patterns	Hydrological studies, Weather and extreme-event forecasting.

4.4. Applications of Remote Sensing in Agricultural Meteorology

Remote sensing has various application in the field of Agricultural meteorology as described below (Figure 4.3).

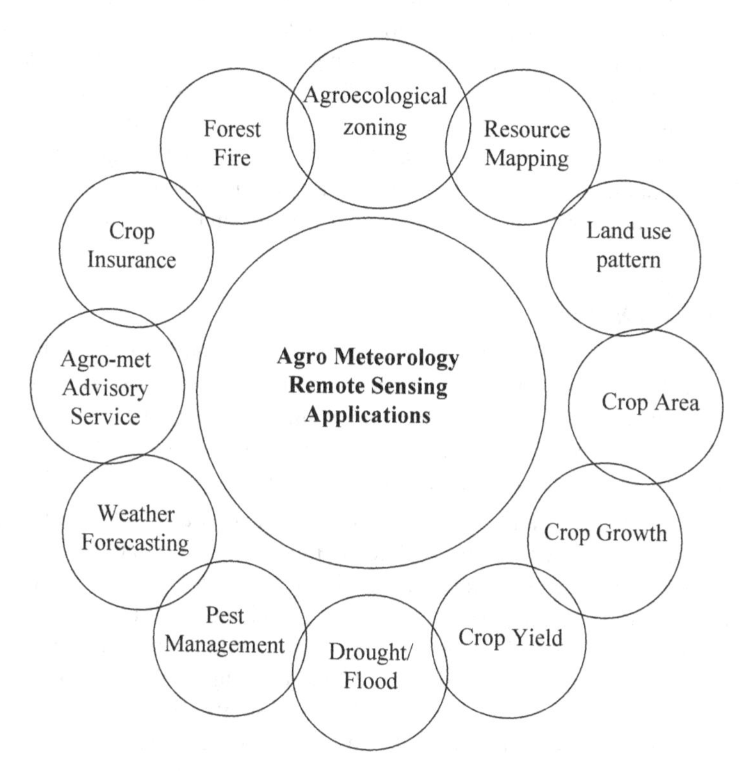

Fig. 4.3. Applications of Remote sensing in Agricultural meteorology

4.4.1. Agro-ecological zoning

Agro-Ecological Zone (AEZ) is a land resource mapping, having a unique combination of land form, soil and climatic characteristics and/or land cover having a specific range of potentials and constraints for land usc (FAO, 1996), and serves as a focus for implementing policies aimed at improving the current situation of land use, either by-development or by reducing land degradation. AEZ is critical for agricultural development planning, since the success and failure of a specific land use or farming system in a given area relies heavily on careful evaluation of agro-climatic resources.

Traditionally, isolines of climate (temperature, rainfall) map was manually superimposed on soil resource map to delineate agro-climate zones. Major limitation in this manual approach is difficulty in treating large quantities of agro-ecological data and necessity for aggregation of information at an early stage itself, which has led to loss of information on spatial variability. The planning commission of India, in its VII Plan (1985-1990) period, divided India into 15 broad agro-climatic zones based on physiography and climate.

New instruments such as satellite remote sensing and the Geographical Information System (GIS) have created new dimensions to track and control land resources effectively for agro-ecological analysis in an integrated manner linking bio-climate, terrain and soil resource inventory information (Patel *et al.,* 2000). Remote sensing provides digital or hard copy data base information on natural resources at zonal scale and this information can be stored and retrieved as and when required and also data can be classified and aggregated for any number of planning exercises. The nature of analysis, involves the combination of layers of spatial information to define zones, lends itself to application of a GIS. The zonal database can also be integrated with non-geographic information such as socioeconomic data, which is relevant for making decision on development priority interventions about the sustainable management of zonal resources.

4.4.2. Resource Mapping

Resources such as water availability, soil types, forest and grasslands, geology, land-use, administrative boundaries and infrastructure (highways, railroads, electricity or communication systems) could be mapped using remote sensing. Possible consequences of human activities that affect the resource availability could also be displayed with combination of remote sensing and GIS technologies.

Identification of the location and extent of salt affected soils are very important to suggest suitable crops and varieties tolerant to salinity and also to plan corrective measures. Saline soils have been shown to have higher reflectance in both the visible and NIR wavelengths than non-saline soils from both ground and satellite radiometric measurements (Rao *et al.,* 1995) allowing for mapping of salinization (Sharma and Bhargava, 1988; Dwivedi, 1992; Wiegand *et al.,* 1996; Yu-liang, 1996). It should be noted, though, that successful mapping of salinization results only when salt scalds are clearly visible on the soil surface. This is not the case for all salt affected soils, which means that salinization is not always easily mapped from remote sensing.

Detection of water from dry land is uncomplicated spectrally leading to the classification of potentially waterlogged areas (Sharma and Bhargava, 1988). However, these areas should be integrated with an accurate land use map as permanent water bodies and natural wetlands could quite easily be included in the classification otherwise. Ground-based electromagnetic (EM-31) surveys (Hume *et al.,* 1999) and airborne radiometrics (Cook *et al.,* 1996) have also proven useful for mapping soil attributes.

Forest is a major resource and play vital role in maintaining the ecological balance and environmental setup. The changes in forest cover are the matter of concern due to its ability of promoting role in carbon cycle. Remote sensing could be potentially applied for assessing and monitoring the changes in forest cover. Advances in the spatial and spectral resolutions of sensors are available for ecologist which is mainly feasible, to study the certain aspects of biological diversity through direct remote sensing. Global and regional scale of temporal remote sensed data is now available for monitoring the changes in forest cover over the last few decades *viz.* Landsat, MODIS,

ASTER, AWiFS and SPOT. Monitoring the changes in forest cover at global and regional scale can contribute to reducing the uncertainties in estimates of emissions of greenhouse gases from deforestation. Remote sensing coupled with Geographical Information System (GIS) as a potential tool, for monitoring the rate of deforestation and quantification of overall forest cover loss at finer scale.

4.4.3. Monitoring the Changes in Land Use Pattern

The land use/cover pattern of a region is an outcome of natural and socio-economic factors and their utilization by man in time and space. Information on land use/cover and possibilities for their optimal use is essential for the selection, planning and implementation of land use schemes to meet the increasing demands for basic human needs and welfare. Invent of the remote sensing, land use/cover mapping has given a useful and detailed way to improve the selection of areas designed to agricultural, urban and/or industrial areas of a region (Selcuk *et al.,* 2003).

Crop area estimation is performed in an attempt to monitor year-to-year land use, it is extremely important to know how much variation is happening under each sector, for example changes in extent of forest area, expansion of urbanization, reduction in potential agricultural land, change in cropping pattern, etc. This is done by comparing the satellite pictures of various time period.

The large areas AVHRR data is very attractive due to its spatial extent and high repeat cycle, however, for Indian condition dominated with small and marginal farmers, the 1 km pixel size is often disproportionately larger than the field sizes. To compensate for mismatched pixel-to-field sizes or to increase the accuracy of area estimations, the spectral characteristics of impure or 'mixed' pixels can be 'unmixed' by linear mixture modelling (Quarmby *et al.,* 1993). As with identification of crop types, ground validation is also very important in area estimation. Traditional estimates of crop area have come from census data over rather large areas, and are usually rather gross and usually contain little specific spatial context. It is important, to make sure that the 'ground truth' data is defensibly more reliable than the observation as well as spatially explicit.

Landsat-TM images represent valuable and continuous records of the earth's surface during the last three decades. El-Asmar *et al.,* (2013) have applied remote sensing indices, *i.e.,* normalized difference water index (NDWI) and the modified normalized difference water index (MNDWI) in the Burullus Lagoon, North of the Nile Delta, Egypt for quantifying the change in the water body area of the lagoon during 1973 to 2011.

In India, Pooja *et al.,* (2012) have quantified land use/cover of Gaga watershed, district Almora using survey of India topographic sheet of the year 1965 and LISS III satellite data for the year 2008 over a period of 43 years. Rawat *et al.,* 2013 carried out study based on 20 years of satellite data from 1990 to 2010 of land use/land cover change, they found that built up area has sharply increased due to construction of new buildings in agricultural and vegetation lands.

Implementation of a Geospatial approach was done for improving the Municipal Solid Waste (MSW) disposal in growing urban environment (Pandey *et al.,* 2012). Biomass estimation of Sariska Wildlife Reserve was carried out using forest inventory and geospatial approach to develop a model based on the statistical correlation between biomass measured at plot level and the associated spectral characteristics (Kumar *et al.,* 2013). Singh *et al.,* (2014) have used recent freely available satellite data of Landsat-8 for assessing the land use pattern and their spatial variation of Orr watershed of Ashok Nagar district, M.P., India.

4.4.4. Crop Area and Crop Type Identification

The most commonly practiced application in remote sensing of agriculture is mapping land cover and to identify crop types. This process primarily uses the spectral information provided in the remotely sensed data to discriminate between perceived groupings of vegetative cover on the ground. The spatial (Atkinson and Lewis, 2000) and temporal information included in single date and time series data, respectively, usually play a secondary role, but can also aid in the classification procedure. Discrimination of crops is usually performed with 'unsupervised' or 'supervised' classifiers.

Unsupervised classification relies upon a computer algorithm to define natural groupings of the spectral properties of the pixels in an image. Problems arising from using this methodology are related and include classification of meaningless groupings, idiosyncratic definition of initial number of classes, and subjectivity involved in combination of similar classes. However, if land cover classes are spectrally well separated, adequate results may be obtained (Barrs and Prathapar, 1994). Supervised classification, which is the most common classification method in agricultural areas, requires *a priori* knowledge. It relies on the analyst to define the perceived groupings by identifying homogeneous areas, called training sites, from either *in situ* collection or directly from the image. These training sites are statistically analysed and then used to assign every pixel in the image to the group in which it has been determined a member (Campbell, 1981; Medhavy *et al.,* 1993).

Common clustering algorithms include maximum likelihood, minimum distance to mean, and parallel piped (Jensen, 1986). In general, classification accuracy assessment involves sampling classes of the thematic map to summarise the amount of overlap between the classified map and what is present on the ground (Richards, 1996). The 'ground truth' information is usually gathered from existing maps, aerial photography, or field surveys using number of sampling schemes such as simple random sample, stratified random sample, systematic sample, *etc.* (Aplin *et al.,* 1998; Barbosa *et al.,* 1996; Le Hegarat-Mascle *et al.,* 2000; McCloy *et al.,* 1987; Medhavy *et al.,* 1993; Kurosu *et al.,* 1997; Panigrahy *et al.,* 1999).

The performances of different sensors for crop identification have been tested over varied geographic areas and crop types. These most commonly include broadband optical (*e.g.,* LANDSAT, SPOT, and AVHRR), and microwave (*e.g.,* ERS, RADARSAT, JERS) used alone or in combination in the form of either single or multiple date imagery. In general, multitemporal microwave imagery results in roughly equal classification accuracies as single date optical imagery when specifically considering rice (*i.e.*, about 90%) (Kurosu *et al.*, 1997) or other crops (Le Hegarat-Mascle *et al.*, 2000). Because of the complementarity of the data, the fusion of both optical and microwave imagery has resulted in higher overall and individual crop classification accuracies (> 95%) than either produce alone (Le Hegarat- Mascle et al., 2000). Higher crop classification accuracies have also been achieved by the combination of GIS and remotely sensed data (Aplin *et al.*, 1998). Similarly, others have improved classification accuracy using the contextual or landscape information inherent in the imagery (Moody, 1997; Stuckens *et al.*, 2000).

4.4.5. Crop Growth and Productivity Monitoring

Crop growth and yield are determined by crop genetic ability, soil, environment, cultivation practices and biotic and abiotic stresses that prevail during the crop growing season. Remote sensing data have a certain benefit over meteorological observations for growth and yield modeling, such as dense observational coverage, direct viewing of the crop and capability to capture the effects of non-meteorological factors such as crop distribution, crop phenology, Leaf Area Index (LAI) and Vegetation Index (VI). With the launch of polar orbiting earth observation satellites (Landsat, Location, IRS, *etc.*), availability of continuous, timely and precise multi-spectral (visible, near-infrared) crop status data is ensured at a range of spatial scales. Remote sensing data, repetitively acquired over agricultural land, assists in the identification and mapping of crops and also in the assessment of crop vigour. Integrating the crop simulation models with remote sensing data under GIS platform can provide an excellent solution for monitoring and modelling crops at a range of spatial scales. Recent developments in GIS technology allow capture, storage and retrieval and visualization and modelling of geographically linked data. LAI derived from remote sensing is integrated and forced in crop simulation model, and this specific parameter is re-initialized or re-calibrated to model and monitor the crop growth on a range of spatial scale under GIS environment.

Re-initialization involves adjustment of initial condition of the particular state variable (LAI) so as to minimize the difference between a remote sensing derived state variable and its simulation. Maas (1988) in his simplified maize model adjusted the initial value of LAI at emergence based on the minimization of an error function between remotely sensed LAI values and simulated LAI values during the course of simulation. Ajith *et al.*, (2017) modified the management parameters such as plant spacing and nitrogen levels in rice crop in the DSSAT model to make changes in the simulated LAI to mimic the remotely sensed LAI and predicted the rice productivity for Tanjore district. For this 30 field level samples were taken as ground truth across Tanjore district and the average yield was underestimated without using the reinitialization of LAI and Use of satellite derived data on LAI for readjustment in DSSAT simulation resulted in only two per cent underestimation of the rice productivity.

This demonstrates that the remote sensing data helps in obtaining a complete and spatially dense observation of crop growth. This complements the information on daily weather parameters that influence crop growth. Remote sensing-crop simulation model linkage captures the crop management and weather on GIS platform, providing a framework to process the diverse geographically linked data.

4.4.6. Crop Yield Forecasting

Crop area measurement is a very common practice in agriculture. Remote sensing is often used for this purpose because of its strengths in relation to spatial extent, temporal density, relative low costs, and potential for rapid assessment of spatial features.

Crop forecasting is essential for various agricultural planning purposes, including pricing, export/import, contingency measures, *etc.* Crop forecasting using remote sensing data, in India, started in late 1980's in Space Applications Centre of ISRO under the Department of Agriculture & Cooperation (DAC)'s sponsored project CAPE (Crop Acreage and Production Estimation). This later on developed into a National level programme, called FASAL (Forecasting Agriculture using Space, Agro-meteorology and Land based observations), which is in operation since August, 2006 (Parihar and Oza, 2006). Remote sensing data, both optical and microwave form the core of crop area enumeration, crop condition assessment and production forecasting. Crop yield is estimated using agrometeorological/spectral yield models and also crop growth simulation models.

Crop yield forecasts can greatly influence farm-level management decisions, such as fertilizer applications and water delivery and farm income assessment. Prevailing method of crop yield estimation includes analysis of crop cuttings at randomly sampled ground plots during harvest (Murthy *et al.,* 1996), or meteorological regression models using rainfall and past yield data (Karimi and Siddique, 1992) that are either not timely nor spatially explicit. Estimation of crop yields using remote sensing because of its ability to produce results quickly and spatially and estimates of yield can be made as early as 1 to 3 months prior to harvest (Quarmby *et al.,* 1993, Rasmussen, 1997), thus impacting management reaction time to yield forecasts.

Both regressions of Vegetation Index (VI) and integration Normalized difference Vegetation Index (∫NDVI) studies generally attempt to estimate yield using remotely sensed data alone. A direct comparison of VI response to yield can result in highly correlated regression equations. Harrison *et al.,* (1984) achieved correlation coefficients (r) as high as 0.9 for rice using the greenness ratio of MSS7/MSS 5. This study developed the yield model based on imagery acquired when the crop was at booting stage. Promising correlation coefficients were also shown between reflective bands of TM imagery and harvested grain yield (up to 0.92 for band 7) (Tennakoon *et al.,* 1992). Likewise, ∫NDVI studies have produced similar results with relation to rice and millet yield in another (Rasmussen, 1992).

Until just recently, the only operational satellite sensor with a repeat cycle sufficient for ∫NDVI studies of crop yield has been NOAA AVHRR (Rasmussen, 1992), TERRA (ASTER and MODIS sensors), and Ikonos satellites. Anticipated sensors include Quickbird-1, and Orb view 3 and 4 could also impact the usefulness of ∫NDVI studies, mostly used in dryland farming.

In an attempt to achieve highly accurate yield estimates, many models have been developed which aim to simulate plant growth. Plant growth models can be validated by comparing simulated and *in situ* measurements of vegetation parameters (*e.g.*, biomass or yield). However, plant growth models may also be validated by comparing simulated reflectance with that measured by remote sensing. This can be achieved by inverting either empirically-based relationships, or physically-based radiative transfer models. This means that a plant growth model estimate of cover can be converted, through a radiative transfer model, to an estimate of reflectance, or recalibrated through the estimation of LAI. This recalibration of crop variables links well to simple growth models because they can be directly related to the fraction of absorbed Photosynthetically Active Radiation (*fAPAR)*, a key variable in most of these models (Inoue *et al.,* 1998).

4.4.7 Monitoring Extreme Weather Events and Their Impacts

Major hydrological related disaster are droughts and floods which affect mainly agriculture and other natural resources like forest vegetation. Due to global warming, increased frequency of rainfall extremes is reported in India. Drought is the single most important weather-related natural disaster often aggravated by human action and has serious impact on food production and food security. The frequency and intensity of droughts is increasing in the recent times in India. In 2016, approximately 33 cores people in 10 Indian states were affected by drought as a result of two consecutive years of poor monsoons that has led to poor harvest. In contrast to droughts, high intensity rainfall in a shorter time period results in flooding due to increased runoff, sometimes landslides, submergence and the related loss in agricultural produces and health hazards. Nearly 40 m ha area in India is flood-prone and every year nearly 8 m ha of land is affected by floods posing serious threat to food security. Reliable and more precise drought and flood forewarning system is need of the hour for effective planning and management at local level to state and central level. Remote sensing techniques make it possible to obtain and distribute information rapidly over large areas in spatial and non-spatial formats and provides tremendous potential for identification, monitoring and assessment of droughts and floods.

4.4.7.1 Drought

Drought is a natural, recurrent climate feature and occurs in all climatic zones, though its characteristics differ greatly from region to region. Drought is classified into Meteorological, Hydrological and Agricultural droughts based on the causative factors such as rainfall deviation; combinations of rainfall with temperature deviations; humidity and or evaporation rates; soil moisture and crop parameter; and climatic indices and estimates of evapotranspiration. There are three phases in disaster management *viz.*, (i) Preparedness (ii) Prevention phase and (iii) Response/Mitigation phase.

Drought preparedness

Identification of drought prone or risk zone area on the basis of historic data analysis of rainfall or rainfall and evaporation and the area of irrigation support. In remote-sensing approach, historical vegetation index data derived from NOAA satellite series are used to get spatial information on drought prone area depending on the trend in vegetation development, frequency of low development and their standard deviations (Jeyaseelan and Chandrasekar, 2002). For drought prediction, initial soil moisture and amount of water available for irrigation, are integrated with the remotely sensed assimilation data on moisture circulation pattern through numerical prediction models or using coupled ocean/atmosphere models.

Drought prevention

Multi-channel and multi-sensor data sources from geostationary platforms (GOES, METEOSAT, INSAT and GMS) and polar orbiting satellites (NOAA, EOS-Terra, DMSP and IRS) are currently used for the measurement, analysis and confirmation of meteorological parameters. Satellites are often used for spatial monitoring of drought conditions through frequent information on synoptic condition on wide-area coverage.

Rainfall monitoring is one of the important steps in drought monitoring and remote sensing provides better spatial estimates of rainfall. Three rainfall estimation procedures are used in the satellite's technique namely 1. Visible and Infrared (VIS and IR) technique - during daytime and IR temperature patterns estimate rainfall over a 10 × 10-pixel array in three categories: no rain, light rain, and moderate/heavy rain. Cloud model techniques aim quantitative improvement of the rain formation processes, 2. Passive microwave technique, precipitation particles are the main source of attenuation of the upwelling radiation and 3. Active microwave system, quantitative measurements of rainfall over land and oceans is obtained.

Satellite based surface temperature estimation is yet another important factor for drought monitoring since it is related to the energy balance between soil and plants on the one hand and atmosphere and energy balance on the other in which evapotranspiration plays an important role. Surface temperature could be quite complementary to vegetation indices derived from the combination of optical bands which is available from MODIS satellite.

Soil moisture in the root zone is a key parameter for early warning of agricultural drought. Remote sensing (RS) has the capability to make frequent and spatially comprehensive measurements of the near surface soil moisture. Passive microwave sensing one among RS methods (radiometry) has shown the greatest potential. Measurements at 1–3 GHz are directly sensitive to changes in surface soil moisture, are little affected by clouds, and can penetrate moderate amounts of vegetation. They can also sense moisture in the surface layer to depths of 2–5 cm (depending on wavelength and soil wetness).

Satellite based monitoring of vegetation condition plays an important role in drought monitoring and early warning. NDVI is calculated by taking the ratio between the difference between the NIR and red bands and their sum is a good indicator of greenness of the area (Deering, 1978).

Drought mitigation

Drought impact assessment using high resolution satellite sensors from LANDSAT, SPOT, IRS, *etc.* can be done by observing changes in land use, persistence of stressed conditions at varied time scale (intra and inter-season), water availability for agriculture, crop yield, etc. Remote sensing is used as decision support for relief management, for example in water management, crop management and for designing mitigation and alternative strategies. Long term drought management action plan maps are being generated at watershed level for implementation.

At Global scale, the NDVI and Temperature Condition Index (TCI) derived from the satellite data are accepted for drought regional monitoring. National Remote Sensing Agency, The Department of Space issues biweekly drought bulletin and monthly reports for India under National Agricultural Drought Assessment and Monitoring System (NADAMS) which uses NOAA AVHRR and IRS WiFS based NDVI with ground-based weather reports.

4.4.7.2 Flood management

Floods can be mapped and monitored with remotely sensed data acquired by aircraft and satellites, or even from ground-based platforms. The sensors and data processing techniques that exist to derive information about floods are numerous. Instruments that record flood events may operate in the visible, thermal and microwave range of the electromagnetic spectrum. Due to the limitations posed by adverse weather conditions during flood events, active radar is invaluable for monitoring floods; however, if a visible image of flooding can be acquired, retrieving useful information from this is often more straightforward. Apart from providing direct information about flooding, remote sensing data can also be integrated with flood models (via model calibration or validation, and data assimilation techniques) or provide floodplain topography data to augment the amount and type of information available for efficient flood management.

Automated Quantification of Surface Water Inundation in Wetlands is done using Optical Satellite Imagery. Fully automated and scalable algorithm for quantifying surface water inundation in wetlands are available and requiring no external training data as this algorithm estimates sub-pixel water fraction (SWF) over large areas and long time periods using Landsat data. This automated algorithm allows for the production of high temporal resolution wetland inundation data products to support a broad range of applications.

Flash floods are considered as earth's most deadly and destructive natural hazards, particularly in arid regions. Use of a hydrological modeling approach to predict the spatial extent, depth, and velocity of the flood waters can be done and hence locate sites at risk of flood inundation. This was accomplished by understanding the characteristics of surface runoff through modeled hydrographs. Here, elevation data were extracted from Shuttle Radar Topography Mission (SRTM) and a two-meter Digital Elevation Model (DEM) derived from WorldView-2 stereo pair imagery would be the result. The land use/land cover and soil properties were mapped from fused ASTER multispectral and ALOS-PALSAR Synthetic Aperture Radar (SAR) data to produce a hybrid image that combines spectral properties and surface roughness respectively.

Satellite Rainfall Estimate (SRE) products provide global-scale spatial data on rainfall intensity. Among them are: Precipitation Estimation from Remotely Sensed

Information Using Artificial Neural Networks (PERSIANN (Sorooshian *et al.*, 2000); the Tropical Rainfall Measuring Mission (TRMM); and the Global Satellite Mapping of Precipitation (GSMaP). These datasets have been applied to numerical hydrological models to simulate floods in various locations of the world (Nguyen *et al.*, 2015). Within South Asia, Nanda *et al.*, (2016) used an SRE dataset to develop a real-time flood-forecasting model for a basin in eastern India. Numerical hydrological models, such as LISFLOOD-FP (Bates and Roo, 2000), HEC-HMS/RAS (2002) and the RRI (Rainfall-Runoff-Inundation) model (2017), have been developed and used to simulate flood inundation. Flood inundation extent and depth have also been simulated by applying solo SREs such as PERSIANN and TRMM to distributional flood models (Schumann *et al.*, 2013). The RRI model has advantages that include: the ability to simultaneously calculate flood inundation and river flow in areas encompassing downstream flood plains and upstream mountain zones; the facility to make calculations for multiple basins in cases where the downstream floodplain can be affected by several rivers; and free availability, making it more accessible to developing countries (Alazzy *et al.*, 2017)

4.4.8 Pest and disease monitoring

A great concern in agricultural systems is the loss of productivity due to crop damage and its negative impact on meeting the increasing demands for food globally (Fox Strand, 2000). Crop damage and subsequent yield reduction can occur for many reasons including, various pathogens, insects, and weeds and often possesses a complex interaction with cropping practices (Ennaffah *et al.*, 1997; Islam and Karim, 1997; Savary *et al.*, 1997).

Many scientists have used remote sensing for disease assessment (Table 4.2)

Table 4.2. Examples of plant patho systems and plant diseases assessed by optical sensors

Sensors	Crop	Disease/Pathogens	References
RGB	Cotton	Bacterial angular Ascochyta blight	Camargo and Smith (2009)
	Sugar beat	Cercospora leaf spot Sugarbeat rust Leaf spot–Phoma leaf spot, bacterial leaf spot	Neumann *et al.*, (2014)
	Grape fruit	Citrus canker	Bock *et al.*, (2008)
	Tobacco	Anthracnose	Wijekoon *et al.*, (2008)
	Apple	Apple scab	Wijekoon *et al.*, (2008)
	Canadian goldenrod	Rust	Wijekoon *et al.*, (2008)

Sensors	Crop	Disease/Pathogens	References
Spectral sensors	Barley	Net blotch Brown rust Powdery mildew	Kushka *et al.,* (2015) Wahabzada *et al.,* (2015)
	Wheat	Head blight Yellow rust	Bauriegel *et al.,* (2011) Bravo *et al.,* (2003) Huang *et al.,* (2007) Moshou *et al.,* (2004)
	Sugar beat	Cercospora leaf spot Sugarbeat rust Powdery mildew Root rot Rhizomania	Bergstrasser *et al.,* (2015); Hillnhutter *et al.,* (2011); Mahlein *et al.,* (2010); Rumpf *et al.,* (2010); Steddom *et al.,* (2003)
	Tomato	Late blight	Wang *et al.,* (2008)
	Apple	Apple scab	Delalieux *et al.,* (2007)
	Tulip	Tulip breaking virus	Polder *et al.,* (2014)
	Sugarcane	Orange rust	Apan *et al.,* (2004)
Thermal sensors	Sugar beat	Cercospora leaf spot	Chaerle *et al.,* (2004)
	Cucumber	Downey mildew Powdery mildew	Berdugo *et al.,* (2014); Oerke *et al.,* (2006)
	Apple	Apple scab	Oerke *et al.,* (2011)
	Rosa	Downey mildew	Gomez *et al.,* (2014)
Fluorescence imaging	Wheat	Leaf rust Powdery mildew	Burling *et al.,* (2011)
	Sugar beat	Cercospora leaf spot	Chaerle *et al.,* (2007) Konanz *et al.,* (2014)
	Bean	Common bacterial blight	Rousseau *et al.,* (2013)
	Lettuce	Downey mildew	Bauriegel *et al.,* (2014) Brabandt *et al.,* (2014)

Some work has been done on forecasting of desert locust. Riley (1989) provided an exhaustive review on the use of remote sensing in the field of entomology. Optical and video imaging in near-infrared and microwave regions were used to quantify even the nocturnal flight behavior of *H. armigera* (Riley *et al.,* 1992). Recently, forecasting methods of potato late blight, apple scab, mango powdery mildew and rice blast are available (Sinha and Banik, 2009).

4.4.9 Satellite Based Weather Forecasting

The specific applications include identification of primary weather system such as low pressure, depression, troughs/ridges, jet streams regions of intensive convection, inter-tropical convergence zones etc. and onset and progress of monsoon system.

Satellite imagery is being extensively used by synoptic network in conjunction with other available conventional meteorological data for analysis and weather forecasting. Zones of cloudiness are identified from the satellite imagery as regions of upward velocity and hence potential areas for occurrence of rainfall is identified.

Satellite imagery is very handy for remote and inaccessible areas such as Himalayas where heavy precipitation usually builds up. Though the characteristic cloud patterns of cold and warm fronts are not seen over India, the western disturbances giving rise to heavy snow fall are well captured in the Satellite imagery (Kalsi and Mishra, 1983). The cloud band ahead of well-marked westerly trough is clearly seen in the satellite imagery. The characteristic structure of snow is easily identified and its areal extent is monitored for estimating run-off and also for long range prediction of monsoon.

Deep penetrative cumulonimbus clouds and thunderstorm complexes are rather easy to be identified in visible and infrared imagery. Squall lines are clearly seen in the satellite loops. Satellite imagery provides powerful signals for forecasting severe weather out break (Purdom, 2003).

The rain bearing southwest monsoon system advances northward usually as an intermittent band of cloudiness called inter-tropical convergence zone (ITCZ). It comprises of numerous rain showers and thunderstorms associated with the convergence in the shear zone. The INSAT and NOAA sounding data have brought out the unique nature of monsoon onset with large scale changes in wind and moisture profiles in lower troposphere prior to monsoon onset. Joseph *et al.*, (2003) have identified conditions leading to onset of monsoon over Kerala using SST, OLR and winds obtained from satellite systems.

The monsoon depressions are the principal rain bearing systems of the southwest monsoon period over India. Satellite imagery shows heavy overcast cloud mass in the southern sector with low level cumulus clouds determining the Low-Level Circulation Centre (LLCC) to the northeast. The LLCC is often free of deep convection. The widespread and heavy rainfall in the southwest sector is often accompanied with deep convection in that sector. Kalsi *et al.,* (1996) have shown from satellite imagery that a few of these depressions acquire structure of marginal cyclones with almost vertical structure up to mid-tropospheric levels. Following Scoffield and Oliver (1977), Mishra *et al.,* (1988) also used the enhanced infrared satellite imagery to compute satellite derived rainfall estimates, which were found to be realistic. These signatures provide a lot of insight into physical and dynamical processes at work in the case of monsoon depressions and are extremely useful for short range forecasting.

The 16 parameters statistical model used by India Meteorological Department has several parameters that are provided by satellite data such as the SST, Snow Cover, EL Nino event etc. Several recent modeling studies show that a significant fraction of the inter-annual variability of monsoon is governed by internal chaotic dynamics

(Goswami, 1998). The numerical weather prediction of monsoon received impetus from the satellite observations. The parameters of SST, cloud motion vector, OLR are found to have impact on model results.

4.4.10 Satellite Data for Agro-met Advisory Service

Remote sensing techniques have been operationally used in many countries to provide basic information on crops, soils, water resources and the impact of drought and flood on agriculture. Procedures for pre-harvest acreage estimation of major crops such as wheat, rice and sorghum, using sampling and digital techniques based on remotely sensed data have improved greatly.

Satellite estimation of rainfall is not likely to be better than rainfall measured through conventional rain gauges, but nevertheless is useful to fill in spatial and temporal gaps in ground reports. Nageswara Rao and Rao (1984) demonstrated an approach for preparing an indicative drought map based on NOAA AVHRR derived rainfall estimation at the seedling stage of crop growth. For drought monitoring, quantitative point-specific rainfall estimates on the daily basis all over the country may not be required. What is needed, however, is a capability to spatially distribute the point rainfall observations over the areal unit in a qualitative manner.

Microwave sensors can provide estimates of soil moisture only in surface layers up to 10 cm thick. This depth is too shallow, compared to the 1-2 m root zone of many field crops in the tropics. Using the water content in the top 10 cm of the surface layer, the moisture content can be calculated within acceptable limits and with minimum error when the surface soil moisture estimation is made just before dawn.

During periods of drought conditions, physiological changes within vegetation may become apparent. Satellite sensors are capable of discerning many such changes through spectral radiance measurements and manipulation of this information into vegetation indices, which are sensitive to the rate of plant growth as well as to the amount of growth. Such indices are also sensitive to the changes in vegetation affected by moisture stress (Das, 2000).

Agrometeorological information is rarely provided as a finished product to the clients. Often it is used to complement the purely meteorological products or delivered in combination with other remotely sensed products, such as information in soil wetness, land or vegetation cover (NDVI), likely presence of pests and/or diseases, estimates of the areal coverage of irrigated or flood-retreat crops, incidence of bush fires, etc. By the nature of their capacity to indicate the probable areal extent of a condition and of the still very rapid evolution of the parameters that can be measured or derived, remotely-sensed data and their derived products will be a growing resource for the supply of agrometeorological products to clients.

This will not only help in planning, advising and monitoring the status of the crops but also will help in responding quickly for taking immediate planning or remedial actions. Planning for seeds distribution, fertilizer supply/requirements, supplying/relocating of sowing/harvesting equipment, procurement of crop from mandis/markets, *etc.*, can be tackled effectively through information derived using these technologies

The use of satellite based remote sensing has proved itself as a strong and unbiased information system at regular intervals of time. While agricultural scientists have shown some interest in developing its usage, there is still a long way to go, as it is only the agricultural scientists who can clearly define what information is actually needed. Besides, they should integrate the remotely sensed information system with their agricultural information system to derive optimum usage, timely recovery of degraded land and refrain from unsustainable activities by use of other advanced technologies to their benefit and to enable increasing productivity through alternate farming system.

4.4.11 Crop Insurance and Remote Sensing

Agricultural insurance is salient, particularly in agrarian societies, like India and China, where the economy is heavily dependent on its agricultural produce. Importance of such schemes in Nations like Kenya which is frequented by rampant fluctuations in weather conditions like 'droughts, floods, hailstorms, excessive rainfall, frost, lightning, and landslides. To add on to the perils, such industrializing Nations rely on food production, not solely for increasing the export shares but for mere subsistence. Such exports play a pivotal role even in industrialized Nations like the U.S, where food products still contribute to a substantial share of total exports produced.

Pradhan Mantri Fasal Bima Yojana (PMFBY), the Government of India's new crop insurance programme, which was launched during Kharif- 2016, is unique in many senses. One of that is promoting the use of technology for better implementation of the crop insurance. The technologies, promoted by PMFBY, are mostly in 3 domains: i) Information Technology, ii) Smartphone Applications, and iii) Satellite and UAV/ drone based Remote Sensing (Ray, 2018).

To address the difficulties that exist in ground-based data collection, the insurance industry had begun utilising satellite-based data either as a supplement to ground-based data or to create novel remote sensing insurance products. Remote sensors provide means to collect data based on specific 'bio-physical' variables like cloud temperature to estimate rainfall. These data accompanying other ground-based data are used to create indices. Hence within these parameters, the risk of loss or damage to crops can be assessed.

There are many possibilities of use of Satellite Remote Sensing (SRS) for crop insurance, includes

Smart Sampling: The Vegetation Index, which is derived from SRS and is representative of crop condition, can be used for designing better sampling plan for CCE

Yield Estimation: With the availability of high-resolution data, there is a possibility of getting crop yield or their proxies using SRS, at village level.

Area Discrepancy: SRS has been operationally used for crop area estimation (under FASAL project), this can be used for checking area discrepancy, especially for major crops of the country.

Yield Discrepancy and Quality Checking: checking yield data through statistical analysis, weather analysis, SRS based vegetation indices and other collateral information.

Loss Assessment and On Account Payment: Various applications have shown that it is possible to assess the severity of impact of drought, hailstorm and flood using satellite data. The NADAMS project regularly assesses the severity of drought at district/sub-district level using an integrated approach. Similarly, microwave remote sensing data are useful to assess the flood inundation of crops.

Prevented Sowing: A comparison of high-resolution SRS data of two years (normal year and the current year) can identify the areas with prevented sowing. This can be overlaid with digitized cadastral maps to identify the farms with prevented sowing condition.

Risk Zone Mapping: In absence of high-quality long-term yield data at block/taluka level, one can use long term satellite-based vegetation indices values for risk zoning. These can be combined with long-term weather data, disaster frequency, etc. for better risk analysis (Bala, 2018).

Furthermore, remote-sensing data archives are readily available, facilitating ease in making comparisons over time and in analysing trends. Remote sensing data are also gathered over large areas of space in a single unit of time and data collection frequently takes place; hence providing users with a more accurate, reliable and real-time set of data.

4.4.12 Forest fire monitoring

Remote sensing has been identified as an effective and efficient tool for monitoring and preventing forest fires, as well as a potential tool for getting an in-depth understanding of how forest ecosystems respond to them.

Through the use of remote sensing and GIS-based technology, forests can be monitored on a daily basis. In comparison to the conventional methods of data collection, satellite-based remote sensing provides the following advantages:

1. Large area coverage
2. Frequent and repetitive coverage of an area of interest
3. Comparably low cost per unit area of coverage
4. Semi-automated computerized processing and analysis via algorithms, etc.
5. Synoptic views in relation to the surrounding environment

Satellite remote sensing is well suited for assessing the extent of biomass burning– which happens to also be a prerequisite for estimating emissions at regional and global scales. This information can help authorities to better understand the effects of fire on climate change. Improved remote sensing techniques also have the potential to accurately date older fire scars and provide estimates of their burn severity.

Through this dashboard, there can be near real-time information on forest fires across a given area. Area Of Interest (AOI) can daily be monitored and reported to officials in charge of forest security and safety. With an end-to-end solution, reporting and taking action can happen with immediate transparency. Implementation is also very simple. Knowledge of these affected areas can aid understanding of the unique structure and dynamics of the vegetated landscape.

4.4.12.1 India

The forest fire is a major cause of degradation of India's forests. According to a paper written by NRSA, it is estimated that the proportion of forest areas prone to forest fires annually range from 33 per cent in some states to over a staggering 90 per cent in others. In recent years, State Forest Departments around the country have been forced to accept the fact that Indian forest fires are not the result of spontaneous combustion due to natural factors, but are rather set by humans. Also, India has recently witnessed a 125 per cent spike (from 15,937–35,888) in such fires in just two years (2015–2017). In three years, forest fires have devoured 875 hectares in the Pune district itself. Now more than ever, India needs to harness the power of GIS and remote sensing capabilities. To prevent these calamities, the Forest Department of India has announced to use GIS especially during the 'Fiery Months'.

The effects of all these fires add up and make a significant impact on the big picture. With populations rising and the amount of farmland decreasing, fires in Southeast Asia are estimated to be responsible for at least 10 per cent of global wildfire emissions. Satellite images have a considerable value in preventing and handling of burnt areas. Explicit knowledge about these areas can aid in the understanding of the structure and dynamics of the vegetation in the landscape. By leveraging the power of GIS and RS, important information on forests can be analyzed and evaluated to make well-informed decisions.

4.5. Problems and Possibilities in Remote Sensing for Agrometeorology

4.5.1 Problems in Remote Sensing

Land cover maps obtained at high spatial and temporal resolutions are necessary to support monitoring and management applications in areas with many smallholder and low-input agricultural systems. Various regional and global land cover products based on earth Observation data have been developed and made publicly available but their application in regions characterized by a large variety of agro-systems with a dynamic nature is limited by several constraints (Mananze *et al.*, 2020).Challenges in the classification of spatially heterogeneous landscapes, include the definition of the adequate spatial resolution and data input combinations for accurately mapping land cover.

Remote sensing-based estimation of woody cover in tropical dry forest ecosystems is challenging due to the heterogeneous woody and herbaceous vegetation structure and the large intra-annual variability in the vegetation due to the occurrence of seasonal rainfall (Passel *et al.*, 2020).

Accurate mapping of winter wheat over a large area is of great significance for guiding policy formulation related to food security, farmland management, and the international food trade. Due to the complex phenological features of winter wheat, the cloud contamination in time-series imagery, and the influence of the soil/snow background on vegetation indices, there remains no effective method for mapping winter wheat at a medium spatial resolution (10–30 m) (Dong *et al.*, 2020)

Progress in the development of sensor technology has increased the speed and convenience of remote sensing (RS) image acquisition. As the volume of RS images steadily increases, the challenge is no longer in producing and acquiring an RS image, but in finding a particular image from numerous RS images that precisely meets user application needs (Hong *et al.*, 2020).

Detailed and accurate information on the spatial variation of land cover and land use is a critical component of local ecology and environmental research. For these tasks, high spatial resolution images are required. Considering the tradeoff between high spatial and high temporal resolution in remote sensing images, many learning-based models (*e.g.*, Convolutional neural network, sparse coding, Bayesian network) have been established to improve the spatial resolution of coarse images in both the computer vision and remote sensing fields. However, data for training and testing in these learning-based methods are usually limited to a certain location and specific sensor, resulting in the limited ability to generalize the model across locations and sensors (Xiong *et al.*, 2020).

Crop species separation is essential for a wide range of agricultural applications—in particular, when seasonal information is needed. In general, remote sensing can provide such information with high accuracy, but in small structured agricultural areas, very high spatial resolution data (VHR) are required (Bohler *et al.*, 2020).

Progress in sensor technologies has allowed real-time monitoring of soil water. It is a challenge to model soil water content based on remote sensing data

4.5.2 Possibilities in Remote Sensing

- The variables comprised spectral bands from Landsat 7 ETM+ and Landsat 8 OLI/TIRS, vegetation indices and textural features and the classification performed within the Google Earth Engine cloud computing could produce reliable land cover maps.
- Partial Least Squares Regression (PLSR) model combined with Sentinel-2 satellite imagery is capable of monitoring woody cover in tropical dry forest regions, which can be used in support of reforestation efforts.
- Phenology-time weighted dynamic time warping (PT-DTW) for identifying winter wheat based on Sentinel 2A/B time-series data can be used to produce winter wheat map with accuracy of 89.98%.
- Recommendations of the INDEX indicator than those of available space (AS) and image extension (IE) and Hausdorff distance for single RS image type selections which is the most common scenario for RS image applications.

- Recently, generative adversarial nets (GANs), a new learning model from the deep learning field, show many advantages for capturing high dimensional nonlinear features over large samples.
- Laser scanning data from unmanned aerial vehicles (UAV-LS) offer new opportunities to estimate forest growing stock volume (VV) exclusively based on the UAV-LS data (Puliti *et al.*, 2020).
- The most accurate crop separation results were achieved using both the Imaging Spectroscopy (IS) dataset and the two combined datasets with an Average Accuracy (AA) of>92 per cent. In addition, in the case of a reduced number of IS features (*i.e.*, wavelengths), the accuracy can be compensated by using additional Near-InfraRed (NIR) Red Green Blue (NIR-RGB) texture features (AA > 90%).
- Surface soil moisture (SSM) can be modelled using Sentinel-1 backscatter data from 2016 to 2018 and ancillary data. However, future work is required to improve the model performance by including more SSM network measurements, assimilating Sentinel-1 data with other microwave, optical and thermal remote sensing products. There is also a need to improve the spatial resolution and accuracy of land surface parameter products (*e.g.*, soil properties and terrain parameters) at the regional and global scales (Chatterjee *et al.*, 2020).

Summarizing the remote sensing in agriculture are associated with its strengths in mapping both in time and space. Seasonal (temporal) and spatial variability of environmental characteristics, mainly crop and soil conditions can be mapped efficiently with remote sensing. These data can then be used to implement timely management actions as the crops develop.

5

Crop Simulation Models

5.1 Introduction

A Crop Simulation Model is a mathematical model which describes crop growth and development processes as a function of environment, soil conditions, and crop management. Simulation model is a schematic representation which works on set of underlying assumptions, equations and algorithms, closely replicating the behaviour of the agricultural systems. A crop growth simulation model predicts the harvestable yield. Apart from this final state of crop production, the model also provides a host of quantitative information about major processes involved in the various stages of crop development. The information covers each component of the crop such as leaves, stems, roots and harvestable products as they change over time, including the changes in soil moisture and nutrient status.

An increase in the population, demands an increase in agricultural production with available resources. Efficient management of available resources under varying weather conditions is important for increasing agricultural productivity. Besides that, the emphasis of agricultural production changes from quantity to quality and sustainability (Aggarwal *et al.*, 2000). Solution of these new challenges lies in understanding the interaction of the numerous elements affecting the plant growth and generating a crop growth model. This extremely complicated information needs to be processed and interpreted on a spatial scale. Information technology which has unleashed new frontiers in data analysis and automation is the driving force behind this model generation. These powerful computer programmes simulate the crop growth or yield of crops under various management regimes and provide near real time information to help farmers to make strategic decision in the management of their crops.

5.2 History and Development of Crop Models

Models were initially intended to improve understanding of crop behavior by illustrating crop growth and development in terms of physiological mechanisms. New ideas and numerous research questions inspired the development of simulation models over the years. In addition to their explanatory role, they quickly recognized the applicability of well-tested models for extrapolation and prediction and created more application-oriented models and now crop simulation models are considered as a potential research and application tool (Boote and Tollenaar, 1994).

The history of agricultural system modeling is characterized by a number of key events and drivers that led scientists from different disciplines to develop and use models for different purposes (Figure. 5.1). Heady (1957) and Heady and Dillon (1964) have done some earliest simulations of agricultural systems to refine decisions on a

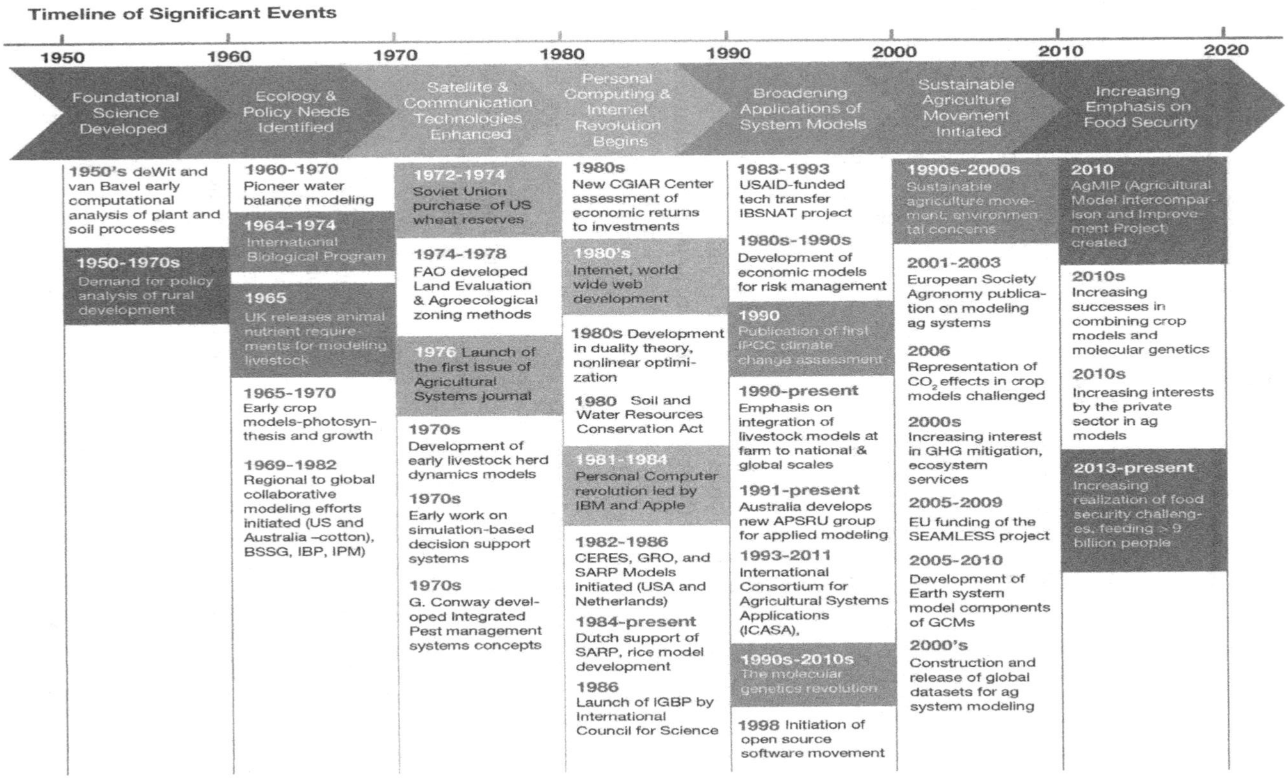

Timeline of Significant Events

1950 | 1960 | 1970 | 1980 | 1990 | 2000 | 2010 | 2020

Foundational Science Developed

Ecology & Policy Needs Identified

Satellite & Communication Technologies Enhanced

Personal Computing & Internet Revolution Begins

Broadening Applications of System Models

Sustainable Agriculture Movement Initiated

Increasing Emphasis on Food Security

1950's deWit and van Bavel early computational analysis of plant and soil processes

1950-1970s Demand for policy analysis of rural development

1960-1970 Pioneer water balance modeling

1964-1974 International Biological Program

1965 UK releases animal nutrient requirements for modeling livestock

1965-1970 Early crop models-photosynthesis and growth

1969-1982 Regional to global collaborative modeling efforts initiated (US and Australia –cotton), BSSG, IBP, IPM)

1972-1974 Soviet Union purchase of US wheat reserves

1974-1978 FAO developed Land Evaluation & Agroecological zoning methods

1976 Launch of the first issue of Agricultural Systems journal

1970s Development of early livestock herd dynamics models

1970s Early work on simulation-based decision support systems

1970s G. Conway developed Integrated Pest management systems concepts

1980s New CGIAR Center assessment of economic returns to investments

1980's Internet, world wide web development

1980s Development in duality theory, nonlinear optimization

1980 Soil and Water Resources Conservation Act

1981-1984 Personal Computer revolution led by IBM and Apple

1982-1986 CERES, GRO, and SARP Models initiated (USA and Netherlands)

1984-present Dutch support of SARP, rice model development

1986 Launch of IGBP by International Council for Science

1983-1993 USAID-funded tech transfer IBSNAT project

1980s-1990s Development of economic models for risk management

1990 Publication of first IPCC climate change assessment

1990-present Emphasis on integration of livestock models at farm to national & global scales

1991-present Australia develops new APSRU group for applied modeling

1993-2011 International Consortium for Agricultural Systems Applications (ICASA),

1990s-2010s The molecular genetics revolution

1998 Initiation of open source software movement

1990s-2000s Sustainable agriculture movement, environmental concerns

2001-2003 European Society Agronomy publication on modeling ag systems

2006 Representation of CO_2 effects in crop models challenged

2000s Increasing interest in GHG mitigation, ecosystem services

2005-2009 EU funding of the SEAMLESS project

2005-2010 Development of Earth system model components of GCMs

2000's Construction and release of global datasets for ag system modeling

2010 AgMIP (Agricultural Model Intercomparison and Improvement Project) created

2010s Increasing successes in combining crop models and molecular genetics

2010s Increasing interests by the private sector in ag models

2013-present Increasing realization of food security challenges, feeding > 9 billion people

(*Source:* Jones *et al.*, 2016)

Fig. 5.1. Summary timeline of selected key events and drivers that influenced the development of agricultural system models.

farm scale and assess the impact of policies on the economic benefits of rural growth, which has inspired additional economic modelling. Dent and Blackie (1979) included models of farming systems with economic and biological components and provided an important source for different disciplines to learn about agricultural systems modeling.

With creation of the International Biological Program (IBP), agricultural economists started modeling farming systems, which has led to the development of various ecological models, including models of grasslands during the late 1960's and early 1970's. As a part of IBP, ecological scientists formed research tools that allowed them to study the complex behavior of ecosystems as affected by a range of environmental drivers (Worthington, 1975; Van Dyne and Anway, 1976). The IBP initiative brought together scientists from different countries, different types of government, and different attitudes toward science (Van Dyne, 1980). Before this program, systems modeling and analysis were not practiced in scientific efforts to understand complex natural systems. IBP left a legacy of thinking and conceptual and mathematical modeling that contributed strongly to the evolution of systems approaches for studying natural systems and their interactions with other components of more comprehensive and managed systems (Coleman *et al.*, 2004).

Models of agricultural production systems were first conceived during 1960's by a physicist, De Wit, of Wageningen University, who believed that agricultural systems could be modeled by combining physical and biological principles. Duncan *et al.*, (1967) published a paper on modeling canopy photosynthesis is an enduring development that has been cited and used by many crop modeling groups and he began creating some of the first crop-specific simulation models for corn, cotton, and peanut (Duncan, 1972).

Some of the first crop models were curiosity-driven with scientists and engineers from different disciplines developing new ways of studying agricultural systems that differed from traditional reductionist approaches and inspiring others to get involved in a new and risky research approach. In 1969, a regional research project was initiated in the USA to develop and use production system models for improving cotton production, building on the ideas of de Wit, Duncan, and Herb Stapleton (Stapleton *et al.*, 1973), an agricultural engineer in Arizona. In 1972, the development of crop models received a major boost after the US government was surprised by large purchases of wheat by the Soviet Union, causing major price increases and global wheat shortages (Pinter *et al.*, 2003). New research programs were funded to create crop models that would allow the USA to use them with newly available remote sensing information to predict the production of major crops that were grown anywhere in the world and traded internationally. This led to the development of the CERES-Wheat and CERES-Maize crop models by Joe Ritchie and his colleagues in Texas (Ritchie and Otter, 1984; Jones and Kiniry, 1986). These two models have continually evolved and are now contained in the DSSAT suite of crop models (Jones *et al.*, 2003; Hoogenboom *et al.*, 2012).

The work started by the early pioneers has continued to evolve throughout the years. Notably, Wageningen University has carried on the legacy of C. T. deWit by training many agricultural system modelers and by developing a number of crop models that are still in use today (Penning de Vries *et al.*, 1991; Bouma *et al*, 1996; van Ittersum *et al.*, 2003). Similarly, some of the early work of Duncan, Ritchie, and others has evolved and contributed to the widely-used DSSAT suite of crop models through collaborative efforts among the University of Hawaii, University of Florida, Michigan State University, the International Fertilizer Development Institute, Washington State University, and others (IBSNAT, 1984; Tsuji *et al.*, 1998; Uehara and Tsuji, 1998; Jones *et al.*, 2003; Hoogenboom *et al.*, 2012).

There were other notable government-funded initiatives in the U.S., Netherlands, and Australia that led to major developments of crop, livestock, and economic models. This includes the 1980 US Soil and Water Conservation Act that led to development of the EPIC model that is still in use today (Williams *et al.*, 1983, 1989), the USAID-funded IBSNAT project that led to the creation of the DSSAT suite of crop models that incorporated the CERES and CROPGRO models (Jones, 1993; Boote *et al.*, 1998, 2010; Jones *et al.*, 2003; Hoogenboom *et al.*, 2012), and the Systems Analysis of Rice Production (SARP) project funded by the Dutch government starting in 1984 that led to the development of the ORYZA rice crop model, now widely used globally (Penning de Vries *et al.*, 1991; Bouman *et al.*, 2001). The establishment of the first fully funded, multidisciplinary crop modeling-oriented research group in Australia in the early 1990s led to the development of the APSIM suite of cropping system models. This was a major milestone and APSIM is currently one of the most widely-used suites of models (McCown *et al.*, 1996; Keating *et al.*, 2003). Another major event was the development of the SEAMLESS project, funded in 2005 and operated for five years. This effort led to major collaboration among agricultural systems modelers and scientists across Europe for development of new data interfaces and models and for the development and integration of models at field, farm, and broader spatial scales, including cropping system and socioeconomic models (van Ittersum *et al.*, 2008).

The concept of Huffaker Integrated Pest Management emerged in the 1970's, on the pests and diseases of plantation crops in Malaysia infused funds for developing insect and disease models of several crops, combined with experimental efforts aimed at reducing pesticide use and more effective use of all measures to prevent economic damage to major crops (Conway, 1987). Coincident with this project was a major increase in the sophistication of population dynamic models in ecology and a growing appreciation of the importance of nonlinearities and the problems for forecasting they imply (Dempster, 1983; Hassell, 1986; Gutierrez *et al.*, 1994; Murdoch, 1994). In 1972, Integrated Pest Management (IPM) project was funded in the USA to address the major problems associated with increasing pesticide use and development of resistance to pesticides by many of the target insects and diseases (Pimentel and Peshin, 2014). Globally, the FAO and various countries were also promoting IPM, with modeling as one of the approaches used to understand how to manage pests and diseases with minimal pesticide use. During this time period, a number of insect and disease dynamic models were developed, and some were coupled with cotton and soybean crop models (Wilkerson *et al.*, 1983; Batchelor *et al.*, 1993), including the

SOYGRO model that is now in DSSAT (Jones *et al.,* 2003). This period of time also led to the development of a general framework for coupling crop models with insect and disease information to estimate impacts on growth and yield (Boote *et al.,* 1983).

The evolution of economic models for different scales and purposes progressed steadily during the last five decades. These developments were fueled by various needs at national and international levels as well as innovations in modeling approaches by the agricultural economics community. The needs included mandates of CGIAR Centers to evaluate returns on investments in research for development, the increased interest in liberalizing global agricultural trade, the evaluation of ecosystem services, and impacts of climate change and adaptation (Rosenzweig and Parry, 1994; Curry *et al.,* 1990; Thornton *et al.,* 2006; Nelson *et al.,* 2009). This steady progress included the development of agricultural risk management analyses, evaluation of national, regional and global policies, and integration of other models with economic models for more holistic assessments, including crop, livestock, grassland, and hydrology models (Havlik *et al.,* 2014; Nelson *et al.,* 2013; Rosegrant *et al.,* 2009; De Fraiture *et al.,* 2007; Elliott *et al.,* 2014a, 2014b).

The advent of personal computers from 1980's and the increasing computing power on the desktop enabled many individual researchers to work with agricultural systems and develop their own models. Another important development in the late 1990's was the open source movement which led to the creation of open source agricultural system models. The access to open source code, enabled community-based development of modules and database interfaces for porting into standard cropping system models like APSIM and DSSAT.

In parallel to funded initiatives, scientists started creating consortia and networks to enhance collaboration for specific purposes, one such example is ICASA- International Consortium for Agricultural Systems Applications formed in 1993 for developing data standards to use with crop models (Ritchie, 1996; Bouma and Jones, 2001) and also constructed and released global datasets of cropping areas, sowing dates, yields, and other management inputs which was used for regional and global analyses of agricultural systems (Ramankutty and Foley, 1999; You *et al.,* 2006; Monfreda *et al.,* 2008; Ramankutty *et al.,* 2008; Fritz *et al.,* 2013).

A global project, International Geosphere-Biosphere Program (IGBP) formed in 1986, led to increasing interest in climate change and the use of models to assess what likely impacts might be under future climate conditions. An early motivation for model use in climate change research was the publication of the first IPCC assessment report on climate change (IPCC, 1990). This led to the use of crop, livestock, and economic models to assess climate change impacts on agriculture as well as agricultural adaptation and mitigation options (Rosenzweig and Parry, 1994). This then prompted crop modelers to incorporate CO_2 effects on crop growth and yield if this effect was missing and to use the models to perform simulation experiments using current and future projected climate conditions (*e.g.,* Curry *et al.,* 1990; Tubiello *et al.,* 2002; Waha *et al.,* 2013). These simulated changes in crop productivity were used in socioeconomic models to evaluate impacts on agricultural trade, food prices, and distribution of impacts (*e.g.*, Rosenzweig and Parry, 1994; Adams *et al.*, 1990; Fischer *et al.*, 1995). Many studies have been conducted since the first work that was led by Rosenzweig, Parry, and others, in particular to provide information for subsequent IPCC assessments as well as various national and regional assessments (Fischer *et al.*, 1995; Rosenzweig and Parry, 1994; Parry *et al.*, 2004).

Long *et al.*, (2006) proved the positive fertilization effects of CO_2 would offset the negative effects of rising temperature and lower soil moisture. Currently, more data are now available from FACE (Free Air CO_2 Experiments) and T-FACE (Temperature FACE) experiments to more comprehensively evaluate and improve the interactive effects of temperature, soil moisture, and CO_2 in current models (Kimball, 2010; Boote *et al.*, 2010). Conducting such evaluations and improvements is one of the goals of the Agricultural Model Inter comparison and Improvement Project (AgMIP) (Rosenzweig *et al.*, 2013a; Thorburn *et al.*, 2014).

The creation of AgMIP in 2010 is a major milestone in the evolution of agricultural models. This initiative created a global community of agricultural system modelers with the goals of comparing and improving crop, livestock, and socioeconomic models, and using the improved models for assessing impacts and adaptation to climate change and climate variability at local to global scales, including evaluating the uncertainties of those assessments (*e.g.*, Asseng *et al.*, 2013; Bassu *et al.*, 2014; Li *et al.*, 2015; Rosenzweig *et al.*, 2013b). Since its start, AgMIP has promoted virtual collaboration among all agricultural modeling groups globally, creating new opportunities for substantially improving abilities to understand and predict agricultural responses to climate, including interacting effects of CO_2, temperature, and water.

Finally, the increasing interest in improving the representation of the earth's land area in regional and global climate models has led to new approaches for modeling agricultural systems (Osborne *et al.*, 2009). This work has led various modeling groups to develop models that represent CO_2, water, and GHG fluxes and also crop growth and yield of grid-cell areas (*e.g.*, Fischer *et al.*, 1995; Rosenzweig *et al.*, 2013b; Elliott *et al.*, 2014a, 2014b). On the livestock side, global gridded models for feed consumption, productivity, manure production, and greenhouse gas emissions for dairy, beef, small ruminants and pork and poultry are now available (Herrero *et al.*, 2013; FAO, 2013).

Major opportunities for advancing agricultural system models, databases, and knowledge products is the evolution of Information and Computer Technologies (ICT). In particular, ICT is likely to lead to more user-driven knowledge products and their linkage to model development. These opportunities are elaborated in the papers by Antle *et al.*, (2017) and Janssen *et al.*, (2017). The continued dedication to develop reliable models has been one of the main features of many agricultural modeling efforts for cropping systems, livestock, and economics (*e.g.*, DSSAT, EPIC, APSIM, STICS, WOFOST, ORYZA, CROPSYST, RZWQM, TOA, IMPACT, SWAP, and GTAP).

5.3 Crop Simulation Models: Paradigm Shift Moving Beyond Individual Crops to Farm Systems

5.3.1 Types of Crop Simulation Models

The crop simulation models are classified into different groups or types depending upon the purpose for which it is designed. These are broad categories that enriches agricultural model development, scientific understanding, decision making and policy support systems (*e.g.*, Loomis *et al.*, 1979; Passioura, 1996; Boote *et al.*, 1996; Bouman *et al.*, 1996; van Ittersum *et al.*, 2003; Ritchie, 1991; McCown *et al.*, 1996).

(a) **Explanatory** models, for improving basic scientific knowledge of agricultural system components or knowledge of interactions leading to overall responses of such systems (Van Ittersum *et al.*, 1998). These models tend to be mechanistic models, as they are usually based on the regulation of known or hypothesized physical, chemical and biological processes occurring in crop or animal production systems, for examples, models of photosynthesis (Farquhar *et al.*, 1980) and water movement in soils (Richards, 1931).

(b) **Descriptive** models, for providing information for supporting decisions and policies, it describes how the agricultural system responds to the external environmental drivers as well as decisions or policies under consideration (Van Ittersum *et al.*, 1998). In this case, independent model-based analysis is used to provide societally relevant knowledge for both public and private decision-making, and analyses are performed to support specific public policy processes and decisions. Users of such models may be interested in prediction of responses that would help guide decisions, or they may be interested in how the system would respond if a particular decision was made.

(c) **Statistical models** to express the relationship between yield or yield components and weather parameters. In these models relationships are measured in a system using statistical technique. Example: Step down regressions, correlation, etc.

(d) **Mechanistic models** explain not only the relationship between weather parameters and yield, but also the mechanism of these models (explains the relationship of influencing dependent variables). These models are based on physical selection.

(e) **Deterministic models** estimate the exact value of the yield or dependent variable. These models also have defined coefficients.

(f) **Stochastic models** have probability element attached to each output. For each set of inputs different outputs are given along with probabilities. These models define yield or state of dependent variable at a given rate.

(g) **Dynamic model** has included time as a variable. Both dependent and independent variables are having values which remain constant over a given period of time.

(h) **Static models** have no time variables. Dependent and independent variables having values remain constant over a given period of time.

(i) **Simulation models** are computer models and in general, are a mathematical representation of a real-world system. One of the main goals of crop simulation models is to estimate agricultural production as a function of weather and soil conditions as well as crop management. These models use one or more sets of differential equations, and calculate both rate and state variables over time, normally from planting until harvest maturity or final harvest.

5.3.2 Spatial and Temporal Scales of Agricultural System Models

Agricultural systems are managed ecosystems comprised of biological, physical and human components operating at various scales. The need for more integrated, farming-system models has been recognized by many researchers for several decades, to carry out the analysis of the trade-offs encountered in improving the sustainability of agricultural systems (Kanter *et al.,* 2018). In the past, crop modelling was built on a single crop-by-crop approach. A new paradigm shift is now emerging that moves beyond 'crops' to 'farming systems.' These new farming system simulation tools incorporate the complexity that comes with many interacting biophysical and socioeconomic components. Users of models or information derived from them and the models themselves vary considerably across spatial and temporal scales as given in Figure 5.2. The scope of the system being modelled and managed varies depending on the questions being asked and the decisions and policies that are being studied.

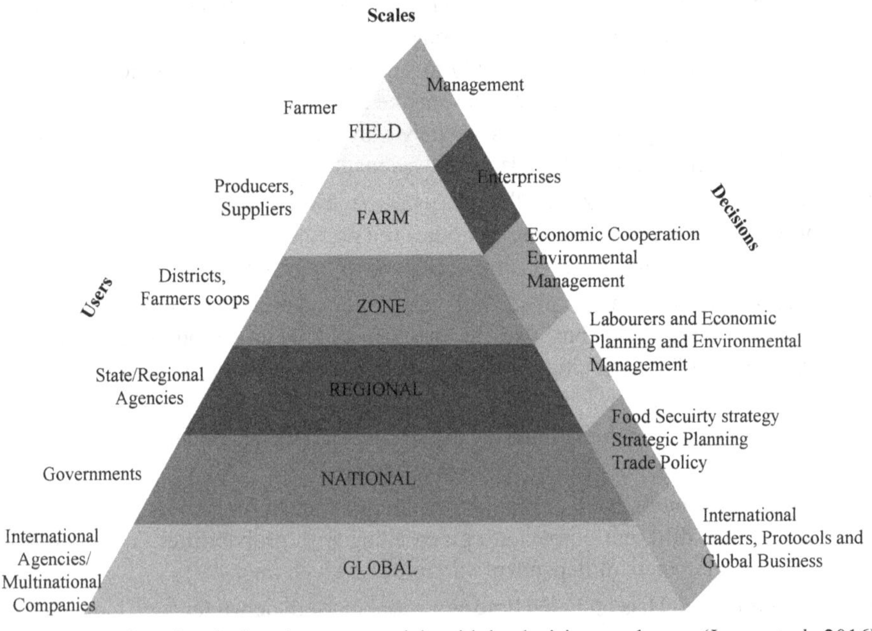

Fig. 5.2. Scales of agricultural system models with its decisions and users (Jones *et al.,* 2016)

5.4. Applications of Crop Simulation Models

5.4.1. Integration of Knowledge Across Disciplines–Standardized Framework

The use of a modular approach in software coding helps the scientist to independently build and test the modules before later incorporating into the main model. The adoption of such a standardized framework enables the integration of basic research that is carried out in different parts of the world. For example, for a particular crop, the modular aspect of the APSIM software allows for the integration of knowledge across crops and across disciplines. This ensures a reduction of research costs by avoiding duplication of research as well as collaboration between researchers at an international level.

5.4.2 Assessing The Yield Gaps and Genetic Gains of Various Crops Towards Crop Intensification

It is predicted that the world population will reach 9 billion by 2050 that will demand an additional 60 per cent more food. This increase means that we need a significant boost in current primary agricultural productivity, and it is estimated that 80 per cent of the increase needed may result from crop intensification. There is increasing concern that "business as usual" may not allow food production to keep up with demand - a situation that could lead to dramatic food price increases, poverty and hunger (FAO, 2003, 2006; Royal Society of London, 2009; Koning and van Ittersum, 2009; Godfray *et al.*, 2010). Achieving the required production target without further large-scale conversion of land to agriculture requires greater increase in crop productivity. However, crop productivity varies greatly from place to place, depending on environment, inputs and practices. Assessing the yield gap of existing cropped lands will indicate the possible extent of yield increase from actual values.

Potential yield is defined as the yield of a crop cultivar when grown in environments to which it is adapted, with nutrients and water non-limiting, and pests and diseases effectively controlled (Evans, 1993). Hence, for a given crop variety or hybrid in a specific growth environment, yield potential is determined by the prevailing weather conditions, mainly the amount of incident solar radiation and temperature. The difference between yield potential and the actual yield represents the exploitable yield gap (Figure. 5.3.)

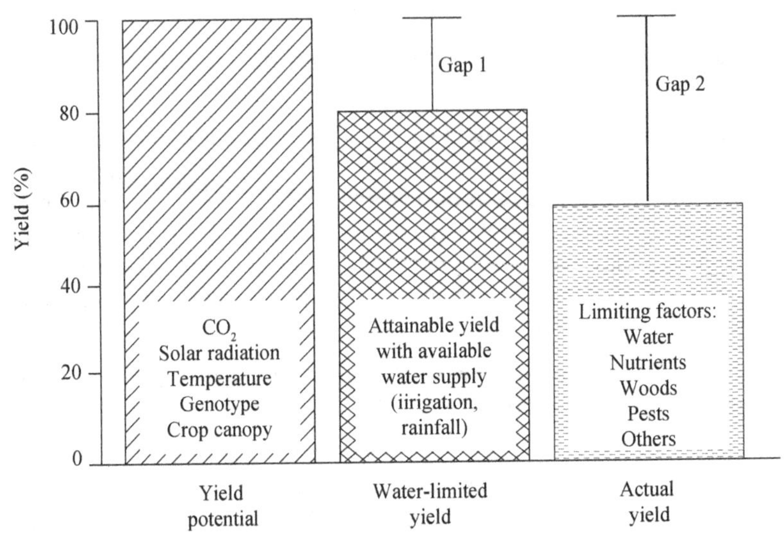

Figu. 5.3. Conceptual framework of potential yield, water limited yield and actual farm yield (Cassman *et al.*, 2010)

Water-limited yield potential is determined by genotype, solar radiation, temperature, plant population, and degree of water deficit. However, the actual farm yields are decided by the yield reduction or loss from factors such as nutrient deficiencies or imbalances, poor soil quality, root and/or shoot diseases, insect pests, weed competition, water limitation, water logging, and lodging.

Crop simulation models are useful tools for estimating regional crop productivity. The mechanistic models such as DSSAT, APSIM and Wheat Grow are constructed based on the processes of crop growth and yield formation, can predict the growth and development dynamics of a given crop or variety under different pedoclimatic conditions. These models have been used extensively to evaluate the effects of environmental, biological, and management changes on crop growth and production.

5.4.3 Application in Crop Breeding Programme- Genomic Selection

Crop simulation models have become more detailed and mechanistic, they closely mimic the agricultural system. More precise information can be obtained regarding the impact of different genetic traits on economic yields and these can be integrated in genetic improvement programs. Researchers use this modelling approach of environmental data for the genomic selection in crop breeding to design crop ideotypes for specific environments.

Analysis of interactions between Genotype, Environment and Management factors requires a deep understanding of their biological basis. Kholova *et al.,* (2014) used APSIM to demonstrate that several genetic traits underlying a stay green phenotype in sorghum increased grain yield across sorghum production regions in India. Some of these trait modifications resulted in trade-offs between grain and stover productivity. For example, breeding for a smaller leaf canopy reduced pre-anthesis water use, which improved water availability for grain filling and ultimately increased yields, but was at the expense of stover productivity, which is unfavorable for farmers, who are interested in crop residues for fodder.

5.4.4 Input Management–Crop Growth Models as Decision Support System

Management decisions regarding cultural practices and inputs have a major impact on yield. Simulation models, which allow the specification of management alternatives, provide a fairly inexpensive means of testing a large range of approaches that would easily become too costly if the conventional approach to experimentation were to be implemented. Proven records are available describing the use of simulation models with respect to cultural management such as selection of variety, optimizing planting date and plant geometry, irrigation optimization and fertilizer scheduling.

Optimizing planting dates

Optimization of planting dates would help the plants to grow in a better environment and also have the potential to offset the negative impacts due to changes in climate (Tingem *et al.*, 2009; Adger *et al.*, 2005; Geethalakshmi *et al.*, 2011). In general, optimum sowing window is decided based on local field experiments conducted with different dates of sowing for few years, few locations and few varieties and final recommendations are extrapolated to other environments. However, extrapolating the results of a limited number of environments is not only difficult but may be misleading (Andarzian *et al.*, 2007; Savin *et al.*, 1995; Timsina *et al.*, 2008). In this context, crop simulation models evaluated using local experimental data could be valuable tools for extrapolating the experimental results to other years and other locations (Mathews *et al.*, 2002).

Crop simulation models have been used to investigate the performance of different cultivars in a range of sowing dates under different soil and climate scenarios (Bannayan *et al.*, 2013; Bassu *et al.*, 2009; Heng *et al.*, 2007; Pecetti and Hollington, 1997). The DSSAT, APSIM, INFOCROP are few comprehensive decision support systems under agro-technology tools (Hoogenboom *et al.*, 2010; Tsuji *et al.*, 1998) that includes model for major crops (Ritchie *et al.*, 1998; Ritchie and Otter, 1985). The CERES-Wheat model can be used to simulate the growth and development of dry land and irrigated wheat across a range of latitudes (Hoogenboom *et al.*, 2010; Nain and Kersebaum, 2007). This model has been evaluated and applied in a range of tropical (Timsina *et al.*, 1995), sub-tropical (Heng *et al.*, 2007) and temperate environments of Asia (Timsina and Humphreys, 2006; Zhang *et al.*, 2013). Similarly, CERES-Rice model is used for rice crop and other respective models are employed for the specific crops for fixing optimum sowing window to enhance the productivity of the crops under current as well as future warmer climate.

Irrigation optimization

In many dry regions of the world, water is the major factor limiting agricultural production. Agricultural, being a major user of water, is expected to use water efficiently to ensure sustainability. Dynamic crop simulation models such as Aqua Crop (Steduto *et al.*, 2009; Raes *et al.*, 2009), DSSAT (Jones *et al.*, 2003), EPIC (Williams, 1995), provide a cost-effective framework for assessing water management strategies to optimize the use of limited water resources in both irrigated and dryland cropping systems. Scheduling deficit irrigation is more complicated compared to scheduling full irrigation because it requires knowledge of crop yield response to water and effects at different growth stages. Linker and Kisekka (2017) demonstrated the potential of model-based optimization of deficit irrigation for determining adequate soil water depletion levels that minimize yield penalty of maize crop using CERES-Maize model.

5.4.5. Risk Mitigation and Management

Using a combination of simulated yields and gross margins, economic risks and weather-related variability can be assessed. These data can then be used as an investment decision support tool.

5.4.5.1. Pest and disease impact assessment and their management

Quantifying the impacts of plant pests and diseases on crop yields is one of the most important research questions for modeling agricultural simulation (Newman *et al.*, 2003, Savary *et al.*, 2006, Esker *et al.*, 2012, Whish *et al.*, 2015). In the past, theoretical models were established to take into account the effect of pests and disease on yield as differentiated by other limiting factors due to the interactions between genotype, environment and management. De Wit and Penning de Vries (1982) introduced the concept of production situation, which includes the combination of yield defining and yield limiting factors, thus determining the attainable yield. Reduction of crop yield due to biotic stresses corresponds to the difference between the attainable and actual yield.

Applied modelling of crop diseases and pests has been dominated by tactical short-term issues such as developing decision support systems to schedule scouting or pesticide applications (Welch *et al.*, 1978, Magarey *et al.*, 2002, Isard *et al.*, 2015). These modelling activities are often focused on specific pest-crop systems, in specific environments, and multi seasonal observations, that allowed building of robust empirical relationships using weather variables and crop phenology (Madden *et al.*, 2007). Two of the most popular examples are phenology models for insect pests (Welch *et al.*, 1978) and SEIR (Susceptible-Exposed-Infectious-Removed) and infection models for plant pathogens (Zadoks, 1971, Magarey *et al.*, 2005). These kinds of model could have application for determining how the changing climate might also alter the frequency of pesticide applications. In some cases, it may be possible to estimate yield impacts by converting forecasts of pest or disease intensity to projections of yield loss (Dillehay *et al.*, 2005).

The process-based insect growth model (GILSYM) is integrated in the process-based crop model EPIC in a modular form, enabling daily feedback between insect growth and consumption and crop growth and stress. If insects are to be simulated, GILSYM is called from the daily loop in EPIC and runs through subroutines simulating insect growth, survival, oviposition, migration, feeding, damage and transition. The damage is compared to a user-specified damage threshold and an insecticide is applied if the threshold is exceeded. Pesticide dosage mentioned in crop model is taken into pest model and efficacy also applied to the pest population (Rasche and Taylor, 2019).

Currently, many decision-making, tactical models for plant diseases rely on collected data on yield losses as reference points, which may no longer be reliable in altered environmental conditions (Donatelli *et al.,* 2017). In EPIC-GILSYM, changing environmental conditions serve as a driver of change for all other processes and cropping systems can be well defined in terms of input intensity, which is an important aspect in the mechanistic simulation of food systems (Savary *et al.,* 2017).

5.4.5.2. Climate change impact assessment

Climate is changing and is expected to change much faster than it was before. Especially with increasing CO_2 emission into the atmosphere, temperatures are increasing with high certainty every year. Importantly the rainfall pattern and quantity are also undergoing wide variations. Changing climate would affect the crop growth and development of all the crops with varying degrees of impact. With growing unpredictability in climate, newer research efforts are being directed to develop adaptation strategies to meet the impact from changing climate.

Mainly for assessing the impact of climate change, climate manipulative experiments such as FACE, FATE and controlled climate chamber facilities are used, which are costly affair and all the research institutes cannot afford to create the facilities. Alternatively, dynamic process-based crop simulation models that are re-calibrated using the newer data sets obtained from manipulated experiment and applied the models across environments and production systems are proved to be successful in assessing the impact of climate change on growth and productivity of crops. Effects of climate change on three major crops–wheat, potato and rice using DSSAT CERES-Wheat, SUBSTOR-Potato and CERES-Rice models under the present and projected future changing climatic conditions with 5.32°C increase in temperature by 2100 indicated 47.6 per cent, 67.8 per cent and 38.6 per cent decrease in yield respectively (Rahman *et al.,* 2018).

However, crop simulation models in assessing the impact should be used with little caution as the models currently being used were developed using control environment experiments and old crop cultivars prevalent during 1970's and 1980's, thus these models may not have incorporated the latest knowledge on crop physiology and response to changing environment, thus may not represent modern crop cultivars and management practices properly (Palosuo *et al.,* 2011; Rotter *et al.,* 2012). Furthermore, most of the modelling studies carried out recently through AgMIP project across the globe indicated the uncertainty that exist with individual models, which limits the model outputs in decision making process (Baranowski *et al.*, 2015, Hoffmann *et al.*, 2015, Persson *et al.*, 2015, Pirttioja *et al.*, 2015, Salack *et al.*, 2015, Semenov & Stratonovitch 2015, Zhao *et al.*, 2015). One such example is simulating the altered management practice in rice crop viz., system of rice intensification (SRI) is not yet possible with DSSAT or any other crop models. Scientists are trying to alter the source code in accordance with the results of the field experiments to mimic the SRI through crop models (Tao *et al.*, 2015).

5.4.5.3. In Socio-economic applications

Climate change has socio-economic impact on individual, society and Nations. Crop simulation models are used in socio-economic studies for foresight analysis of agricultural systems under global change scenarios and the consequences of potential food system shocks (Reynolds *et al.*, 2018). The predicted effects of climate change call for a multi-dimensional method to assess the performance of various agricultural systems across economic, environmental and social dimensions. Climate Smart Agriculture (CSA) recognizes that the three goals of climate adaptation, mitigation and resilience must be integrated into the framework of a sustainable agricultural system. New simulation-based method was developed based on the Regional Integrated Assessment (RIA) methods developed by the Agricultural Model Inter-comparison and Improvement Project (AgMIP) for climate impact assessment. This method combines available data, field- and stakeholder-based surveys, biophysical and economic models, and future climate and socio-economic scenarios. It features an integrated farm and household approach and accounts for heterogeneity across biophysical and socioeconomic variables as well as temporal variability of climate indicators. This method allows for assessment of the technologies and practices of an agricultural system to achieve the three goals of CSA.

In crop insurance applications

Crop Insurance is risk mitigation effort to support the farmer in times of distress. The insurers use these models to run probabilistic worst-case scenarios to decide on the quantum and scope of insurance cover. For example, shock scenarios were developed for wheat stem rust spread based on Strong El-Nino conditions and explored using the **IFPRI IMPACT model**. Lloyd's model was constructed by de-trending globally aggregated FAO data from 1961 to 2013 to identify probable changes in crop area and yield. The estimated production response to the shock was deemed drastic but plausible (Lunt *et al.*, 2016). The effects emulate recent country-level effects on major crops, such as maize in southern Africa, where major yield losses have been observed. According to the IMPACT bio-economic model, production shocks would generate global losses of 7 per cent for wheat, 10 per cent for maize, 11 per cent for soybean and 7 per cent for rice.

5.4.6 Application in Policy Decisions

Yield estimation is done for making policy decisions on import or export of a crop produce, price fixation and also for crop insurance claim settlement purpose. Crop simulation models play a major role in yield prediction and estimation at varied scales.

a. Introduction of a new crop

Agricultural research is linked to the prevailing cropping system in a particular region. Hence, data concerning the growth and development of a new crop in that region would be lacking. Developing a simulation model based on scientific data collected elsewhere and a few datasets collected in the new environment helps in the assessment of temporal variability in yield using long- term climatic data. Running the simulations with meteorological data in a balanced network of locations also helps in introducing new crops in a location.

b. Linking with remote sensing for yield estimation

Remote sensing information when linked with crop simulation models can act as an effective tool to predict the yield in advance at district, State and National level. Spatial analysis tool built in crop simulation model is used for this purpose. In AgMIP programme an interface has been created to handle large quantity of data with respect to management, soil, genetics and weather to perform the spatial analysis. When a model with a sound physiological background is adopted, it is possible to extrapolate to other environments. Several simulation models are used to assess climatically-determined yield in various crops and in separating yield gain into components due to changing weather trends, genetic improvements and improved technology. For example, the CANEGRO model has been used along the same lines in the South African sugar industry. Through this modelling approach, quantification of yield reductions caused by non-climatic causes (*e.g.*, delayed sowing, soil fertility, pests and diseases) becomes possible.

5.5 Case Studies of Application of Crop Simulation Model for Farm Management Decisions

Simulation modelling is increasingly being applied in research, teaching, farm resource management, policy analysis and production forecasts. Some case studies are presented below.

5.5.1 Application of APSIM Model in Deciding the Best Cropping System

The Agricultural Production Systems Simulator APSIM is one of the dynamic crop growth simulation models, capable of dealing with water and N dynamics under different fertility management conditions (mineral and organic amendments). An example of how model outcome can be used to guide the field experimentation, given that testing all options in field is not feasible and are too expensive. Irrigated rice (*Oryza sativa* L.)-wheat (*Triticum aestivum* L.) in northwest India is entirely dependent upon groundwater pumped from the underlying aquifer, resulting in alarming rates of reductions of the groundwater table and the potential for saline groundwater intrusion into fresh groundwater (Humphreys *et al.,* 2010). There is therefore great interest in finding sustainable solutions to reduce the decline in groundwater table. As the rice crop is highly water demanding, rice equivalent yield (REY), irrigation requirement and evapotranspiration of a rice-wheat system and maize-wheat system in central

Punjab, India was evaluated (Figure 5.4a-d).

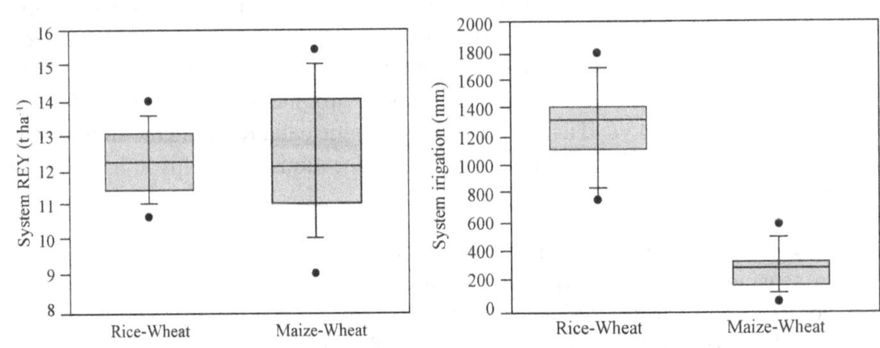

(a) Probabilistic distribution of rice equivalent yield

(b) Probabilistic distribution of irrigation

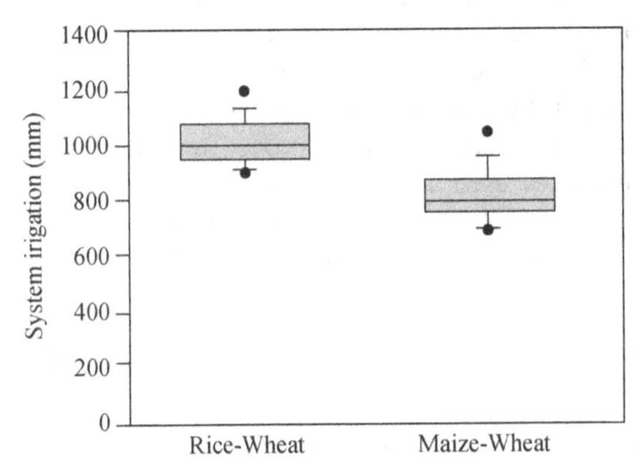

(c) Probabilistic distribution of evapotranspiration of a rice-wheat system and maize-wheat system

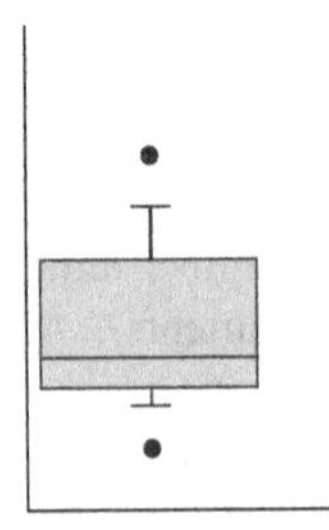

Shaded bars are 25th–75th percentiles; whisker caps are 10th and 90th percentile and black dots 5th and 95th percentile, horizontal line is median

(d) Details about shaded bars

(*Source*: Humphreys *et al.*, 2010)

Fig. 5.4 (a-d) Evaluation of rice-wheat system and maize-wheat system in Central Punjab in India over 40 years (1970–2010)

APSIM model was applied to evaluate the effects of different crop management practices such as rice variety, duration, sowing date, replacing rice with other crops, and the use of conservation agriculture on the land and water productivity of the total system, and on the components of water balance (Balwinder-Singh *et al.,* 2015). From the Figure 4, it is clear that maize-wheat system is best suited compared to rice-wheat system in Punjab, in terms of using less water and for producing more rice equivalent yield. This signifies the usage of dynamic crop simulation model in making important decision on cropping system. If minimizing permanent loss of water from the system is the objective, the best method is a partial conservation agriculture rice-wheat system with a short duration rice variety. If maximizing productivity is the sole objective, a full conservation agriculture system with a medium duration variety is the best option. If the objective is to produce the maximum output per unit of evapotranspiration, this can be achieved using a short duration variety planted in mid-July under full conservation agriculture (Balwinder-Singh *et al.,* 2015). Replacing rice with rainy season maize (*Zea mays* L.) in the system can maintain yields while reducing the total irrigation amount by 80 per cent, which can contribute to huge reductions in pumping costs and energy use. Maize-based systems also reduce permanent water loss from the system by 200 mm. This approach can be extended to rainfed rice systems worldwide as a risk reducing strategy in a variable monsoon.

5.5.2. Application of DSSAT Model in Programming Irrigation

In Bangladesh's southern delta, farmers mostly grow grass pea (*Lathyrus sativus* L.) and mungbean (*Vigna radiata* L.) during the *Rabi* (dry) season (December–April) (Krupnik et al., 2017). These crops are generally not irrigated, instead drawing on residual soil moisture and the fact that the water table is close to the surface, fluctuating between 1–2.5 m depth in the dry season (Shamsudduha *et al.,* 2009). In this environment, tactical irrigation is a useful and cheap option to increase water use efficiency, especially as these crops can be a major source of cash income for smallholder farmers, while significant yield losses are often caused by sub-optimal plant establishment due to dry conditions around sowing. The International Maize and Wheat Improvement Center (CIMMYT) conducted several experiments on maize, mung bean, and wheat in the delta region to better understand irrigation needs and develop a decision support app for farmers and irrigation service providers.

Simulated soil water content, with and without capillary up flow (a function of soil type, rooting depth, and distance to the water table), based on work by Talsma (1963) and an algorithm developed by Meyer for the farm-level water balance model BASINMAN (Figure. 5.5). The soil-water balance used in the smart phone application is based on the widely used CERES model, of which several specific crop variants are included in DSSAT.

The soil-water balance is based on the widely used CERES model, embedded in DSSAT. The green line represents the drained upper limit, while the red indicates the target stress level (50% of drained upper limit until anthesis and 30% thereafter). With a water table present, irrigations were required 7 and 32 days after planting, whereas without a water table, two more irrigations were required at 77 and 92 days after planting.

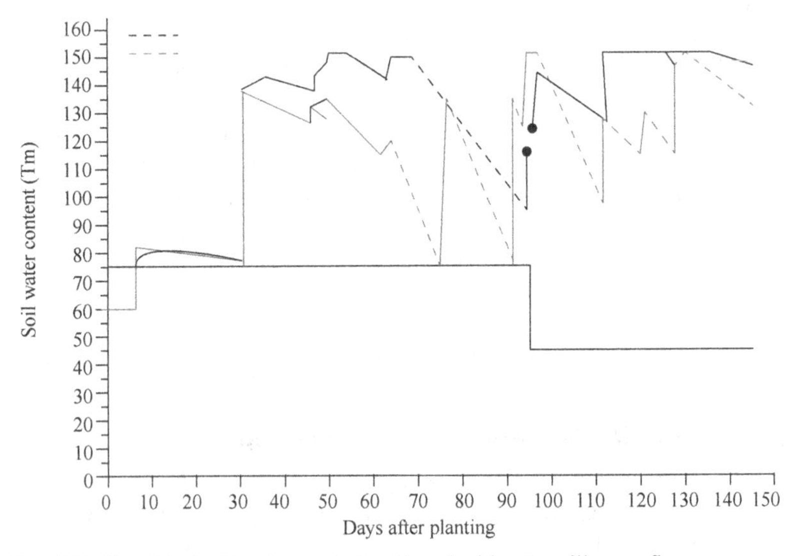

Fig. 5.5. Simulated soil water content with and without capillary upflow.
(*Source*: Reynolds *et al.*, 2018)
***Note*:** Bars at the bottom indicate daily upflow from the water table.

The simulation model has been integrated into a smartphone application called PANI (Program for Advanced Numerical Irrigation), which runs on a daily time step and uses forecasted daily maximum and minimum temperatures to predict irrigation needs one week in advance. PANI addresses the needs of irrigation service provider and farmers, as both receive a weekly SMS informing them whether a field needs to be irrigated or not. PANI demonstrates that it is possible to bring crop simulation models, or components thereof, to the fingertips of smallholder farmers. It is possible to automatically run the model with representative soil and weather information based on the GPS location; the only inputs required from the user are sowing and irrigation dates and ground cover photos taken every 10–15 days.

5.5.3 Application of APSIM Model in Nitrogen Management Decision

A study conducted in Sahelian centre of ICRISAT tested the performance (calibration and validation) of the APSIM model in terms of pearl millet response under Sahelian climatic conditions, to the three main fertility management practices: manure (M1: 300 kg/ha; M2: 900 kg/ha; M3: 2700 kg/ha), crop residue (R1: 300 kg/ha; R2: 900 kg/ha; R3: 2700 kg/ha) and fertilizer (F1:0 N + 0 P; F2: 15 kg/ha N + 4.4 kg/ha P; F3: 45

kg/ha N + 13.1 kg/ha P) and used the validated model to identify N application rates that are better adapted to subsistence small-holder millet farming in the Sahel through scenario analyses based on long term (23 years) simulations (Akponikpè *et al.*, 2010). Comparison between observed and APSIM simulated grain (b) yield is presented in Figure. 5.6.

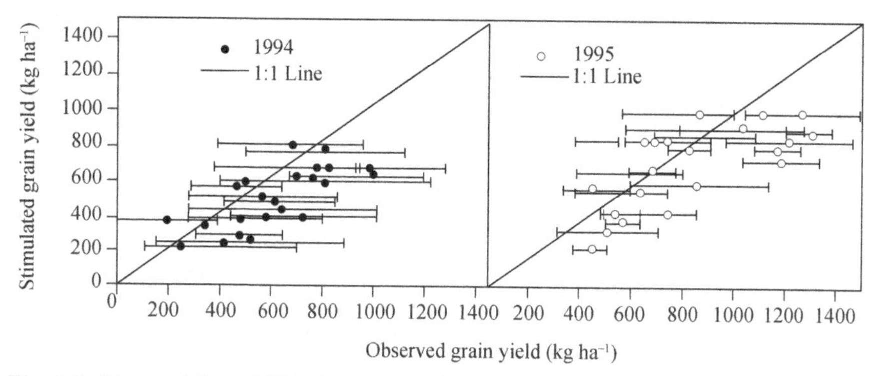

Fig. 5.6. Observed Vs. APSIM simulated grain yield as affected by combined application of cattle manure, crop residue and mineral fertilizer in 1994 and 1995.
(*Source*: Akponikpè *et al.*, 2010)

Note: Error bars denote standard deviation of observed means, n = 3.

For the total above ground biomass at harvest, the RMSE was 615 and 1294 kg ha^{-1} in 1994 and 1995, respectively. The d index of agreement was 0.84 and 0.67 for these two years, respectively. APSIM model performance was relatively good in 1994 but biomass yield was slightly over predicted in 1995. A 23-year, long term simulation with the validated APSIM model showed that moderate N application (15 kg N ha^{-1}) improves both the long-term average and the minimum guaranteed yield without increasing inter-annual variability compared to no N input. Although it does imply a lower average yield than the current fertilizer recommendation (30 kg N ha^{-1}), the application of 15 kg N ha^{-1} appears more appropriate for small-holder, subsistence farmers as it guarantees higher minimum yield in worst years, thereby reducing their vulnerability (Akponikpè *et al.*, 2010).

5.5.4 Python-based Environmental Policy Integrated Climate (PEPIC) Model for Nitrogen Loss Assessment

Spatially explicit and crop-specific information on global N losses into the environment and knowledge of trade-offs between N losses and crop yields are largely lacking. Crop growth model, Python-based Environmental Policy Integrated Climate (PEPIC) was used to determine global N losses from three major food crops: maize, rice, and wheat. Rice accounts for the highest N losses, followed by wheat and maize. The N Loss Intensity (NLI), defined as N losses per unit of yield, presents high variation among different countries (Figure. 5.7). Simulations of mitigation scenarios indicate that redistributing global N inputs and improving N management could significantly abate N losses and at the same time even increase yields without any additional total N inputs (Liu *et al.*, 2016).

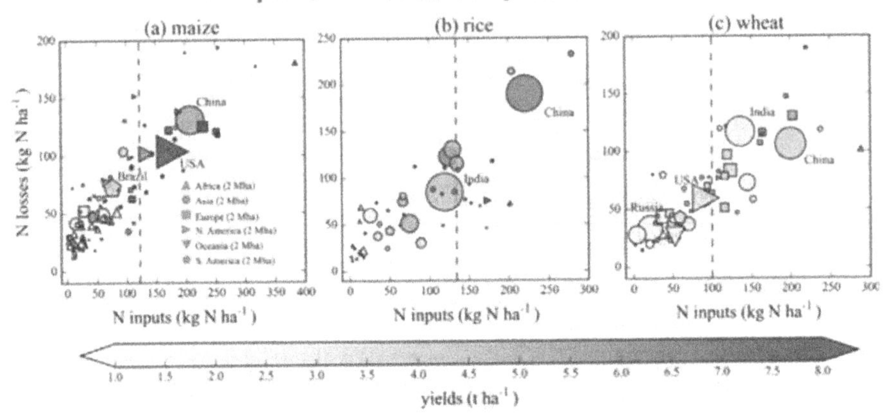

Fig. 5.7. Annual loss of Nitrogen to environment and water for major countries around the world
(*Source*: Liu *et al.*, 2016).

The global TN_t (environment loss) and TN_w (water loss) were 14,253 and 8584 Gg N yr-1. Asia has about 45 per cent of total global wheat yield but 60 and 69 per cent losses in nitrogen applied to the field. Default parameters of the EPIC model were used for global application, as it is difficult to adjust model parameters at specific regions on such a large scale (Liu *et al.*, 2016).

5.5.5 AQUACROP- Economic Model for Irrigation Management

A model at farm scale was developed and applied to an area in South-western Spain to assist farmers in pre-season decision making on cropping patterns and on irrigation strategies. Yield predictions were obtained from the Aqua Crop model which was validated for four different crops. The results were fed into the economic model. The model simulated the impact on farm income of: (a) irrigation water constraints (b) variations in agricultural policies (c) changes in product and water prices and (d) variations in the communication to farmers of the specific level of irrigation water allocation. The model predicted a strong negative impact on farm income of delaying a decision on the level of seasonal water allocation by the water authority, reaching up to 300 € ha^{-1} in the study area (García-Vila and Fereres, 2012). The results in Figure. 5.8 emphasize the need to improve the decision-making process by water authorities to attain the desired goals of properly managing water scarcity without compromising the sustainability of farming.

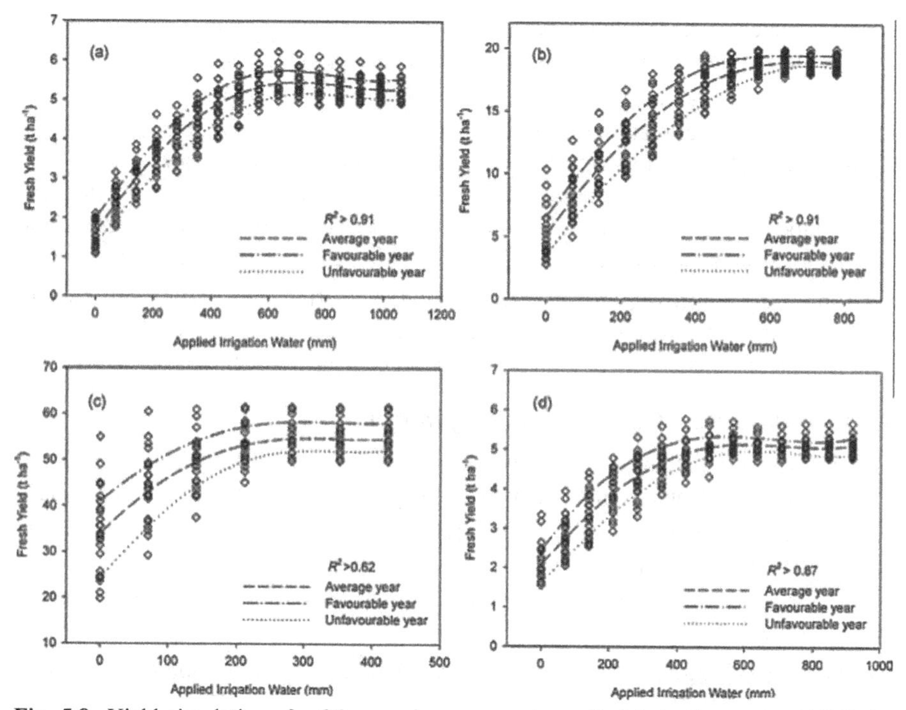

Fig. 5.8. Yield simulations for 26 years in response to applied irrigation water (AIW) for: (a) cotton; (b) maize; (c) potato; and (d) sunflower
(*Source*: García-Vila and Fereres, 2012).

Note: showing the water-production functions for an average, a favourable (80 percentil of Pc), and an unfavourable year (20 percentil of Pc). Pc = Normal price of produce obtained by the farmer.

5.6 Challenges of Crop Modelling

- Crop simulation models exist for major crops, but their use in backstopping crop improvement remains marginal. This is, in part, due to the convention of breeding 'mega-varieties,' cultivars thought to be adaptable to many different farming contexts. Crop simulation brings a new paradigm of breeding for specific conditions, focused at smaller geographical scales, to deliver larger overall returns.
- Looking ahead, modelling analysis can be embedded in the scope of breeding programs. Good modelling tools exist (*e.g.*, Simple Simulation Modeling, SSS), but the community of scientists using these tools is still small and we need to do more to improve their use.
- Key frontiers include expanding the number of pulse species and cultivars modeled as well as developing the coefficients and datasets needed to validate models for different pulses in specific regions.
- More work is also needed to simulate the beneficial effects of pulse crops in farming

systems such as the amount of residual nitrogen available to crops following pulses in a rotation and the nutritional (*e.g.* protein) benefits produced on a farm when pulses are grown.

- Global gridded climate–crop model ensembles are increasingly used to make projections of how climate change will affect future crop yield. However, the level of certainty that can be attributed to such simulations is unknown. Parameterizing crop models with grid-specific information on wheat cultivars tended to decrease the crop model uncertainty, particularly for low latitudes (Xiong et al., 2019).
- Crop model improvements and better-quality spatial input data more closely representing the wide range of growing conditions around the world will be needed to reduce the uncertainty of climate change impact assessment of crop yields

5.6.1 Challenges of Crop Modelling in India

1. Accurate prediction of climate extremes in the simulation models and projection at finer resolutions so that it helps to develop appropriate preparedness policies
2. Human resource development to perform ground-level data collections and modeling in a regional scale
3. Conducting long-term field experiments to understand fundamental climate processes
4. Land use changes are the major driving agent for the water resources compared to climatic threats.
5. Basin-scale water modeling should be processed through spatiotemporally downscaling.
6. Location-specific indigenous model for simulating country-specific climatic impact is essential.
7. Biophysical and socioeconomic determinant of the crop productivity should be included in the simulation model.
8. Data repository development and meta-data reliability assurance
9. Knowledge gaps should be filled, and information must be reached in intact form by the end users in time.

Finally, the development of crop simulation models is both an art and a science. The complex process of the understanding all agro inputs, visualization of all the inter-related parameters and delivering the model in time to the farmers is exciting and challenging. Fast changing climate, perceptible change in temperature and rainfall pattern means that the past data needs to be revalidated in greater frequency.

These crop simulation models can be used for a wide range of applications. As research tools, model development and application can contribute to identifying gaps in our knowledge, thus enabling more efficient and targeted research planning. Models that are based on sound physiological data are capable of supporting extrapolation to alternative cropping cycles and locations, thus permitting the quantification of temporal and spatial variability. Over a relatively short time span and at comparatively low costs, the modeler can investigate a large number of management strategies that would not be possible using traditional methodologies. Despite some limitations, the modelling approach remains the best means of assessing the effects of future global climate change, thus helping in the formulation of national policies for mitigation purposes. Other policy issues, like yield forecasting, industry planning, operations management, consequences of management decisions on environmental issues, are also well supported by modelling.

6

Weather Codes and Their Management

Under normal weather situation, agriculture depends on timely occurrence of rainfall for crop management activities irrespective of different farming situations namely, irrigated and unirrigated agriculture. Changes in temperature and rainfall pattern together with frequent occurrence of extreme weather events are major anticipated threats from the climate change that occurs now around the world. This threat in long run, if not addressed properly, would affect food security of a country. Pratap Narain *et al.,* (2000) observed an increase in moderate and severe drought occurrence numbers in the arid zone of India during last decade (1991-2000) as compared to the earlier decades of the last century. Similarly, Pasupalak (2008) reported daily maximum rainfall of 400.3 mm in the last decade against earlier record of 256.4 mm (1969–78) in a location at Odisha State. Further the inter- annual variation within the decade did increase and also the number of rainy days with very heavy rainfall > 125 mm also found increased in the location against the daily mean rainfall of 20 mm/day in India.

A day may be with heavy rainfall or with bright sunshine or windy or in combination of these. When heavy rainfall continues for more than a week with some system like cyclone, floods may be the result. Similarly, if dry spell continues during a cropping season for more than 15 days, drought would initiate in that region. This is again depending upon the nature of the soil water holding capacity and nature of soils, topography and green cover of that region. In between these two extremes, we have normal season, where we get normal seasonal rainfall during a cropping season. Hence like dress code, in agro-climatology, there are three weather codes, namely floods code, drought code and normal code. The jargon weather code was developed by Dr.M. S Swaminathan in India. Across the countries of the World, if these weather codes occur, the impact would be universal. The strength of adaptation capacity to maneuver these extreme events by the farming community rests with their past experience made for similar situations, financial support and also the technology options that are available from the scientific institutions.

The India Meteorological Department(IMD) has developed a common and scientific rules that whenever the seasonal rainfall of a particular year of a particular region falls between (+) 19 to (–) 19 per cent of the long period average, then that season is called as normal season, while the seasonal rainfall falls between (–) 19 to (–) 59 per cent from the long period average, then it is called as deficit seasonal rainfall, while if the seasonal rainfall falls between (–) 59 to (–) 99 per cent, then it is classified as drought. On the other side, the excess season rainfall and wet seasonal rain fall are brought under (+) 19 to (+) 59 per cent and (+) 59 to (+) 99 per cent respectively of the long period average seasonal rainfall. For computing normal seasonal rainfall, a minimum of thirty years past seasonal rain fall data are required. Prof. Dr. M. S Swaminathan used to spell out always that farmers must be empowered to meet the challenge from these different weather codes namely floods code, and drought code, which are common to be seen under monsoon rainfall situations. In India all the three weather codes namely normal, drought and floods codes are common to be seen. Under climate variability situation which is triggered by natural balance, these weather codes would come in rhythm or come under mathematical function. This means that out of five years in Tamil Nadu, one year would be with drought code, one year with floods code and three years with normal weather code. But under "Climate change" scenario, the frequency occurrence of both floods and drought codes would be more and they may not follow any fixed rotation as noticed under climate variability scenario. Balasubramanian (2008) in his concept paper on Farmer and weather, indicated that community must be prepared with skill to meet the challenges from floods and droughts based on their past experience as cited elsewhere in this chapter. He further suggested that selection of a well-trained climate manager from the community at the village level to manage the climate for crop and animal production is a need of the hour and it is a must for sustaining society welfare.

For managing drought and flood weather codes in agriculture, the information's from long range weather forecast or seasonal climate forecast are to be considered. In a nut shell weather based farm decisions have to be taken whether it is tactical or strategical decisions. But for normal weather code, the existing normal agricultural practices can be continued. Let us look on these weather codes in detail.

6.1 Normal Weather Code

Since normal seasonal rainfall is anticipated based on the long-range weather forecast, farmers can go with their existing plan of crop and animal managements, by taking weather-based farm decisions especially for medium range weather forecast and short-range weather forecast information.

6.2 Flood Weather Code

This situation occurs when excess rainfall occurs during a cropping season. It may occur at any time during cropping season and hence suitable crop selection is very important along with planning for water drainage from the fields. Flood weather information can be known before the on-set of the season based on long range/seasonal climate forecast. For field and crop management during cropping season, both medium range and short-range weather forecast information's are necessary. Pre seasonal workshop meeting along with preparation of contingency planning is required before the on-set of seasons based on long range/seasonal climate forecast information.

6.3. Drought Weather Code

This situation occurs when lesser than 25 per cent of mean seasonal rainfall is received during a cropping season. Drought may occur during early period of the season or mid period of season or late period of the season. The drought weather information can be known before the on-set of the season based on long range/seasonal climate forecast. Whenever drought occurs, the crop's productivity would be below normal or sometimes there is high probability for crop's failure. Aridity index map given by

IMD or satellite imageries of remote sensing satellite would provide enough support to give real time information on the occurrence and spread of the drought over a region. For field and crop management, during cropping season, both medium range and short-range weather forecast information are necessary for taking weather-based farm decisions. If the occurrence of drought is known through seasonal climate forecast or long-range forecast, the wise decision is to reduce the strength of animals in the farm as well as reducing the cultivable area. If the anticipated drought intensity is found be lesser, selection of drought tolerant crops along with recommended technologies for drought management could be followed and no doubt, the productivity would be lesser than normal weather code. Pre seasonal workshop meeting along with preparation of contingency planning is required based on the long range or seasonal climate forecast information received during the on-set of a season.

6.4 Contingency Plan

Contingency plan is a process that prepares action plan to respond coherently to aberrant weather situation like drought or flood or to prepare agricultural plan to an emergency situation to sustain productivity. In literature it is indicated that contingency plan is a plan devised for an outcome other than in the usual plan. It is often used for risk management. The crop management activities meant for normal weather situation cannot be practiced either during floods or droughts situation. Hence technologies suitable for crop production to manage floods and droughts must be tailored in advance to sowing season for each village in consultation with farming community and agricultural extension personnel. Past experiences also could be considered for this process. Preparation of contingency plan mainly depends on forecasted long range weather forecast or seasonal climate forecast information and under Indian condition, these forecast on rainfall information's are available 15 days in advance of seasonal sowing. The India Meteorological Department provides south west monsoon rainfall forecast (June–Sept) in advance. Similarly, the Agro- climate Research Centre of TNAU provides seasonal climate forecast for south west monsoon and as well as for north east monsoon seasonal rainfall for different districts of Tamil Nadu in advance. From the weather forecast, we can infer that whether the coming season is droughty season or flood season or normal season. Accordingly, contingency plan can be prepared for droughty and flood weather codes. It is not necessary to prepare plan for normal weather code. It is suggested to prepare the contingency plan for a village level rather than for a block level, since variability is getting enlarged at block level. Precious plan could be obtained at the village level. The contingency plan preparation includes:

- Preparation of contingency plan for agricultural activities
- Preparation of contingency plan for animal activity including fodder management
- Preparation of plan for drinking water and other natural resources management

The steps to prepare the contingency plan includes

1. Let us understand what is long range weather forecast or seasonal climate forecast information for the ensuing crops season (Whether it is droughty season or flood season or normal season)
2. Arrange pre seasonal village level meeting at village level and inform the anticipated weather code and discuss the options to be implemented with the villagers.
3. Prepare a detailed plan covering crop and animal managements and discuss with selected elderly farmers and get their views and get refined. The indigenous knowledge's that are already available may also be considered for inclusion as best bet technologies or community practices. Technical knowledge from extension wing also can be obtained and incorporated in the plan.
4. Procurement of inputs and fixing responsibilities with the villagers on the division of work to be attended
5. Mass communication of the plan prepared with in the village and keep ready for implementation.

The following points must be considered for preparing contingency planning

(a) Existing cropping pattern/cropping system
(b) Ground/surface water availability
(c) Forage/fodder requirement
(d) Crops to be cultivated
(e) Variety to be selected for the selected crop
(f) Technology to be selected including best bet technology
(g) Input requirement and procurement
(h) Pest and disease load at the village level
(i) Labor availability
(j) Normal agricultural practices of the village
(k) Farmer's previous experience at the village.

As indicated already elsewhere in this book, weather is defined as the day to day change in the atmosphere. The anticipated change in the weather for three to five days in advance in terms of change in rainfall, temperature, wind speed, wind direction, cloud amount and relative humidity will be communicated to the farmers by the scientific body through weather forecast. The scientific body predicts change in weather elements though Global circulation model or by using Regional climate model as the case may be. This forecast will be communicated to the client farmers for making weather-based farm decisions. Especially in agriculture weather forecast information would be very useful to change or shift the proposed farm operations to reduce crop management risks. Weather-based farm decisions is defined as altering the proposed farm operations/decisions to reduce climate and weather-related risks based on weather forecast information received.

Some examples are as follows

- If heavy rain is forecasted on the proposed day of rice harvest, postpone the rice harvests.
- If rain more than 25 mm is received during dry land sowing season, sowing of dry land crops can be done successfully.
- Anticipating rain proposed irrigation schedule can be postponed to 10–15 days later based on the soil water holding capacity

  There are two types of weather-based farm decisions and those are.

  (a) Farm decisions based on the weather forecast

  (b) Farm decisions based on the prevailing weather

To understand contingency plan for three weather codes let us go for on example.

Name of the village: A

Total area under cultivation: 500 ha (dry land/upland-200 ha; low land- 100 ha; irrigated upland-200 ha)

Cattle strength: 1000 (cows-100; buffaloes-400; goats-500)

Forest cover of the village: 25 per cent of the total village area

Soil type and AWHC: Sandy clay loam; 1.5 mm/cm depth of soil

Normal weather code:

Rainfall anticipated based on long range weather forecast during SWM season; 600mm (normal seasonal rainfall: 650 mm); NEM season; 400 mm (normal seasonal rainfall 450mm).

Cropping pattern/system followed at the village level based on animal fodder requirement under normal situation;

Dry land/upland: Maize (June-Sept)- chickpea (Oct- Dec)-200ha

Low land: Rice (June- Oct)-rice (Nov-Feb)- rice fallow black gram- 100ha

Irrigated upland: Maize (June-Sept)-groundnut (Oct-Jan)-sesame (Feb-May)-200ha

Crop production and animal production technologies are followed as per the recommendations of State Department of Agriculture.

Floods weather code

Rainfall anticipated based on long range weather forecast during SWM; 1200 mm (normal seasonal rainfall: 650 mm) and 600 mm during NEM season against seasonal average rainfall of 450 mm.

Cropping pattern/system to be followed at the village level based on animal fodder requirement as per the discussion made by the farmers during pre-seasonal climate workshop conducted at village level in the presence of technical people;

Dry land: Rice (June-Sept)- ragi (Oct- Jan)-200 ha under compartmental bunding, ridges and furrows land managements with enough draining facilities.

Low land: Rice (June- Oct)-rice (Nov-Feb)- rice fallow black gram (March -May)-100 ha

Irrigated upland: Sugar cane (June-March) -200ha

Crop production and animal production technologies are based on contingency crop plan as decided by the villagers during pre-seasonal climate workshop conducted at village level in the presence of technical people.

Drought weather code
Rainfall anticipated based on long range weather forecast during SWM; 150 mm (normal seasonal rainfall: 650 mm) and 100 mm during NEM season against seasonal average rainfall of 450 mm.

Cropping pattern/system to be followed at the village level based on animal fodder requirement as per the discussion made by the farmers during pre-seasonal climate workshop conducted at village level in the presence of technical people;.

Dry land: Fallow -200 ha

Low land: Green manure (June- Oct)- Fallow- 100 ha

Irrigated upland: Sorghum (June-Sept)–Minor millets (Oct-Jan) 200 ha

Crop production and animal production technologies are based on contingency crop plan as decided by the villagers during pre-seasonal climate workshop conducted at village level in the presence of technical people

Animal strength will be reduced to 50 per cent for sustainable maintenance.

6.5 General Management During Floods and Drought

6.5.1 Floods

Normally flood occurs when the intensity of rainfall is more than 100 mm per hour against Indian average of 20 mm per day followed by its duration of rain intensity that occurs. In addition, the nature of topography of a region, its soil type and ground cover also are very important factors to trigger flood situation. When the runoff of water from heavy rainfall is > 30 per cent and the infiltration capacity of the soil is lesser than the run-off water, then there is every opportunity to get floods situation. Normally flood occurs with cloud bursting and also during monsoon season with the onset of cyclones. In an average, from the genesis of cyclone to its land fall, it lives for 7–10 days. This duration along with heavy rainfall from the cyclone would sufficient to make foods situation in an area. Floods may be slow or fast rising but in general it develops over a period of days based on the factors responsible for floods occurrence. Many types of floods could be seen across literature and among them are river floods, coastal floods, urban floods and flash floods. Whatever may be the types of flood, its management in general is very important considering the human and animal loss and also property losses.

6.5.1.1 Reasons for floods

- Intensity of rainfall: > 80 mm/hour
- Duration of rainfall intensity: > four hours continuously of rainfall >80mm
- Topography of the area
- Ground cover
- Amount of accumulated debris including plastic materials in drainage channels, stream and river course
- Silting in water bodies
- Converting water bodies for habitation
- Land slides
- Dam or tank bursts

6.5.1.2 Managing floods

How to manage floods through overall planning and their execution in advance includes;

- Desilting water bodies
- Strengthening the bunds of water bodies especially tanks, ponds etc.,
- Repairing the sluices of the tanks
- Providing contour bunding in cultivable area
- Make the cultivable area in to good tilth before start of the season for easy and higher infiltration
- All drainage channels in the inhabited area may be free from debris including polyethylene bags
- All runoff water may be directed to abandoned open wells
- Getting information's from early warning centre

6.5.2 Droughts

There is no universal definition for drought. Each country in the world has its own definition for its application purposes. In India the IMD defines drought as If a meteorological division receives seasonal rainfall < 75 % of normal seasonal rainfall, then drought occurs.

Rainfall quantity to be received	Category of drought
If a meteorological division receives seasonal rainfall between 25 and 50 % of normal seasonal rainfall......	Moderate drought
If a meteorological division receives seasonal rainfall < 25 % of normal seasonal rainfall......	Severe drought

There are different types of drought namely. meteorological drought, hydrological drought, agricultural drought, sociological drought and socio-economic drought. Whatever may the droughts, the main point is non availability of needy water for use. The meteorological drought is nothing but receipt of deficit rainfall to level of < 75 per cent of normal seasonal rainfall. The hydrological drought indicates non availability of water in the water bodies. The agricultural drought occurs when soil moisture is inadequate to meet crops ET. The sociological drought is nothing but the confusion and fearness of the society to meet the challenges from incoming drought. Under socio-economic drought, there is lesser purchasing capacity, zero cash flow, poverty, foodless day and so on. As a result, sometimes revolution also may happen. Or a situation, where water shortage ultimately affects the established economy of the region adversely

Since we are interested in agro-climatology, under agricultural droughts, there are umpteen number of classifications and those are;

- Early season drought
- Mid-season drought
- Late season drought
- Apparent drought-rainfall may be sufficient for one crop (pulse) and not sufficient for the other crop (cotton) raised in the same season and same area under rainfed/dry land
- Permanent drought- Over years the soil moisture is insufficient to meet crops ET

Based on periodicity of drought occurrence, there are two regions and those are;

1. Drought affected area-Drought occurs once in a way
2. Drought prone area -Permanent drought; the solution is to provide river water by linking.

Many causes have been observed for the occurrence of drought and among them the following are for worthy to note:

- Geographical position of the region
- Global warming and Climate change
- Deforestation
- Atmospheric pollution
- Early withdrawal of seasonal rainfall or late onset of seasonal rainfall or mid-season dry spell
- ENSO (El-Nino and Southern Oscillation)
- Higher surface albedo would affect radiation balance

6.5.2.1 Understanding drought

The drought occurrence can be known through many ways in advance and those are;

1. Understanding the long-range weather forecast information given for the season or understanding seasonal climate forecast information given for the season.
2. Analyzing historical rainfall data for the probability of occurrence.
3. EL-Nino signal.
4. Seasonal initial probability rainfall analysis.
5. Satellite derived data.

6. Agricultural Rainfall Index (ARI). This can be computed as ARI = Present monthly rainfall (mm)/monthly mean PET (mm); if the worked value is < 40, then it is interpreted that drought situation starts.
7. Aridity Anomaly Index: Aridity Index (IA) = (PE- AE)/PE, wherein IA = Weekly aridity index, AE = actual evapotranspiration from water balance study, PE = Potential evapotranspiration. The departure of real time IA from its long-term mean is expressed in per centage as aridity anomaly. From the computed aridity anomaly, the drought intensity can be measured as follows:

Aridity anomaly	Drought intensity
< 25 %	Mild drought
26-50%	Moderate drought
> 50 %	Severe drought

8. Getting early warning information from the centre.

6.5.2.2 Impact of drought

The Impacts from drought are many and the important ones are as follows;

- Crops failure and sometimes productivity loss
- Change in land use pattern
- Depletion in usable water
- Non availability of fodder
- Distress sale of farm animals
- Imbalance in rural economy
- Social disharmony
- Political instability
- Spread of poverty
- Ecological imbalance
- Conflicts for resources

6.5.2.3 Drought management

- Preparing contingency plan
- Establishing village knowledge centre similar to the concept of M.S. Swaminathan Research Foundation, Chennai.
- Developing climate manager at the village level.
- Opening night school for climate understanding at the village level.
- Increasing organic carbon content of the soils.
- Implementing afforestation programme.
- Literacy enhancement
- Maintaining water bodies well
- Society cooperation
- Pre-seasonal climate workshop

7

Integrated Weather Forecast and Agro-Advisories

7.1 Importance of Weather Forecast in Agriculture

In India, agriculture is in reality to be a "gamble activity" with the monsoon rain, that means the productivity and profit from agriculture depend upon the goodness of the season. If any malevolent weather situation that happens during a crop season, whether it is at the beginning of the crop season or at the mid of the crop season or at the terminal of the crop season or at the whole of the crop season, the productivity of the crops raised would be under stake. In practice the irrigated crop area is lesser than the area under unirrigated crop in India in any region except the regions where irrigation is routed through perennial rivers. Normally the production from unirrigated crop is 30 to 40 per cent of the same crop grown under irrigated condition in an environment in the same season.

In India, in a region or in a taluk, if the irrigated area is lesser than 30 per cent of total cultivable area, there is every opportunity for the occurrence of food shortage. That means assured production depends upon irrigation rather than managing the crops with seasonal rainfall alone. This clearly indicates that the higher probability of occurrence of crop production risks falls under rainfed/dry land situation. The crop production risk is also common to irrigated area like unirrigated crop but the intensity gets varied. In this context one example that happened in India can be quoted. During December- January of 1982, with the occurrence of heavy rain/snow as a result of prevalence of western disturbance weather system, the wheat crop stood for harvest at Punjab got damaged with a loss of 124 crore of rupees. If this was anticipated earlier through weather forecast development, (which is common in all developed countries) and communicated to the farmers in advance, then the yield loss would have been reduced by following some tactical decisions to be taken either by the farmers or by the concerned Government. At that time, India did not have medium range weather forecast system and hence Indian farmers met with greater losses.

In India there are three weather systems and those are: (i) South west monsoon season (June-Sept-the rain processes is orographic- trade winds), (ii) North east monsoon season (Oct-Dec–the rain process is due to cyclone, vortex and convection process) and (iii) western disturbance (the rain process is passing of jet streams of low pressure from west to east of Central India). Eighty per cent of the geographical area of entire country, receive 80 per cent total annual rainfall from South west monsoon season and only Northeast monsoon benefits 20 per cent of the total area in India. The western disturbance covers only a central strip portion of the central part of India from west to east. Both the South west and North east monsoon have their own inherited problems of both low and uneven distribution of seasonal rainfall. The crop production risks due to uneven distribution of rainfall and also with low rainfall quantity could be seen under rainfed/dry land situations during both south west and north east monsoon seasons. This situation definitely warrants weather forecast system for agriculture in order to address the issues from uneven rainfall distribution. No doubt with the weather forecast given with higher amount of accuracy and lead time would definitely reduce the production risks to a greater extend, if proper tactical and strategical farm decisions are taken. But here another point is to be considered. The accuracy of weather forecast is about 90 per cent in temperate countries, where the weather system is slow and steady. But countries like India, which is situated nearer to the equator, the weather system is always turbulent and as a result, the accuracy is getting down though same model of that is used in temperate countries was used to develop weather forecast in India. However presently the accuracy of weather forecast issued to agriculture is of good and farmers tend to put their confidence on this for taking farm decisions. Still it is a long way to go.

Let us ask a question that what is weather forecast?

- Forecasting future weather change in spatial and temporal scales by employing different scientific tools
- Predicting future weather change in spatial and temporal scales by employing different scientific tools
- Foretelling future weather change in spatial and temporal scales by employing different scientific tools

7.2 Genesis of Weather Forecast in India

Historically the science of meteorology in India was written in Upanishads as early as 3000 BC, which provides the information such as formation of clouds, rain and seasonal cycle as caused by the movement of earth around the sun. During 700 AD, Honorable Kalidasa has written Meghdoot. The content provides the information on on-set of monsoon over Central India and traces the path of the monsoon clouds. The British East India Company established the meteorological observatory in Calcutta during 1785 and in Madras during 1796 to study weather and climate. In 1875 the India Meteorological Department (IMD) was established. Mr. H. F Blandford was appointed as meteorological reporter to the Government of India. Sir. John Eliot was appointed as Director General of observatories in May 1889. The Head quarter of IMD was latter shifted to Shimla and then to Poona (Pune) and finally to New Delhi. Mr. H.F. Blandford initiated the system of long-range forecasting for monsoonal rainfall in India. The long-range forecast was latter improved through evolution of phases by eminent pioneers like Sir. John Eliot and Sir. Gilbert Walker. Sir. Gilbert Walker has identified a phenomenon of linking monsoon rainfall with global meteorological situations and discovered the Southern Oscillation and this is popularly known presently to control south west monsoon rainfall in India to a greater extent. The first farmers weather bulletin was issued by the IMD in collaboration with All India Radio on daily basis and broadcasted in 26 languages. The forecast was at macro level and did not contain any agro-advisory. With wheat loss in Punjab during 1982, the IMD, with the establishment of National Centre for Medium Range Weather Forecasting started giving weather forecast for agriculture for three days from1991 and this forecast with agro advisory was given to the farmers weekly twice through its Agro-Met Field units at the agro climatic scale and latter shifted weather forecast with agro-advisories for five days under district level as a unit. The crop yield forecast was also issued by the Drought Research Unit of IMD by using statistical–empherical model for rice and wheat crops.

Now IMD through its 127 agro-climatic sub zones wants to give weather forecast at block level along with agro-advisories through its Agro-Met Field units. Presently the Agro-Climatic Research Centre of Tamil Nadu Agricultural University, Tamil Nadu, India, is operating automated weather forecast for five days along with agro-advisories through mobile phone for block level by employing Regional Climate Model.

7.3 Weather Forecast

Before understanding the weather forecast, we must know about the scales of meteorological motions and those are given in Table 7.1.

Table 7.1 Scales of meteorological motion system

Scale type	Horizontal scale (km)	Vertical scale (km)	Time scale (hours)
Planetary scale-macro climate	> 2000	10	200–400
Synoptic scale-macro climate	500–2000	10	100
Meso climate	1–100	1-10	1–16
Micro climate	< 100 metre	200 metre	6-12 minutes

Different methods for forecasting are available in the science and those are:

1. Synoptic method
2. Statistical method and
3. Numerical weather prediction.

The relevant information on these methods is given in Table 7.2.

Table 7.2 Methods of weather forecasting

Methods of forecasting	Characters
Synoptic method	Using weather elements observation at surface and upper level from different locations and preparation of weather charts, satellite information also used along with analogue analysis
Statistical method	Multiple regression, Arithmetic Regression Integrated Moving Average (ARIMA)
Numerical Weather Prediction (NWP)	Equations drawn from Coriolis force, frictional force, pressure gradient force and gravitational force are integrated in to multiple equations and it becomes a model. Weather elements observations are taken at different height from ground level to troposphere and those are included in the model for solution. Model is run from present to future. Initial weather condition is very important.

There are different types weather forecast in India with different temporal dimensions and those are:

1. Nowcast- a short -time weather forecast issued generally for the next few hours
2. Short range weather forecast- This is issued for next 24 hours with an outlook for another 24 hours
3. Medium range weather forecast- This is issued for next three to 10 days
4. Long range weather forecast- This is issued from 10 days to a season of more than three months
5. Seasonal climate forecast- This is used for a season especially for rainfall and farm decisions making

The further details of the forecast are given in Table 7.3 and 7.4.

Table 7.3 Details of different weather forecast

Name of the weather forecast	Issued by	Forecasted weather elements	Methods used in India	Lead time	Accuracy (%)
Nowcasting	IMD	Thunder storm, dust storm, cold and heat waves	Synoptic and weather map, NWP	One to two hours earlier to the events to occur	90 to 98
Short range	IMD	Cloud spread, rainfall, temperature, cyclone warning	Synoptic and weather map, NWP	One day	80–90
Medium range	IMD	Rainfall, temperature, RH, wind speed and direction, cloud cover	GNWPM and RCM	3–10 days	70–75
Medium range	Tamil Nadu Agricultural University (TNAU)	Rainfall, temperature, RH, wind speed and wind direction, and cloud cover	Regional climate model	7 days	70–75
Long range	IMD	Seasonal rainfall alone	Statistical regression model	30–40 days	60
Seasonal climate	TNAU	Seasonal rainfall	Rain man software-statistical	15–20 days	60

In the Table 7.4 for different weather forecast, the information on nature of clients, mode of transmission and resolution are given.

Table 7.4 Information on clients and other details for different weather forecast.

Name of the weather forecast	Clients	Mode of communication	Usefulness for agro-met advisory preparation	Resolution
Now casting	Publics and farmers	Radio, television, weather warning centres	Nil	State level
Short range	Publics and farmers	Radio, television, dailies,	Nil	State level
Medium range	Farmers	Television. Mobiles, web site	Highly useful for preparation	District/block level
Long range	Farmers and others	Radio, television and dailies, web page	Highly useful for preparation	North, south, western and eastern part of India
Seasonal climate	Farmers	Television, dailies, radio, web page	Highly useful for preparation	District and monthly level

7.4 Integration of Weather Forecast

Integrated weather forecast means all forecast information along with agro-advisories are to be given to the farmers for taking weather-based farm decisions one by one from long range to short range from Agro-Met Feld units of IMD in temporal dimension. But presently they give only medium range weather forecast information along with agro-advisory. Let us discuss the weather-based farm decision to be taken from each forecast when they are given to the farmers. Since the list is exhaustive, some few examples are given. Under long range weather forecast, based on information given, area to be cultivated, crops and technology to be selected, taking action to purchase required inputs, computation of labour requirement and time of sowing, etc can be decided before the on-set of season under crop planning. Under medium range weather forecast information, land preparation, refined date of sowing, time of fertiliser application and plant protection and other crop-based activities can be planned successfully. The information from short range forecast could be effectively used for water management, harvesting and mid correction of decision taken under medium range weather forecast information. In a nutshell seasonal crop plan can be decided before the onset of season based on the information from long range. Hence the first information from Agro Met Field unit is to give long range weather forecast information. After the season is set in, the Agro-Met Field unit must give both medium range and short-range weather forecast information, so that the farmers can modify the decision taken already under medium range weather forecast information to short range weather forecast information which has higher accuracy value. This is single window delivery system. One by one weather information has to be given to the farmers for taking weather-based farm decisions

7.5 Automated Weather Forecast System

As indicated earlier the first step taken in India to introduce automated weather forecasted agro- advisory was done by Balasubramanian *et al.*, (2014) at Agro-climate Research Centre (ACRC), Tamil Nadu Agricultural University, Coimbatore. With the objective of providing weather windows at block level of Tamil Nadu for developing agro- advisories in respect of dominant crops, an exploratory research was conducted with four levels of rainfall, three levels of maximum temperature, three levels of minimum temperature, three levels of mean day relative humidity and three levels of wind speed. Adopting the permutation and combination methodology, a total of 324 combinations were noted from the identified levels of five weather elements and out of these combinations, 54 weather windows each with range of weather values were selected based on the validation of real-time weather data collected covering both temporal and spatial weather dimensions of the State Tamil Nadu. During 2016 again Balasubramanian *et al.*, (2016), with the objective of developing agro-met advisories for rice crop for 54 independent selected weather windows (SWW) covering eight rice growing seasons/systems of Tamil Nadu, one research was carried out. This was done to provide weather-based agro-met advisory to the farming community. Weather sensitive rice agro-met advisories were developed for its nine stages through group discussions based on rice crop sensitiveness to SWW. The available literature on interaction between rice crop and weather elements at their different threshold levels was also properly considered during this exercise. Proto-types were run for one block of Tamil Nadu and thus problems identified for seeking solutions. Now these 54 weather windows have been used along with agro-advisories for 104 crops at ACRC, Coimbatore for issuing automated weather agro advisories to the farmers.

7.6 Weather Forecast Validation/Verification

Validation or verification is an important tool to understand the value of forecast information given for taking farm decisions. This topic carries more weight to put confidence on the information of forecast given to the farmers. The verification is required to improve the method of forecasting to model developers and also to develop confidence on the forecast received by the farmers as indicated earlier. From simple formulae to complex mathematic equations are available to validate the forecast values received and some are presented hereunder.

1. Forecast Accuracy (ACC) or Ratio score or Hit score

$ACC = YY + NN/YY + NN + YN + NY$

First letter in the pair: Forecasted one

Second letter in the pair: Events occurred

Y = yes; N = No

2. Heidke Skill Score (HSS)

HSS = ZH–FM/{(Z + M) (M + H) + (Z + F) (F + H)}/2

Z = No of correct predictions of no rain (neither predicted nor observed)

F = No of false alarms (predicted but not observed)

M + No. of misses (observed but not predicted)

H = No of hits (predicted and observed)

3. Root Mean Square Error (RMSE)

RMSE = $[\ 1/n \sum (f_1 - 0_1)^2]^{1/2}$

F_1 = Forecast value

0_1 = Observed value

N = Total number of observations.

Correlation also can be used for validation between forecasted value and observed value.

For this a minimum of 30 pairs must be there for analysis. Chi square test also can be used for this purpose.

7.7 Traditional Knowledge on Weather Forecast

This is based on observations made over generations and many of them found scientifically valid. These can be considered as a tool along with scientific weather forecast for taking farming decisions.

a. Short Range Traditional knowledge

1. There will be heavy rain pour, if the rain bow appears on the eastern sky at the time of raining. If it is on the western sky, there will be no rain
2. When crab comes to the bund, it may rain
3. Red clouds in the east sky, it may rain in the next day
4. Frogs crocking in chorus then it is followed by rain
5. When dragon flies, fly down (low), it may rain that there will be rainfall within a day or two.
6. Increased mosquito bite predicts rain
7. Dense fog in the early morning indicates no rain
8. If there is an accumulation of clouds in the South-East direction in a layered form accompanied by winds blowing from the southern direction then it is claimed to get rain
9. If there is a swelling on the lower portion of the camel's legs then rainfall is predicted by the farmers. The swellings are probably caused due to higher relative humidity.

b. Medium range traditional knowledge

1. The closer circle to the moon, the nearer is the shower and vice versa
2. If a snail climbs certain trees, there will be no rain

c. Long range traditional knowledge

1. If the Tatihari bird (Lap wing) lays eggs on the higher portion of the lake bunds or on the top of any structure, the coming season carries heavy rainfall. If the same bird lays eggs on the lower side of the tank or on the floor of the water body structure, drought is anticipated. Further it is also believed that if a single egg is laid by the Lap wing bird, then there will be rainfall only for one month out of four months of the rainy season. If two eggs are laid then rainfall will occur for two months and similarly four eggs indicate there will be rainfall during all the four months of the rainy season.
2. If the clouds thunder on the first day of mid-April, there shall be no rain for 72 days.
3. If crow cries during night and fox howls during the day, then there would be severe drought.
4. If the "Tillbohara" (Dragon fly), which appears generally in the rainy season, are observed to swarm in a large group over a water surface (Pond) then dry weather is predicted but if they swarm over open dry lands or fields then early rainfall is predicted by the farmers.
6. If the "Khejri" tree bears good fruit in a particular year, then farmers predict good rainfall during the next rainy season and vice versa less rain is predicted in the event of a poor fruit crop.

7.8 Weather Thumb Rules Developed by Community in the Absence of Weather Forecast Information

This is based on farmers weather observation combined with farm activities done over years and found validated. This becomes thumb rule over time. It varies with villages and taluks and districts.

a. Rainfall:

Rainfall amount (mm)	Farm activities to be done
5–10	Plant protection
10–12	Hand weeding and hoeing
12–15	Fertiliser application
25–30	Sowing, postponement of irrigation, harvesting of tuber and groundnut

b. Temperature

Temperature (°C)	Farm activities to be done
Day temperature > 32 and continued for a week	Sucking pest menance
Night temperature < 20 and continued for a week	Germination of disease spores
Day temperature more than 34 and continues	Start irrigation

c. Wind

Wind speed (Kmph)	**Farm activities to be done**
<5	Plant protection, hand weeding and hoeing
10–15	Winnowing
15–20 + rainfall >25mm	Propping sugarcane of more than eight months and banana of more than five months
> 20	Propping sugarcane of more than eight months and banana of more than five months
Wind speed >20 with minimum temperature <20°C + evening RH > 60%	Epidemic spread of diseases

d. Relative humidity

RH (per cent)	**Farm activities to be done**
Evening RH is more than 60 and continued for a week with minimum temperature of < 20 °C	Disease initiation and planning for plant protection
Evening RH is more than 60 and continued for a week with minimum temperature of <20 °C + wind speed > 15 Kmph	Epidemic spread of diseases

8

Climate Change

8.1 Climate Change

Climate change is a broad term of global negative phenomena as created predominantly by burning of fossil fuels, which add heat-trapping gases to earth's atmosphere. The over accumulation of greenhouse gases in the atmosphere lead to increased temperature and this is called as global warming. As a result, sea-level rise; ice mass loss in Greenland, Antarctica, Arctic and mountain glaciers worldwide; shifts in flowering/ plant blooming; and extreme weather events occur.

In simple terms, climate change of a particular place is nothing but the change or shift from the average mean temperature, rainfall, *etc.*, (Figure. 8.1).

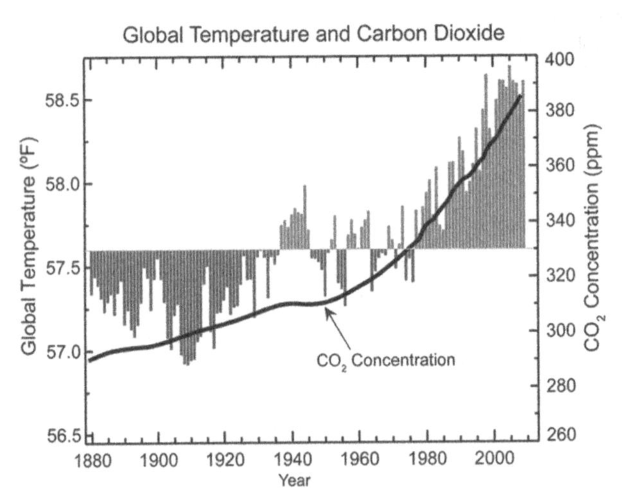

Fig. 8.1 Increasing of global temperature over the years (*Source*: Mota-Babiloni, 2016)

8.1.1 Climate Variability

Climate variability is defined as a short time variation in the mean state and other statistics of the climate on all temporal and spatial scales.

8.1.2 Climate Change vs Variability

In essence, climate variability looks at changes that occur within smaller time frames, such as a month, a season or a year, and climate change considers changes that occur over a longer period of time, typically over decades or longer (Figure. 8.2).

Variations are a natural component of the climate caused by changes in the systems that influence the climate, such as the general circulation system. At times systems are so strong or weak to give rise to extreme climate events. Whereas, climate change refers to permanent shift in the traditional space-time patterns of climate. The example is change from one climate mode to another climate mode, which is outside the normal range.

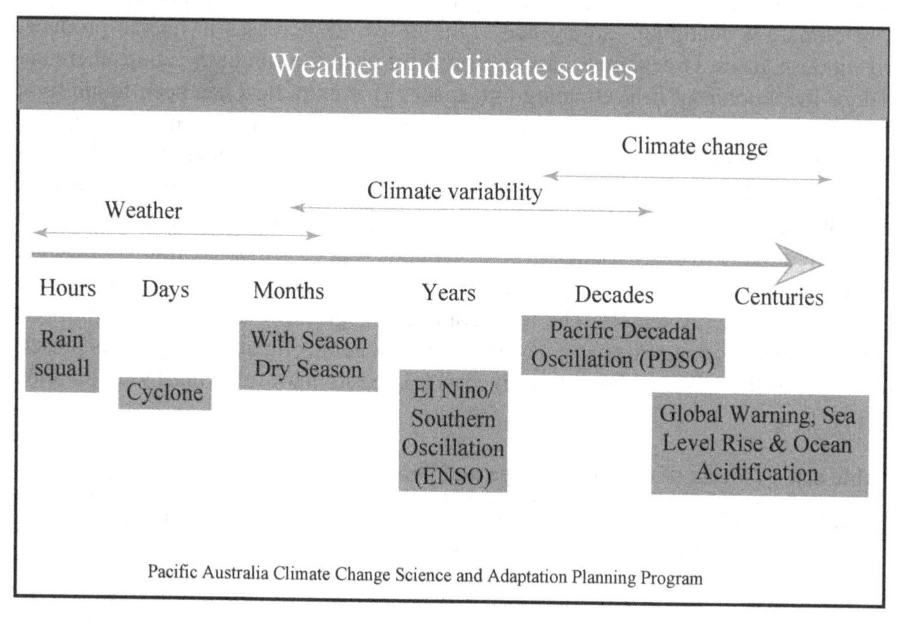

Fig. 8.2. Climate change vs variability
(*Source*: Mohamed *et al.*, 2002)

In the period since the industrial revolution, emissions of greenhouse gases from fossil fuel combustion, deforestation and agricultural practices have led to global warming and climate change. Observed and anticipated changes in the climate include higher temperatures, changes in rainfall patterns, changes in the frequency and distribution of weather events such as droughts, storms, floods and heat waves, sea level rise and consequent impacts on human and natural systems. Many scientists argue that the impacts of climate change will be devastating for natural and human systems and that climate change poses a severe threat to human civilisation. However, action to respond to climate change has been slow. Climate change draws attention to the relationship between science and society, challenges global governance institutions, and triggers new social movements.

8.1.3 Characterizing and Historicizing Climate Change

Literally 'climate change' denotes to long-term change in the statistical distribution of weather patterns (*e.g.*, temperature, precipitation *etc.*) over decades to millions of years of time. Climate on earth has changed on all time scales (IPCC, 2007). But UNFCCC (UNFCC, 1994) defined climate change as "a change of climate which is attributed directly or indirectly to human activity that alters the composition of the global atmosphere and which is in addition to natural climate variability observed over comparable time periods". However, the IPCC definition of climate change includes change due to natural variability alongside human activity (UNFCC, 2013). Australian Government's DCCEE (DCCEE, 2012) in its website described climate change - 'our climate is changing, largely due to the observed increases in human produced greenhouse gases. Greenhouse gases absorb heat from the sun in the atmosphere and reduce the amount of heat escaping into space. This extra heat has been found to be the primary cause of observed changes in the climate system over the 20th century'. Thus, in the environmental discourse different stakeholders have characterized climate change as mainly the change in modern climate augmented by human activities and the adverse human activities for example burning fossil fuel, deforestation etc. are considered likely to bring change in some climatic aspects which are briefly presented in the Table 8.1. Also, while talking about climate change, these are the features that the term mainly entails. The term climate change however through definitions, policy propagation qualifies as a negative anthropogenic climate change in its present meaning, at the onset it did not really appreciated its harmful brunt (Vlassopoulos, 2012).

Table 8.1. Aspects of Climate Change and perceived implications

Global warming	GHG concentration	Emission of Green House Gases through industrialization, travelling etc. is increasing the GHG concentration in the atmosphere. At this moment CO_2 concentration is at its highest concentration1 in 650 000 years–393 ppm (Buis, 2019)
	Change in world temperature	GHG concentration along with some other issues leads to warming the world. Earth has warmed since 1880. Most of this warming has occurred since the 1970s, with the 20 warmest years having occurred since 1981 and with all 10 of the warmest years occurring in the past 12 years (Buis, 2019) Being central to the issue predominantly, Global warming brings about change in following different features of the human environment
Ozone layer depletion		Slow, steady decline of about 4 per cent per decade in the total volume of ozone in earth's stratosphere (the ozone layer) since the late 1970s is estimated which is likely to bring health implications (different cancerous diseases), augmenting extreme weather events (desertification, drought) through opening the curtain that was protecting earth from hazardous sun rays
Shrinking ice sheets		Greenland lost 150 km^3 to 250 km^3 (36 mi^3 to 60 mi^3) of ice per year between 2002 and 2006 and Antarctica lost about 152 km^3 (36 mi^3) of ice between 2002 and 2005 (Buis, 2019). This on the other hand contributing to the next problem sea level rise.

Rise in Sea Level	Global sea level rose about 17 cm (6.7 inch) in the last century (Buis, 2019). Continual increase is very likely to inundate many island states, low-lying delta regions leaving their population having no land to inhabit.
Ocean Acidification	Since 1750 the CO_2 content of the earth's oceans has been increasing and it is currently increasing about 2 billion tons per year which has increased ocean acidity by about 30 per cent (Buis, 2019).
Warming Oceans	With the top 700 m (about 2300 ft) of ocean showing warming of 0.16°C since 1969 due to absorbed increased heat of the earth (Buis, 2019). These two changes are likely to bring massive change/destruction in ocean habitations

When the author described greenhouse effect in his article published at the 'Annales de la Chimieet de Physique' (Vlassopoulos, 2012). About half a century later Arrhenius (1896) published first calculation of global warming from human emissions of CO_2 though Keeling was the first to measure accurately CO_2 in the earth's atmosphere in 1960 (Weart, 2008). Vlassopoulos (2012) noted that climatic variations were perceived precisely as a scientific issue until 1970, hence the debate was mostly confined to the scientific community of climatologists and relevant research was fragmented into different university endeavors only.

AIP (Weart, 2010) maintains that the rise of environmentalism in the early 1970s raised public doubts about the benefits of human activity for the planet which in other way turned the curiosity about climate into anxious concern. Since then, concern about anthropogenic global degradation spreads which ignited numbers of international cooperation, programs and meeting to date of concerned stakeholders including representatives from interested community other than the scientist only (Vlassopoulos, 2012, Weart, 2010).

Programs and meetings in 1970's appear to take place to explore and acknowledge the extent of anthropogenic Climate Change. Global Atmospheric Research Program (GARP) organized by World Meteorological Organization (WMO) and the International Council of Scientific Unions (ICSU) in 1974 is an example in this regard. It was aimed to examine the highly complex problem of the physical basis of climate (Flohn, 1977). Another example could be the first World Climate Conference (WCC) organized by the WMO in 1979. Whereas, some significant events in the 1980's and 1990's, can be said, were inclined to devise methods to address it. For example, Montreal Protocol of the Vienna Convention in 1987 imposes international restrictions on emission of ozone-destroying gases. Two major events in 1990's, one, 1992 conference in Rio-de-Janeiro produces UN Framework Convention on Climate Change and another 1997 International conference produces Kyoto Protocol (came into effect in 2005) that set targets for industrialized nations to reduce greenhouse gas emissions (Weart, 2008).

Kyoto Protocol is regarded as the most significant commitment in addressing global climate change so far. That's why as it expires at the end of 2012, through different conventions, from UNFCCC to the Conference of the Parties (COP-17) to the convention, held in November–December 2011 in Durban, South Africa, the world nations are continuing to strive to negotiate what may become the post-Kyoto (Olmstead *et al.,* 2012).

8.2. Definition of the Term "Global Warming"

Global warming is the ongoing rise of the average temperature of the earth's climate system and has been demonstrated by direct temperature measurements and by measurements of various effects of the warming (Figure. 8.3).

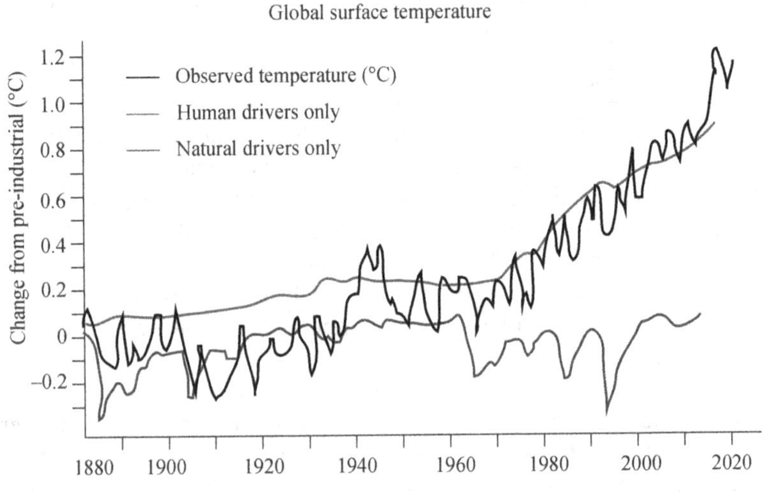

Fig. 8.3. Global warming over the years
(*Source*: NASA, 2020)

The earth's climate continuously changes following a natural process. It is as a consequence of such natural process that the ice-cold climate of the earth became favourable for the evolution of mankind. The solar energy that passes through the earth's atmosphere is absorbed by the surface of the earth while a major portion of it is reflected into space. This process, if continued naturally, would not have disturbed the atmosphere of the earth and hence would not have become a cause of concern. However, with industrialization, overpopulation, land-use change, deforestation and change in lifestyles, the concentration of the different gases present in the atmosphere have undergone drastic changes. Some of such gases like carbon dioxide, nitrous oxide, methane, etc., create a partial blanket over the earth's atmosphere and do not allow the outgoing infra-red radiations to travel back into space. They, in turn, trap the infra-red radiations and reflect them into the earth's atmosphere. This mechanism of trapping gases is mostly adopted in greenhouses to take aid of the warmth generated from such gases in the quicker growth of plants present therein. As the mechanism involved is similar to what is deliberately done in a greenhouse, such an effect in the earth's atmosphere is known as the 'Greenhouse effect'.

As a consequence of the greenhouse effect, the global mean temperature of the earth increases. Such increase in temperature continues over a long period of time and is the main cause of extensive warming of the planet and hence termed as 'Global Warming'. The fallacy here is that warming of the earth is necessary to make it habitable otherwise the earth would be a big ball of ice, non-conducive to life, but such warming should be following the natural cycle. When the increase in the concentration of the greenhouse gases happens irrationally due to human induced activities, the enhanced greenhouse effect occurs and hence global warming.

The emission of the greenhouse gases has been on the rise since pre-industrial times but such increase has become extensively rapid in recent times with the increase in the burning of fossil fuels and land use change. It may be observed herein that the Fourth Assessment Report of the Intergovernmental Panel on Climate Change published in 2007 has stated that the increase in the global greenhouse gas emissions has been up to 70 per cent between 1974 and 2006.

Apart from the naturally occurring greenhouse gases namely carbon dioxide, methane and nitrous oxide, there are certain other man-made greenhouse gases like chlorofluorocarbons (CFC's), per fluorocarbons (PFC's), hydrofluorocarbons (HFC's) and sulphur hexafluoride (SF_6). The increase in the concentration of such man-made gases further enhances the average temperature of the earth and causes the enhanced greenhouse effect. When the increase in the stock of the greenhouse gases goes beyond the carrying capacity or the assimilative capacity of the earth, there occur manifestations of the same through changes in the weather regime and impacts are also observed on the diverse ecosystems of the globe.

8.2.1 Global Warming - Relationship with Climate Change

The Synthesis Report of the Fourth Assessment Report on Climate Change published by the Intergovernmental Panel on Climate Change in 2007 clearly establishes the fact that global warming is occurring at an alarming rate which is observable from the increase in the average air temperature and ocean temperature, the widespread melting of ice-sheets and the rise in the sea-levels. The Third Assessment Report of the Intergovernmental Panel on Climate Change, published in 2001, established the fact of climate change and stated that it is beyond the 'normal'. Climate change refers to the changes brought about in the earth's climate system over a wide period of time taking place mainly by human intervention more specifically called, human induced global warming. Article 1 of the United Nations Framework Convention on climate change defines 'climate change' as "a change, which is attributable directly or indirectly to human activity that alters the composition of the global atmosphere and which is in addition to natural climate variability observed over comparable time periods".

Interpretation of such definition implies the following:

(a) That human induced activities are primarily responsible for climate change;

(b) That climate change brings about a change in the composition of the global atmosphere, *i.e.,* a change in the acceptable concentrations of gases in the atmosphere;

(c) That it is not naturally induced.

It has however to be kept in mind that all the impacts observed are not due to the current emissions of greenhouse gases but a result of the cumulative impact of such emissions that have happened since pre-industrial times. The temperature of the earth has already increased by 0.8 degrees Centigrade since the Industrial Revolution. With the current trend of global warming, it is expected that there would be a further increase of about 0.7°C.

8.3 IPCC and Its Reports

The Intergovernmental Panel on Climate Change (IPCC) was created to provide policymakers with regular scientific assessments on climate change, its implications and potential future risks, as well as to put forward adaptation and mitigation options.

Through its assessments, the IPCC determines the state of knowledge on climate change. It identifies where there is agreement in the scientific community on topics related to climate change, and where further research is needed. The reports are drafted and reviewed in several stages, thus guaranteeing objectivity and transparency.

The IPCC has published five comprehensive assessment reports reviewing the latest climate science as well as a number of special reports on particular topics.

The IPCC published its First Assessment Report (FAR) in 1990, a supplementary report in 1992, a Second Assessment Report (SAR) in 1995, Third Assessment Report (TAR) in 2001, a Fourth Assessment Report (AR4) in 2007 and a Fifth Assessment Report (AR5) in 2014. The IPCC is currently preparing the Sixth Assessment Report (AR6), which will be completed in 2022.

Each assessment report is in three volumes, corresponding to Working Groups I, II, and III. It is completed by a synthesis report that integrates the working group contributions and any special reports produced in that assessment cycle.

8.3.1 First Assessment Report

The IPCC First Assessment Report (FAR) was completed in 1990, and served as the basis of the UNFCCC.

The executive summary of the WG I Summary for Policymakers report says, they are certain that emissions resulting from human activities are substantially increasing the atmospheric concentrations of the greenhouse gases, resulting on average in an additional warming of the earth's surface. They calculated with confidence that CO_2 has been responsible for over half the enhanced greenhouse effect. They predicted that under a "Business As Usual" (BAU) scenario, global mean temperature will increase by about 0.3 °C per decade during the 21st century. They judged that global mean surface air temperature has increased by 0.3 to 0.6 °C over the last 100 years, broadly consistent with prediction of climate models, but also of the same magnitude as natural climate variability. The unequivocal detection of the enhanced greenhouse effect is not likely for a decade or more.

Supplementary report of 1992

The 1992 supplementary report was an update, requested in the context of the negotiations on the UNFCCC at the Earth Summit (United Nations Conference on Environment and Development) in Rio de Janeiro in 1992.

The major conclusion was that research since 1990 did "not affect our fundamental understanding of the science of the greenhouse effect and either confirm or do not justify alteration of the major conclusions of the first IPCC scientific assessment". It noted that transient (time-dependent) simulations, which had been very preliminary in the FAR, were now improved, but did not include aerosol or ozone changes.

8.3.2 Second Assessment Report

Climate Change 1995, the IPCC Second Assessment Report (SAR), was finished in 1996. It is split into four parts:

i. A synthesis to help interpret UNFCCC article 2.
ii. The Science of Climate Change (WG I)
iii. Impacts, Adaptations and Mitigation of Climate Change (WG II)
iv. Economic and Social Dimensions of Climate Change (WG III)

Each of the last three parts was completed by a separate Working Group (WG), and each has a Summary for Policymakers (SPM) that represents a consensus of national representatives.

The SPM of the WG I report contains headings:

- Greenhouse gas concentrations have continued to increase
- Anthropogenic aerosols tend to produce negative radiative forcing
- Climate has changed over the past century (air temperature has increased by between 0.3 and 0.6 °C since the late 19th century; this estimate has not significantly changed since the 1990 report).
- The balance of evidence suggests a discernible human influence on global climate (considerable progress since the 1990 report in distinguishing between natural and anthropogenic influences on climate, because of: including aerosols; coupled models; pattern-based studies)
- Climate is expected to continue to change in the future (increasing realism of simulations increases confidence; important uncertainties remain but are taken into account in the range of model projections)
- There are still many uncertainties (estimates of future emissions and biogeochemical cycling; models; instrument data for model testing, assessment of variability, and detection studies)

8.3.3 Third Assessment Report

The Third Assessment Report (TAR) was completed in 2001 and consists of four reports, three of them from its Working Groups:

i. **Working Group I:** The Scientific Basis

ii. **Working Group II:** Impacts, Adaptation and Vulnerability
iii. **Working Group III:** Mitigation
iv. Synthesis Report

A number of the TAR's conclusions are given quantitative estimates of how probable it is that they are correct, *e.g.*, greater than 66 per cent probability of being correct. These are "Bayesian" probabilities, which are based on an expert assessment of all the available evidence.

Robust findings of the TAR Synthesis Report include

- Observations show earth's surface is warming. Globally, 1990's very likely warmest decade in instrumental record.
- Atmospheric concentrations of anthropogenic (*i.e.,* human-emitted) greenhouse gases have increased substantially.
- Since the mid-20th century, most of the observed warming is "likely" (greater than 66% probability, based on expert judgement) due to human activities.
- Projections based on the Special Report on Emissions Scenarios suggest warming over the 21st century at a more rapid rate than that experienced for at least the last 10,000 years.
- Projected climate change will have both beneficial and adverse effects on both environmental and socio-economic systems, but the larger the changes and the rate of change in climate, the more the adverse effects predominate.
- Ecosystems and species are vulnerable to climate change and other stresses (as illustrated by observed impacts of recent regional temperature changes) and some will be irreversibly damaged or lost.
- Greenhouse gas emission reduction (mitigation) actions would lessen the pressures on natural and human systems from climate change.
- Adaptation [to the effects of climate change] has the potential to reduce adverse effects of climate change and can often produce immediate ancillary benefits, but will not prevent all damages. An example of adaptation to climate change is building levees in response to sea level rise

8.3.4. Fourth Assessment Report

The Fourth Assessment Report (AR4) was published in 2007. Like previous assessment reports, it consists of four reports:

i. **Working Group I:** The Physical Science Basis
ii. **Working Group II:** Impacts, Adaptation and Vulnerability
iii. **Working Group III:** Mitigation
iv. Synthesis Report

People from over 130 countries contributed to the IPCC Fourth Assessment Report, which took 6 years to produce. Contributors to AR4 included more than 2500 scientific expert reviewers, more than 800 contributing authors, and more than 450 lead authors.

Robust findings of the Synthesis report include

- Warming of the climate system is unequivocal, as is now evident from observations of increases in global average air and ocean temperatures, widespread melting of snow and ice and rising global average sea level.
- Most of the global average warming over the past 50 years is "very likely" (greater than 90% probability, based on expert judgement) due to human activities.
- Impacts [of climate change] will very likely to increase due to increased frequencies and intensities of some extreme weather events.
- Anthropogenic warming and sea level rise would continue for centuries even if GHG emissions were to be reduced sufficiently for GHG concentrations to stabilise, due to the time scales associated with climate processes and feedbacks. Stabilization of atmospheric greenhouse gas concentrations is discussed in climate change mitigation.
- Some planned adaptation (of human activities) is occurring now; more extensive adaptation is required to reduce vulnerability to climate change.
- Unmitigated climate change would, in the long term, be likely to exceed the capacity of natural, managed and human systems to adapt.
- Many impacts [of climate change] can be reduced, delayed or avoided by mitigation.

Global warming projections from AR4 are shown below

- The projections apply to the end of the 21st century (2090–'99), relative to temperatures at the end of the 20th century (1980–'99).
- Descriptions of the greenhouse gas emissions scenarios can be found in Special Report on Emissions Scenarios.

Special Report on Emissions Scenarios (SRES)

The Special Report on Emissions Scenarios (SRES) is a report published by the IPCC in 2000. The SRES contains "scenarios" of future changes in emissions of greenhouse gases and sulphur dioxide. One of the uses of the SRES scenarios is to project future changes in climate, *e.g.*, changes in global mean temperature. The SRES scenarios were used in the IPCC's Third and Fourth Assessment Reports.

The SRES scenarios are "baseline" (or "reference") scenarios, which means that they do not take into account any current or future measures to limit greenhouse gas (GHG) emissions. SRES emissions projections are broadly comparable in range to the baseline projections that have been developed by the scientific community.

AR4 global warming projections

Best estimates and *likely* ranges for global average surface air warming for six SRES emissions marker scenarios are given in this assessment and are shown in Table8.2. For example, the best estimate for the low scenario (B1) is 1.8°C (*likely* range is 1.1°C to 2.9°C), and the best estimate for the high scenario (A1FI) is 4.0°C (*likely* range is 2.4°C to 6.4°C) (Table 8.2).

Table 8.2. AR4 global warming projections

Emissions scenario	Likely (°C)	Best estimate range (°C)
B1	1.8	1.1–2.9
A1T	2.4	1.4–3.8
B2	2.4	1.4–3.8
A1B	2.8	1.7–4.4
A2	3.4	2.0–5.4
A1FI	4.0	2.4–6.4

"Likely" means greater than 66 per cent probability of being correct, based on expert judgement.

8.3.5 Fifth Assessment Report

The IPCC's Fifth Assessment Report (AR5) was completed in 2014. AR5 followed the same general format as of AR4, with three Working Group reports and a Synthesis report. The Working Group I report (WG1) was published in September 2013.

Conclusions of AR5 are summarized below

Working Group I

- Warming of the climate system is unequivocal, and since the 1950s, many of the observed changes are unprecedented over decades to millennia.
- Atmospheric concentrations of carbon dioxide, methane, and nitrous oxide have increased to levels unprecedented in at least the last 800,000 years.
- Human influence on the climate system is clear. It is extremely likely (95–100% probability) that human influence was the dominant cause of global warming between 1951–2010.

Working Group II

- Increasing magnitudes of [global] warming increase the likelihood of severe, pervasive, and irreversible impacts.
- A first step towards adaptation to future climate change is reducing vulnerability and exposure to present climate variability.
- The overall risks of climate change impacts can be reduced by limiting the rate and magnitude of climate change.

Working Group III

Without new policies to mitigate climate change, projections suggest an increase in global mean temperature in 2100 of 3.7 to 4.8 °C, relative to pre-industrial levels (median values; the range is 2.5 to 7.8 °C including climate uncertainty).

The current trajectory of global greenhouse gas emissions is not consistent with limiting global warming to below 1.5 or 2 °C, relative to pre-industrial levels. Pledges made as part of the Cancún Agreements are broadly consistent with cost-effective scenarios that give a "likely" chance (66–100% probability) of limiting global warming (in 2100) to below 3 °C, relative to pre-industrial levels.

8.3.6. Representative Concentration Pathways of AR5

A **Representative Concentration Pathway (RCP)** is a greenhouse gas concentration (not emissions) trajectory adopted by the IPCC. The RCPs; originally RCP 2.6, RCP 4.5, RCP 6, and RCP 8.5; are labelled after a possible range of radiative forcing values in the year 2100 (2.6, 4.5, 6, and 8.5 W/m^2, respectively) (Fig. 8.4).

Projections in AR5 are based on "Representative Concentration Pathways" (RCP's). The RCP's are consistent with a wide range of possible changes in future anthropogenic greenhouse gas emissions. Projected changes in global mean surface temperature and sea level are given in the main RCP article.

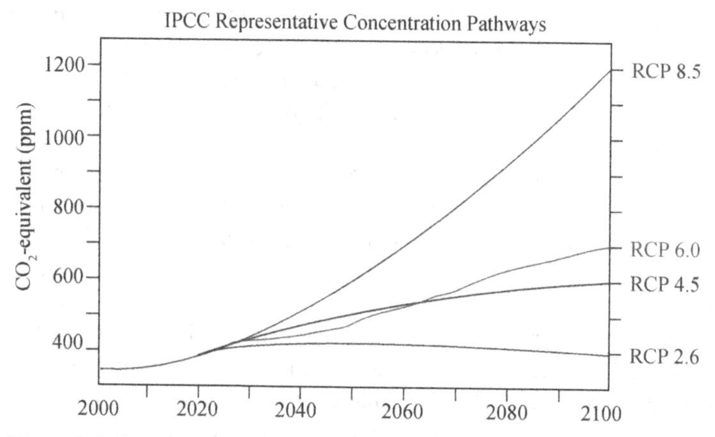

(*Source*: IPCC Assessment Report, 2007)

Fig. 8.4. Representative Concentration Pathways of AR5

8.4. Observed and Projected Changes in Climate

8.4.1 Observed and Projected Changes in Climate in India

In the last 100 years the mean annual surface air temperature of India has increased by 0.4–0.6°C (Rupakumar, 2002). Annamalai *et al.,* (2007) has reported decreasing rainfall tendency in both southwest and northeast monsoon seasons in most parts of central and northern India. In contrast, peninsular parts of India, particularly over the region 9–16°N latitude encompassing the rice growing areas showed an increasing rainfall tendency. This increase was particularly strong during the northeast monsoon season. Lal *et al.,* (1995) projected the climate change scenario for the Indian subcontinent and the results indicated an increase in annual mean maximum and minimum surface air temperatures by 0.7°C and 1.0°C respectively over land in the 2040's with respect to the 1980's level.

Shakeel *et al.,* (2009) have reported a projected increase in temperature by 0.4–2.0°C in *Kharif* and 1.1–4.5°C during *Rabi* by 2070 over Indian region. The IPCC (2007) in its fourth assessment report has predicted 2.7–4.3°C temperature rise over India by 2080's. Rupakumar *et al.,* (2006) projected a temperature rise of 2.9° and 4.1°C for India under the B2 and A2 scenarios of SRES respectively, in 2080's relative to

1970's. The CMIP3-based model ensemble mean projects a warming of 3.19°C under the A1B scenario for 2080's compared to the 1970's baseline. Krishnakumar *et al.*, (2010) projected a warming of 3.5°–4.3°C over the same period for the A1B scenario. The CMIP3-based model ensemble mean projects a warming of 3.19°C under the A1B scenario for 2080's compared to the 1970's baseline (Rajiv Kumar Chaturvedi *et al.*, 2012). All India annual mean temperature increases by 1.7°C–2.02°C by 2030's under different RCP scenarios and by about 2°C–4.8°C by 2080's, relative to the pre-industrial base (1880's). The CMIP5-based model ensemble projects a warming of 2.8°C and 4.3°C under the RCP 6.0 and RCP 8.5 scenarios respectively, for 2080's compared to the 1970's baseline (Rajiv Kumar Chaturvedi *et al.*, 2012).

According to Khan *et al.*, (2009), the mean rainfall of India for the SRES is projected to increase by 10 per cent during *Kharif* and *Rabi* seasons during 2070 from the reference year 2010. The study conducted by Rupakumar *et al.*, (2003) revealed that marked increase in rainfall in the 21st century is likely to be evident after 2040's in India. They have also inferred that the number of rainy days is likely to be increased by 5 - 10 days in the foot hills of Himalaya and Northeast India. All-India annual precipitation increases by 1.2–2.4 per cent by 2030's under different RCP scenarios and by 3.5–11.3 per cent by 2080's, relative to the pre-industrial base. Precipitation is projected to increase almost all over India except for a few regions in short-term projections (2030's). RCP 2.6 experiences the least increase in precipitation with the projected precipitation change varying from 0 to 15 per cent, while and RCP 8.5 is associated with the largest changes in precipitation, with the projected precipitation changes varying from 5 to 45 per cent (Rajiv Kumar Chaturvedi *et al.*, 2012).

In the northern parts of India, the warming would be more pronounced. The extremes in maximum and minimum temperatures are expected to increase under changing climate; some places may remain dry whereas few places are expected to get more rain. Leaving Punjab and Rajasthan in the North West and Tamil Nadu in the South, which show a slight decrease, on an average a 20 per cent rise in all India summer monsoon rainfall over all states is expected. Number of rainy days may come down (*e.g.* MP) but the intensity is expected to rise at most of the parts of India (*e.g.* North East). Gross per capita water availability in India will decline from 1820 m^3/yr. in 2001 to as low as 1140 m^3/yr. in 2050 (Mahato, 2014).

The observations on the mean sea level along the Indian coast revealed a long-term (100 year) rising trend of about 1.0 mm/year. However, the recent data suggests a rising trend of 2.5 mm/year in sea level along Indian coastline. The sea surface temperature adjoining India is likely to warm up by about 1.5–2.0°C by the middle of this century and by about 2.5–3.5°C by the end of the century. A 1meter sea-level rise is projected to displace approximately 7.1 million people in India and about 5764 km^2 of land area will be lost, along with 4200 km of roads (NATCOM, 2004).

8.4.2 Observed Changes in Climate in Tamil Nadu (case study)

Maximum temperature

Normal average annual maximum temperature of Tamil Nadu varies both in space and time. Annual maximum temperature of Tamil Nadu (Figure. 8.5) varied from 29.0°C (Cuddalore) to 33.9°C (Karur). To understand the seasonal fluctuations across the region, data were segregated into seasons. Among the seasons, cold weather period (CWP) had a range between 29.3°C (Cuddalore) to 32.1°C (Kanyakumari) while in hot weather period (HWP) it ranged between 32.1°C (Cuddalore) to 37.0°C (Chennai). During southwest monsoon Tamil Nadu witnessed maximum temperature from 27.5°C (Cuddalore) to 35.3°C (Ariyalur, Villupuram, Karur, Vellore, Nagapattinam, Trichy, Perambalur and Thanjavur) while northeast monsoon witness 27.7°C (Cuddalore) to 31.2°C (Kanyakumari) variation.

Minimum temperature

Normal average annual minimum temperature of Tamil Nadu has both spatial and temporal variability (Figure. 8.5). Annual minimum temperature ranged between 18.8°C (Cuddalore) and 24.4°C (Vellore and Nagapattinam). Among the seasons, cold weather period (CWP) had a range from 16.5°C (Cuddalore) to 22.1°C (Kanyakumari) while hot weather period (HWP) it ranged from 20.2°C (Cuddalore) to 25.7°C (Vellore and Nagapattinam).

Maximum temperature distribution over Tamil Nadu

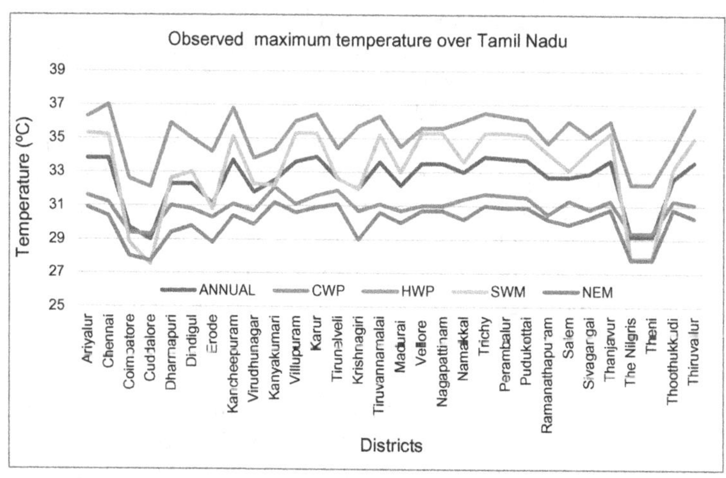

Minimum temperature distribution over Tamil Nadu

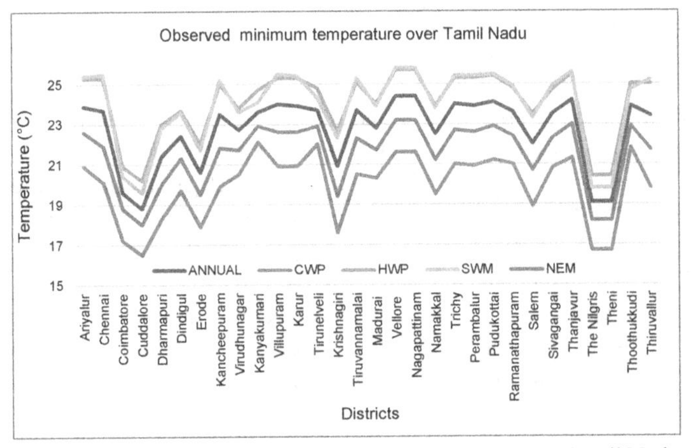

Fig. 8.5 Temperature distribution in space and time over Tamil Nadu (*Source*: Gowtham *et al.*, 2017)

During southwest monsoon, Tamil Nadu witnessed maximum temperature from 19.6°C (Cuddalore) to 25.8°C (Vellore and Nagapattinam) while northeast monsoon witnessed 18.0°C (Cuddalore) to 23.2°C (Vellore and Nagapattinam).

Rainfall

Normal annual rainfall was studied grid wise with data from 183 points that covered the entire State of Tamil Nadu. The per cent of grids segregated under the devised rainfall category well indicates the spatial variability while its Coefficient of Variation (CV) explains its temporal variability. A maximum of 38.8 per cent (71 grids) of grids receive 801 to 1000 mm of rainfall annually with CV ranging between 20–52 per cent while 30.1 per cent (55 grids) of grids receive 601–800 mm with 22–50 per cent CV. Thus a majority of (68.8 %) grids receive 601–1000 mm. Above 1000 mm of rainfall was received over 27.9 per cent of grids, 14.2 per cent (26 grids) of grids receive 1001–1200 mm with 22–44 per cent of CV, 9.8 per cent (18 grids) of grids receive 1201–1400 mm of rainfall with 22 to 39 per cent CV and 2.2 per cent (4 grids) receives 1401 to 1600 mm (25 to 103 % CV) and 0.5 per cent or 1 grid each receive 1601–1800mm (35 % CV), 2001–2200 mm (54 % CV) and above 2200 mm (26 % CV) respectively. The least of 401–600 mm of rainfall was received over 3.3 per cent (6 grids) of grids with CV ranging from 31–57 per cent.

To understand the seasonal variation, cumulating rainfall for the monsoon months did similar grid wise analysis. During SWM, a maximum of 60.7 per cent (111 grids) of grids receive 200–400 mm of rainfall with CV varying between 27–124 per cent followed by 19.7 per cent of (36 grids) grids that receive 0–200 mm of rainfall with CV varying from 42–126 per cent. About 16.4 per cent (30 grids) of grids receive 400–600 mm of rainfall with CV ranging from 31–79 per cent while 1.6 per cent (3 grids) of grids receive 801–1000 mm with 51–159 per cent of CV. One grid each or 0.5 per cent of grids receives 601–800 mm, 1401–1600 mm and 2001–2200 mm with a CV of 29, 65 and 33 per cent respectively.

In case of NEM, a maximum of 43.7 per cent (80 grids) of grids receive 201–400 mm of rainfall followed by 37.2 per cent (68 grids) receive 401–600 mm of rainfall with CV varying from 36–74 and 33–67 per cent respectively. Around 14.2 per cent of grids (26 grids) receive 601–800 mm of rainfall with CV varying from 29–57 per cent. Least of 4.9 per cent (9 grids) of grids receives 801–1000 mm of rainfall with 28–43 per cent of CV (Figure. 8.6).

Rainy days

Similar to rainfall, the normal annual rainy days was also studied grid wise with 183 points that covers entire Tamil Nadu. A maximum of 64.5 per cent (118 grids) of grids distributes the annual rainfall through 41 to 60 rainy days with CV varying between 15–72 per cent while 26.2 per cent (48 grids) of grids distribute through 61–80 rainy days with 12–42 per cent of CV. About 4.9 per cent (9 grids) of grids had rainy days from 21–40 with a CV from 26–54 per cent. Least of 2.2 (4 grids), 1.6 (3 grids) and 0.5 (1 grid) per cent of grids distribute through 101 to 120, 81 to 100 and 121– 140 rainy days. The CV varies about 19–40, 14–41 and 10 per cent respectively.

Seasonal variation in rainfall distribution was studied grid wise aggregating the rainy days for the monsoon months. During SWM, a maximum of 50.3 per cent (92 grids) of grids distributes the SWM rainfall through 21–40 rainy days with CV ranging between 18 and 130 per cent followed by 47 per cent (86 grids) of grids that distribute through 1–20 rainy days with CV from 26 to 125 per cent. Least of 0.5 (1 grid), 1.6 (3 grids) and 0.5 (1 grid) per cent of grids distributes their SWM rainfall through 41–60, 61–80 and 81–100 rainy days with CV 41, 26 to 51 and 15 per cent respectively. During NEM, maximum of 84.2 per cent (154 grids) of grids over Tamil Nadu distributes its NEM rainfall through 21–40 rainy days with 19–55 per cent CV followed by 15.3 per cent (28 grids) of grids that distribute through 1–20 rainy days with 31–54 per cent of CV. Least of 0.5 per cent or one grid distributes through 41–60 rainy days with 22 per cent of CV (Gowtham *et al.,* 2017).

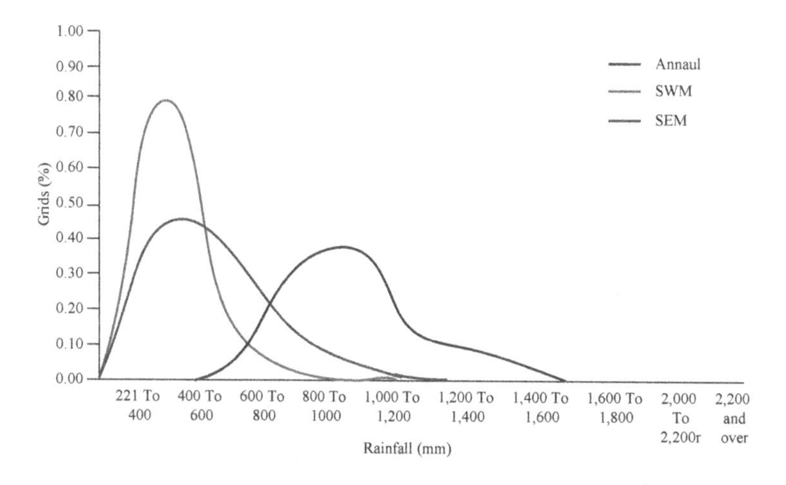

(a) Rainfall distribution over Tamil Nadu

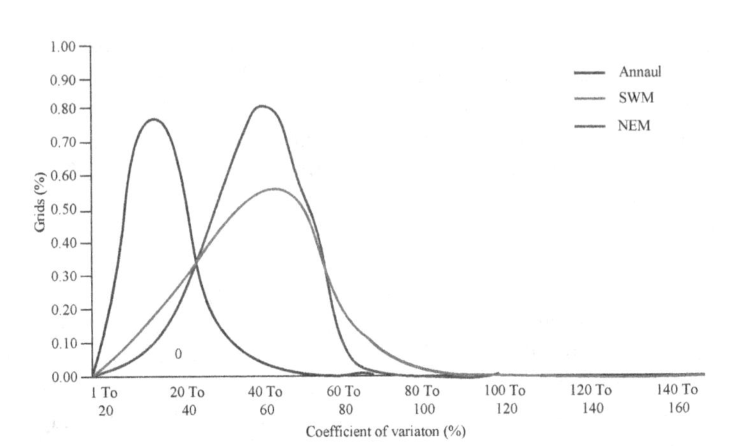

(b) Rainfall Variaiton over Tamil Nadu

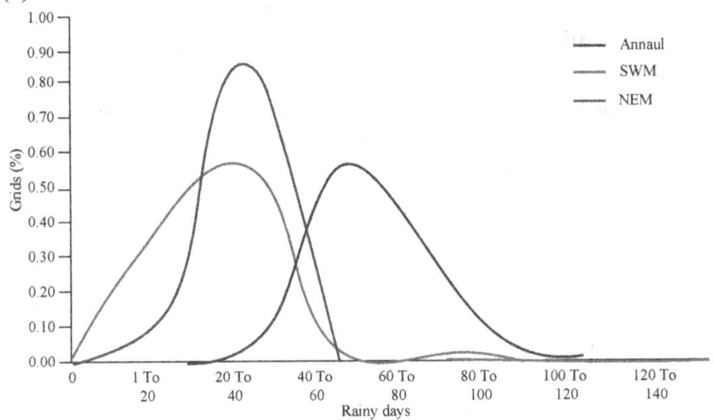

(c) Rain days over Tamil Nadu

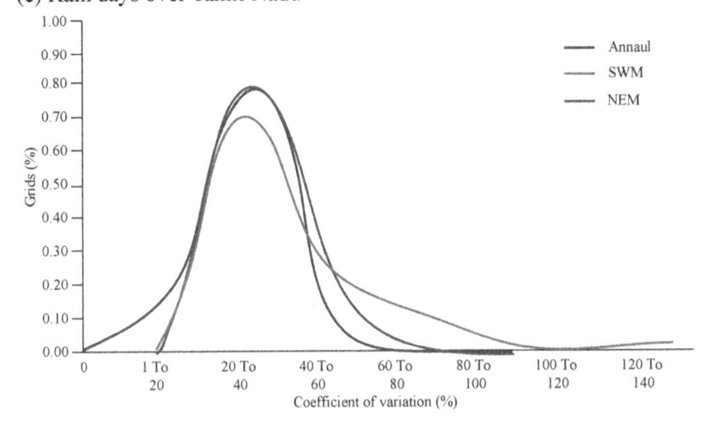

(a) Rainy days Variation over Tamil Nadu

Fig. 8.6. Variation in annual and seasonal rainfall and rainy days over Tamil Nadu (*Source*: Gowtham *et al.*, 2017)

8.4.3 Future Climate Projections Under RCP 4.5 and RCP 8.5 Scenarios Over Tamil Nadu

The maximum and minimum temperatures over Tamil Nadu are projected to increase irrespective of the models and scenarios studied. Maximum temperature over Tamil Nadu are expected to increase up to 3.7°C and 4.7°C by the end of mid-century as projected through stabilization (RCP 4.5) and overshoot emission (RCP 8.5) pathways. Minimum temperature over Tamil Nadu are expected to increase up to 4.9°C and 5.7°C by the end of mid-century as projected through stabilization and overshoot emission pathways. The rate of increase in minimum temperature is higher than that of maximum temperature. Among the monsoon seasons, SWM is projected to have a higher increase in both maximum and minimum temperatures than NEM over Tamil Nadu. Annual rainfall over Tamil Nadu is expected to vary between a decrease of -17.1 per cent to an increase of 34.6 per cent and –33.0 per cent to 45.1 per cent as projected through RCP 4.5 and RCP 8.5 scenarios. Rainfall during the SWM is expected to have wider variation in future than NEM (Gowtham *et al.*, 2017).

Maximum Temperature

Annual maximum temperature: All the models projected an increase of temperature for all the locations over Tamil Nadu. The selected five climate models BNU-ESM, CanESM2, CCSM4, CMCC-CM, CMCC-CMS under RCP 4.5 projected 0.7°C (BNU-ESM and CCSM4) to 3.7°C (CanESM2) increase in annual temperature by mid-century. The lowest increase was projected over Vellore by both the models while highest increase was over Krishnagiri. The Climate models MIROC5, CanESM2, GFDL-ESM2, CMCC-CM, HadGEM2-AO selected under RCP 8.5 projected an increase of 0.7°C (GFDL-ESM2) to 4.7°C (CanESM2). Interestingly through RCP 8.5 also, the lowest increase was projected over Vellore and highest increase over Krishnagiri (Figure. 8.7).

Seasonal Maximum temperature: Seasonal variation in climate projections was also studied. The changes projected for Cold Weather Period (CWP) through RCP 4.5 is from 0.9°C (BNU-ESM, CMCC-CM and CMCC-CMS) to 3.1°C (CanESM2) by mid-century. The lowest increase was projected over Ariyalur, Theni and Tiruvannamalai while the highest increase was projected over Krishnagiri. Climate simulations through RCP 8.5 projected an increase of 0.3°C (GFDL-ESM2) to 4.5°C (CanESM2). The lowest increase in CWP is projected over Vellore and highest increase over Krishnagiri. During Hot Weather Period (HWP), RCP 4.5 projected an increase ranging from 0.5°C (BNU-ESM) to 3.2°C (CanESM2). The lowest increase was projected over Tiruvannamalai while the highest increase was projected over Vellore. RCP 8.5 projected an increase ranging from 0.8°C (GFDL-ESM2) to 4.2 °C (CanESM2). Interestingly, RCP 8.5 also projected the lowest increase in HWP over Tiruvannamalai while the highest increase was over Vellore.

During SWM, 0.6°C (BNU-ESM) to 4.8°C (CanESM2 and CMCC-CM) was projected through RCP 4.5. The lowest increase was projected over Villupuram while the highest increase was projected over Krishnagiri by both the models (CanESM2 and

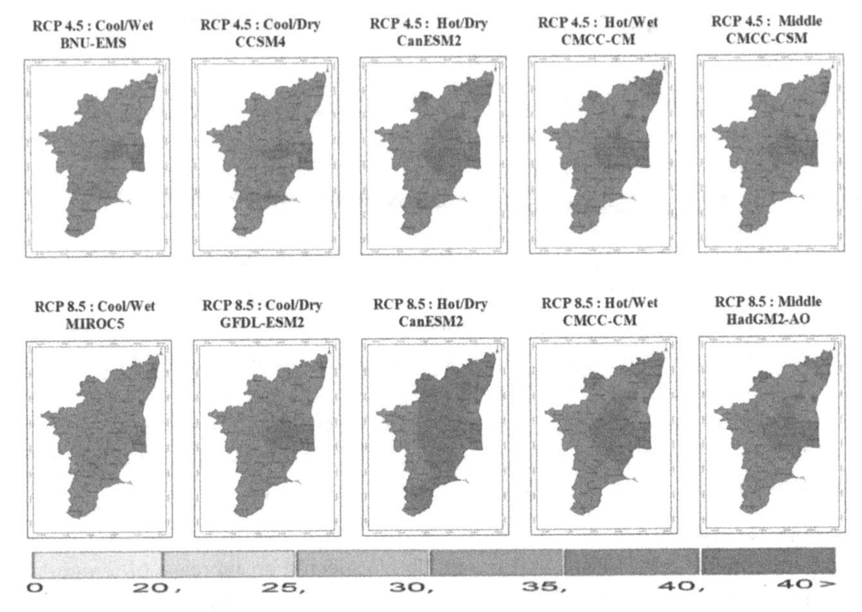

Fig. 8.7. Projected annual Maximum temperature (°C) for RCP 4.5 and RCP 8.5 scenarios (*Source*: Gowtham *et al.*, 2017)

CMCC-CM). RCP 8.5 projected an increase of 0.3°C (MIROC 5) to 5.7°C (CanESM2) and the lowest increase was projected over Villupuram while the highest increase was projected over Krishnagiri similar to RCP 4.5. During NEM, 0.1°C (CCSM4) to 4.7°C (CMCC-CM) was projected through RCP 4.5. The lowest increase was projected over Vellore while the highest increase was projected over Dharmapuri. RCP 8.5 projected an increase of 0.3°C (MIROC 5) to 5.5°C (CanESM2) and the lowest increase was projected over Vellore while the highest increase was projected over Dharmapuri similar to RCP 4.5.

Minimum Temperature

Annual minimum temperature: It was projected to increase by all the models studied for all the locations over Tamil Nadu. The selected five climate models BNU-ESM, CanESM2, CCSM4, CMCC-CM, CMCC-CMS under RCP 4.5 projected 0.8°C (CCSM4) to 4.9°C (CanESM2 and CMCC-CM) increase in annual temperature by the end of mid-century. The lowest increase was projected over Villupuram by both the models while highest increase was over Krishnagiri. The RCP 8.5 projected an increase of 1.3°C (MIROC 5 and GFDL-ESM2) to 5.7°C (CanESM2). The lowest increase was projected over Villupuram and Thiruvallur, and highest increase over Krishnagiri (Fig 8.8).

Seasonal minimum temperature: Seasonal variation in climate projections was also studied. The changes projected for Cold Weather Period (CWP) through RCP 4.5 is an increase ranging from 0.3°C (BNU-ESM) to 5.3°C (CanESM2) by the end of mid-century. The lowest increase was projected over Villupuram while the highest increase was projected over Krishnagiri. RCP 8.5 projected an increase of 1.2°C (MIROC 5) to 6.3°C (CanESM2). The lowest increase in CWP is projected over villupuram and highest increase over Krishnagiri. During Hot Weather Period (HWP), RCP 4.5 projected no change (BNU-ESM) to an increase of 5.1°C (CanESM2). The lowest increase was projected over Vellore while the highest increase was projected over Krishnagiri. The RCP 8.5 projected an increase ranging from 0.3°C (GFDL-ESM2) to 6.0 °C (CanESM2). Interestingly, in RCP 8.5 also the lowest increase in HWP is projected over Vellore while the highest increase was over Krishnagiri.

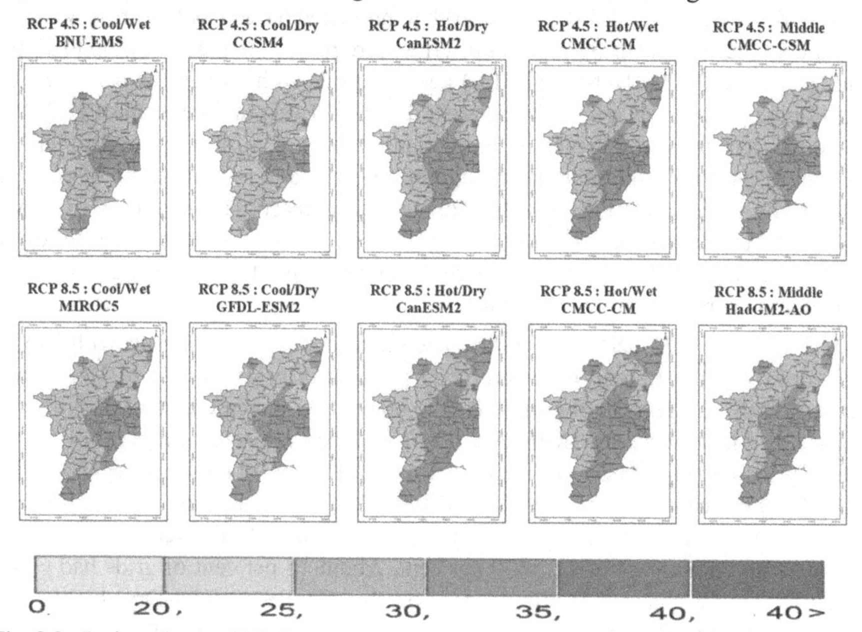

Fig. 8.8. Projected annual Minimum temperature (°C) for RCP 4.5 and RCP 8.5 scenarios *Source*: Gowtham *et al.*, 2017)

During SWM, 0.8°C (CCSM4) to 4.8°C (CanESM2) was projected through RCP 4.5. The lowest increase was projected over Villupuram while the highest increase was projected over Krishnagiri by both the models. The RCP 8.5 projected an increase of 1.1°C (MIROC 5) to 5.6°C (CanESM2). Similar to RCP 4.5, lowest increase was projected over Villupuram while the highest increase was projected over Krishnagiri similar to RCP 4.5. During NEM, 0.5°C (BNU-ESM) to 4.9°C (CMCC-CM) was projected through RCP 4.5. The lowest increase was projected over Villupuram while the highest increase was projected over Krishnagiri. The RCP 8.5 projected an increase of 1.3°C (MIROC 5) to 5.5°C (CMCC-CM) and the lowest increase was projected over Chennai, Villupuram, Tiruvannamalai, Vellore and Thiruvallur while the highest increase was projected over Krishnagiri.

Rainfall

Annual Rainfall: It was projected to vary between a decrease of -17.1 per cent (CMCC-CM) to an increase of 34.6 per cent (CMCC-CMS) by RCP 4.5 while RCP

8.5 projected a variation of -33.0 per cent (CanESM2) to 45.1 per cent (GFDL-ESM2). The range projected by RCP 8.5 is higher than that of RCP 4.5. In RCP 4.5, annual rainfall projected by all the five climate models had almost similar distribution of rainfall over the grids of Tamil Nadu. Most of the grids of Tamil Nadu was projected to witness a change between–20 to + 20 per cent. More than 50 per cent of the grids were projected to have 1 to 20 per cent increase in rainfall except CanESM2 (Hot-dry), which projected a decrease from 1 to 20 per cent. The model CCSM4 (Cool-Dry) had 92 per cent of grids with 1 to 20 per cent increase and 8 per cent of grids with 0 to 20 per cent decrease. BNU–ESM (Cool-wet) had projected 63 per cent of grids with 1 to 20 per cent increase while 37 per cent of grids were projected to decrease up to 20 per cent. The CMCC-CM (Hot-wet) had 54 per cent grids with 1 to 20 per cent increase and 43 per cent of grids with 0–20 per cent decrease. Least of 3 per cent of grids had 21–40 per cent increase. Typically, Projections through CanESM2 (Hot-dry) had 89 per cent of its grids with either no change or a decrease in rainfall up to 20 per cent followed by 11 per cent of grids with 1–20 per cent increase. The model CMCC-CMS (mid of the quadrant) had 77 per cent of grids with 1–20 per cent increase followed by 21 per cent of grids with 21–40 per cent increase. Only two per cent grids had no change or a decrease up to 20 per cent.

In RCP 8.5, GFDL-ESM2 (Cool-dry) had highest of 74 per cent grids with 1 to 20 per cent increase in rainfall followed by 25 per cent of grids with 21 to 40 per cent increase. Only one per cent of grids had 41 to 60 per cent increase in rainfall. Not even a single grid had a decrease. MIROC5 (Cool-wet) had projected a highest of 55 per cent of grids with 1–20 per cent increase in rainfall followed by 45 per cent of grids with 21–40 per cent increase. CMCC-CM (Hot-Wet) had a maximum of 58 per cent of grids distributed with 1 to 20 per cent increase in rainfall followed by 24 per cent of grids with an increase of 21 to 40 per cent. About 18 per cent of grids had either no rainfall or a decrease was projected. Typically, CanESM2 (Hot-Dry) had 97 per cent of grids with either no change or decrease up to 20 per cent followed by 8 per cent of grids with 21–40 per cent decrease in rainfall. Least of 3 per cent of grids had 1 to 20 per cent increase in rainfall. The HadGEM2-AO (middle of four quadrants) had maximum of 50 per cent of grids over 21 to 40 per cent increase followed by 26 per cent of grids with 1–20 per cent increase. About 21 per cent of grids had either no change or a decrease up to 20 per cent. Least of 3 per cent of grids had 41–60 per cent increase in rainfall (Figure. 8.9).

Rainfall projection for southwest monsoon and its spatial distribution: Rainfall during SWM was projected to vary between a decrease of -15.8 per cent (CMCC-CM) to an increase of 60.1 per cent (CMCC-CM) by RCP 4.5, while RCP 8.5 projected a variation of -19.9 per cent (CanESM2) to 75.4 per cent (MIROC5). The spatial spread of SWM rainfall projected by all models under RCP 4.5 showed major number of grids with increase in rainfall except CanESM2. CCSM4 (Cool-Dry) had projected a maximum of 98 per cent of grids with 1–20 per cent increase in rainfall ,while 2 per cent of grids had 0–20 per cent decrease in rainfall. Interestingly, BNU-ESM (Cool-Wet) had projected all the grids of Tamil Nadu to have increase in rainfall. Maximum of 78 per cent of grids had 21 to 40 per cent increase in rainfall followed by 22 per cent grids with 1–20 per cent increase. The CMCC-CM (Hot-Wet) had projected 90 per cent of grids with increase in rainfall while 10 per cent of the grids were projected to have reduction up to 20 per cent. About 54, 31 and 5 per cent of grids were projected to have 1–20, 21–40 and 41–60 per cent increase in rainfall. The CanESM2 (Hot-Dry) had projected 55 per cent of the grids with decrease in rainfall up to 20 per cent followed by 45 per cent of grids with increase in rainfall up to 20 per cent. Only one per cent grids had increase in rainfall from 21 to 40 per cent. The CMCC-CMS (middle of four quadrants) had projected 85 per cent of grids to have increase in rainfall from 1 to 20 per cent followed by 9 per cent of grids with 21–40 per cent increase. About 6 per cent of grids had decrease in rainfall up to 20 per cent.

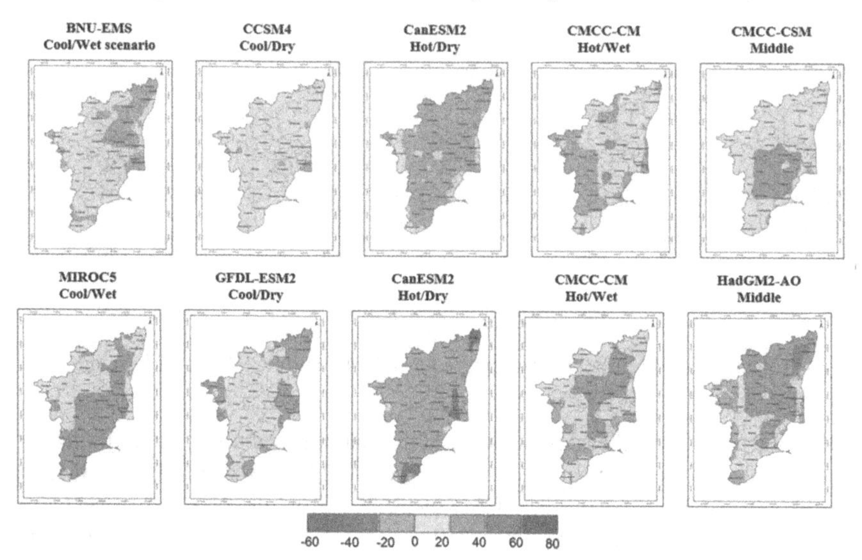

Fig. 8.9. Projected annual rainfall deviation (%) for RCP 4.5 and RCP 8.5 over baseline (*Source*: Gowtham *et al.*, 2017)

Except GFDL-ESM2, all other models under RCP 8.5 had projected an increase in rainfall for most of the grids. The GFDL-ESM2 had projected a maximum 69 per cent of grids with no change or a decrease up to 20 per cent. About 27 and 4 per cent of grids had 1–20 and 21–40 per cent increase in rainfall respectively. The MIROC5 (Cool-Wet) had projected all its grids to have an increase in rainfall. About 63 per cent of grids were projected to have an increase in rainfall of 21–40 per cent followed by 29 per cent of grids with 41–60 per cent increase. Minimum of 4 and 3 per cent of grids had 1–20 and 61–80 per cent increase in rainfall. The CMCC-CM (Hot-Wet) had projected 96 per cent of grids to have increase in rainfall and only 4 per cent of grids alone had decrease in rainfall up to 20 per cent. The CanESM2 (Hot-Dry) had also projected 96 per cent of grids with increase in rainfall followed by 4 per cent of grids with decrease in rainfall up to 20 per cent. The HadGEM2-AO (middle of four quadrants) projected 99 per cent of grids an increase in rainfall while only one per cent grids was projected with decrease of 20 per cent. About 45 per cent of grids had increase in rainfall from 1–20 per cent followed by 46 per cent of grids with 21–40 per cent increase. Minimum of 2 and 6 per cent of grids had 41–60 and 61–80 per cent increase in rainfall respectively (Figure. 8.10).

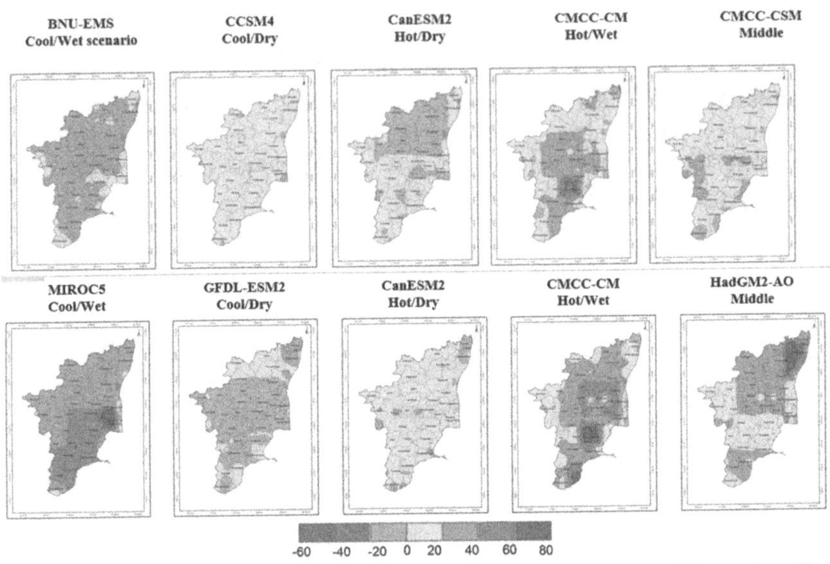

Fig. 8.10. Projected SWM rainfall deviation (%) for RCP 4.5 and RCP 8.5 over baseline (*Source*: Gowtham *et al.*, 2017)

Rainfall projection for northeast monsoon and its spatial distribution: Rainfall during NEM was projected to vary between a decrease of -28.4 per cent (BNU-ESM) to an increase of 57.5 per cent (CMCC-CM) by RCP 4.5, while RCP 8.5 projected a variation of -31.6 per cent (CanESM2) to 60.2 per cent (GFDL-ESM2). All the models had projected major number of grids with increase of NEM rainfall except BNU-ESM and CanESM2. The CCSM4 (Cool-Dry) had projected 92 per cent of grids to have increase in rainfall up to 20 per cent, while 8 per cent of the grids had decrease in rainfall up to 20 per cent. The CMCC-CM (Hot-Wet) projected 58 per cent of grids with 1–20 per cent increase in rainfall while 35 per cent of grids were projected to have decrease in rainfall up to 20 per cent. Least of 7 and 1 per cent of grids had 21–40 and 41–60 per cent increase in rainfall respectively. The CMCC-CMS (middle of four quadrants) projected 86 per cent of grids with increase in rainfall from 1–20 per cent followed by 11 and 1 per cent of grids with 21–40 and 41–60 per cent increase in rainfall respectively.

Typically, the model BNU-ESM (Cool-Wet) had projected 85 per cent of grids to have decrease in rainfall up to 20 per cent followed by 13 per cent of grids to have increase in rainfall up to 20 per cent. Least of one per cent of grids each had 21–40 per cent increase and decrease respectively. The CanESM2 (Hot-Dry) had projected 93 per cent of the grids to have decrease in rainfall up to 20 per cent while 1 per cent of grids had decrease from 21–40 per cent. Least of 6 and 1 per cent of grids was projected to have increase in rainfall from 1 to 20 per cent and 21 to 40 per cent respectively.

In RCP 8.5, all the representative models projected increase in rainfall over maximum number of grids over Tamil Nadu except CanESM2. The GFDL-ESM2 projected an increase in rainfall for all the grids. About 26 per cent of grids were projected to have increase in rainfall up to 20 per cent and maximum of 63 per cent of grids were projected to have 21 to 40 per cent increase. Least of 11 per cent had 41 to 60 per cent increase in rainfall. The MIROC5 (Cool-Wet) projected 99 per cent of grids to have an increase in rainfall while only one per cent had decrease in rainfall from no change to 20 per cent. About 85 per cent of the grids were projected to have 1 to 20 per cent increase in rainfall followed by 14 per cent of grids with 21 to 40 per cent increase in rainfall. The CMCC-CM (Hot-Wet) had projected 97 per cent of grids to have increase in rainfall while 3 per cent of grids were projected to have decrease up to 20 per cent. About 65 per cent of the grids had 1–20 per cent increase followed by 31 and 1 per cent of grids with 21–40 and 41–60 per cent increase in rainfall. The HadGEM2-AO (middle of four quadrants) projected 98 per cent of grids to have an increase in rainfall while 2 per cent grids were projected to have decrease up to 20 per cent. About 64 per cent grids had 21–40 per cent increase followed by 33 per cent grids with 1–20 per cent increase and least of 1 per cent grid had 41–60 per cent increase in rainfall. Typically, CanESM2 (Hot-Dry) alone had projected 98 per cent of grids to have decrease in rainfall, while one per cent of grids each were projected to have 1–20 and 21–40 per cent increase in rainfall. Out of 98 per cent, 60 per cent of grids had 0 to 20 per cent decrease while 38 per cent grids were projected to have 21–40 per cent decrease in rainfall (Figure. 8.11).

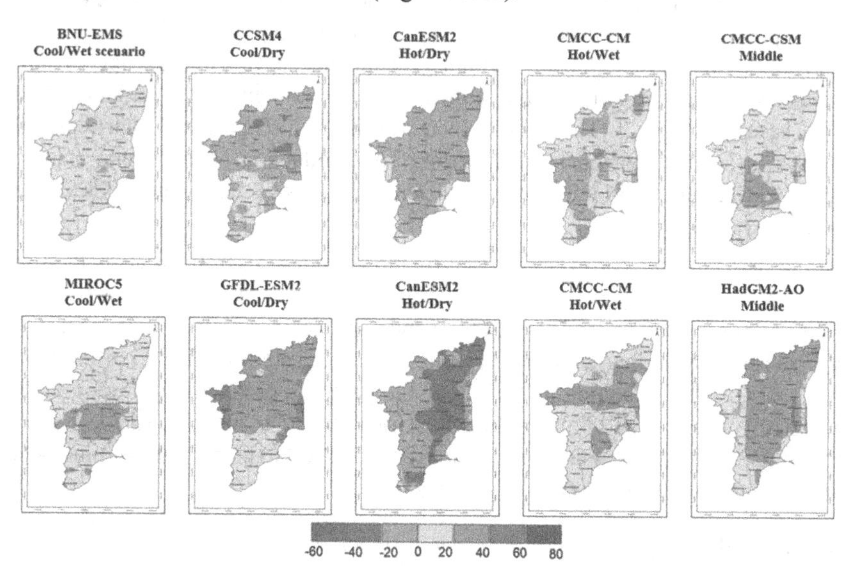

Fig. 8.11. Projected NEM rainfall deviation (%) for RCP 4.5 and RCP 8.5 over baseline (*Source*: Gowtham *et al.*, 2017)

8.5 Climate Change Impact on Agriculture

Agriculture is one of the largest employers in the world and the most weather dependent of all the human enterprises. India has to support about 17 per cent of the world's human population and 15 per cent of the livestock with only 2.4 per cent of the world's geographical area and 4 per cent of world's water resources. Agriculture is one of the important sectors of the Indian economy and it contributes 14 per cent of the Nation's GDP, about 11per cent of its exports, about half of the population still relies on agriculture as its principal source of income and it is a source of raw material for a large number of industries. Latest projections suggest that global population would grow from 7 billion to more than 9 billion people in 2050 (UNESCO, 2012). To feed this growing population, it is expected that a 60 per cent increase in global agricultural production will be required by 2050 (FAO, 2012).

The challenge of rapidly boosting productivity is deepened by the current and expected impacts of climate change. Direct weather impacts *viz.,* droughts, floods, untimely rain, frost, hail, heat and cold waves and severe storms contribute to the annual crop losses in the world agriculture. Global agriculture will be under significant pressure to meet the demands of rising populations using finite, often degraded soil, and natural, manmade resources that are predicted to be further stressed by the impact of climate change. Hence, agriculture in developing countries must undergo significant transformation to meet the growing and interconnected challenges of food insecurity and climate change (FAO, 2010).

8.5.1 Climate Change Implication on Agriculture

Temperature, solar radiation, water, and atmospheric CO_2 concentration are the climate and atmospheric variables of importance to plant productivity, these parameters are changing under the current global environment. Climate change will most likely result in new combinations of soil, climate, atmospheric constituents, solar radiation, and pests, diseases, and weeds. Thus, the expected changes in climate will alter regional agricultural systems, with consequences for food production. The climate change impact on agriculture will be different in different regions of the world and multiple effects are expected.

Agricultural impacts, for example, will be more adverse in tropical areas than in temperate areas. Developed countries will largely benefit since cereal productivity is projected to rise in Canada, northern Europe and parts of Russia. In contrast, many of today's poorest developing countries are likely to be negatively affected in the next 50–100 years, with a reduction in the extent and potential productivity of crop land. Most severely affected will be sub-Saharan Africa due to its inability to adequately adapt through necessary resources or through greater food imports. Potential impacts of climate change on agriculture are as follows,

i. Geographical shifts

Agricultural crop distribution and production is largely dependent on the geographical distribution of thermal and moisture regimes. Global warming is significantly increasing the area with temperature regimes not conducive to growth and production of agricultural crops. Losses due to pest, disease and weed will increase. The range of many insects will change or expand and new combinations of pest and disease may emerge as natural ecosystems respond to shift in temperature and precipitation profiles.

ii. Desertification

Increase in desertification of land is expected due to changes in soil property due to loss of soil organic matter, leaching of soil nutrients, salinisation and erosion (all this might be accelerated due to climate change).

iii. Loss of land through sea level rise and associated salinisation

Sea level rise will affect agriculture in low-lying coastal areas. Flood frequency increases in the already flood insecure areas. Increase in frequency and intensity of extreme weather events are expected. All these events will have a drastic negative impact on the crop growth and productivity.

iv. Effects of changes in temperature

The most adverse agro-ecological impacts of global climate change are in the thermal regime (temperature) and moisture regime. In regards to tropical and subtropical agriculture, increase in temperature and decrease in available moisture are the fundamental changes limiting plant/crop production. Increase in air temperature can accelerate crop growth and consequently shorten the growth period. In cereal crops for example, such changes can lead to poor vernalization (*e.g.*, hastened flowering) and reduced yield. Increased heat stress to crop and livestock; *e.g.*, higher night-time temperatures, which could adversely affect grain formation and other aspects of crop development; increased evapotranspiration rates caused by higher temperatures, and lower soil moisture levels.

v. Effects of changes in concentrations of CO_2

Higher atmospheric concentrations of CO_2 will improve the water use efficiency of all crops (by reducing the evapotranspiration) and increase the rate of photosynthesis of most crops. Increase in the carbon dioxide concentrations are expected to increase the biomass in many crop species. Order of response will be C_3 plants associated with nitrogen fixing microbes > C3 plants > C4 > CAM. Annuals plants respond more to elevated CO_2 levels compared to perennials. The direct effects of CO_2 however will be small in regions where fertilizer usage is low.

vi. Changes in precipitation

Possible decline in precipitation is expected in some food-insecure areas such as southern Africa and the northern region of Latin America. Changes in precipitation level and increase in temperature leads to reduced soil moisture level. Along with reduced water availability for irrigation, impairment of crop growth is expected in non-irrigated areas of many regions. Concentration of rainfall into a smaller number of rainy events with increases in the number of days with heavy rain, increasing erosion and flood risks; Changes in seasonal distribution of rainfall, with less falling in the main crop growing season.

Indirectly, climate change may have considerable effects on land use pattern due to availability of irrigation water, frequency and intensity of inter- and intra-seasonal droughts and floods, and availability of energy. All these can have tremendous impact on agricultural production and hence, food security of any region.

Analysis of the food grains production/productivity data for the last few decades reveals a tremendous increase in yield, but it appears that negative impact of vagaries of monsoon has been large throughout the period. In this context, a number of questions need to be addressed as to determine the nature of variability of important weather events, particularly the rainfall received in a season/year as well its distribution within the season. These observations need to be coupled to management practices, which are tailored to the climate variability of the region, such as optimal time of sowing, level of pesticides and fertilizer application. Hence, the research on the impact of climate change and vulnerability on agriculture is a high priority in India as the changes in CO_2, temperature and precipitation are more evident.

8.5.2 Crops Response to Climate Change

An atmosphere with higher CO_2 concentration would result in higher net photosynthetic rates (Cure and Acock, 1986, Allen *et al.*, 1987). Higher CO_2 concentrations may also reduce transpiration (*i.e.*, water loss) as plants reduce their stomatal apertures, the small openings in the leaves through which CO_2 and water vapor are exchanged with the atmosphere. The reduction in transpiration could be 30 per cent in some crop plants (Kimball, 1983). However, stomatal response to CO_2 interacts with many environmental (temperature, light intensity) and plant factors (*e.g.* age, hormones) and, therefore, predicting the effect of elevated CO_2 on the responsiveness of stomata is still very difficult (Rosenzweig and Hillel, 1995).

The effect of an increase in carbon dioxide would be higher on C3 crops (such as wheat) than on C_4 crops (such as maize), because the former is more susceptible to carbon dioxide shortage. Moreover, the protein content of the grain decreases under combined increases of temperature and CO_2. For rice, the amylase content of the grain-a major determinant of cooking quality-is increased under elevated CO_2. With wheat, elevated CO_2 reduces the protein content of grain and flour by 9-13 per cent. Concentrations of Fe and Zn, which are important for human nutrition, would be lower.

Although increase in CO_2 is likely to be beneficial to several crops such as rice, wheat and pulses, associated increase in temperatures and increased variability of rainfall would considerably impact food production. The IPCC (2007) and few other global studies indicate considerable probability of loss in crop production in India with increase in temperature. Some of these projected loss estimates for the period 2080-2100 are 5–30 per cent (Rosenzweig and Parry, 1994; Fischer *et al.*, 2002; Parry *et al.*, 2004; IPCC, 2007). The IPCC (2007) projects that cereal yields in seasonally dry and tropical regions such as India, are likely to decrease for even small local temperature increases (1–2°C). As a result of these changes, globally the potential for food production is projected to increase with increase in local temperature over a range of 1–3°C, but above this range it is projected to decrease.

There are few Indian studies on these themes and they generally confirm similar trend of agricultural decline with climate change (Aggarwal and Sinha, 1993; Aggarwal and Mall, 2002). More recent studies done at the Indian Agricultural Research Institute (IARI), New Delhi, indicate the possibility of loss of 4–5 million tons in wheat production with every rise of 1°C temperature throughout the growing period even after considering benefits of carbon fertilization. This analysis assumes that irrigation would remain available in future at today's levels and there is no adaptation. The study carried out at IARI also indicates that losses in wheat production at 1°C increase in temperature can be reduced from 4–5 million tons to 1–2 million tons if a large percentage of farmers could change to timely planting and changed to better adapted varieties. These adaptation benefits become smaller as temperature increases further. Wheat growth simulator (WTGROWS) developed at IARI, New Delhi, has been extensively tested for different agro-environments (Aggarwal and Kalra, 1994). Using WTGROWS, a strong linear decline in wheat yield was noticed with the increase in January temperature. For every degree increase in mean temperature, grain yield decreased by 428 kg/ha.

All the yield components in rice viz. panicle number (effective tillers), filled grains per panicle and grain weight responded positively to enhanced carbon dioxide levels without considering the increase in temperature. Increased photo-assimilate supply possibly increased the maturity percentage of the seeds. For every 75-ppm increase in CO_2 concentration, rice yields will increase by 0.5 t/ha, but yield will decrease by 0.6 t/ha for every 1°C increase in temperature (Sheehy *et al.,* 2006).This is presented in Fgure 8.12. The CO_2 enrichment has generally shown significant increases in rice biomass (25–40 per cent) and yields (15-39per cent) at ambient temperature, but those increases tended to be offset when temperature was increased along with rising CO_2 (Ziska *et al.,* 1996; Moya *et al.,* 1998). The yield losses as caused by concurrent increases in CO_2 and temperatures are primarily caused by high-temperature-induced spikelet sterility (Matsui *et al.,* 1997). Increased CO_2 levels may also cause a direct inhibition of maintenance respiration at night temperatures higher than 21°C (Baker *et al.,* 2000).

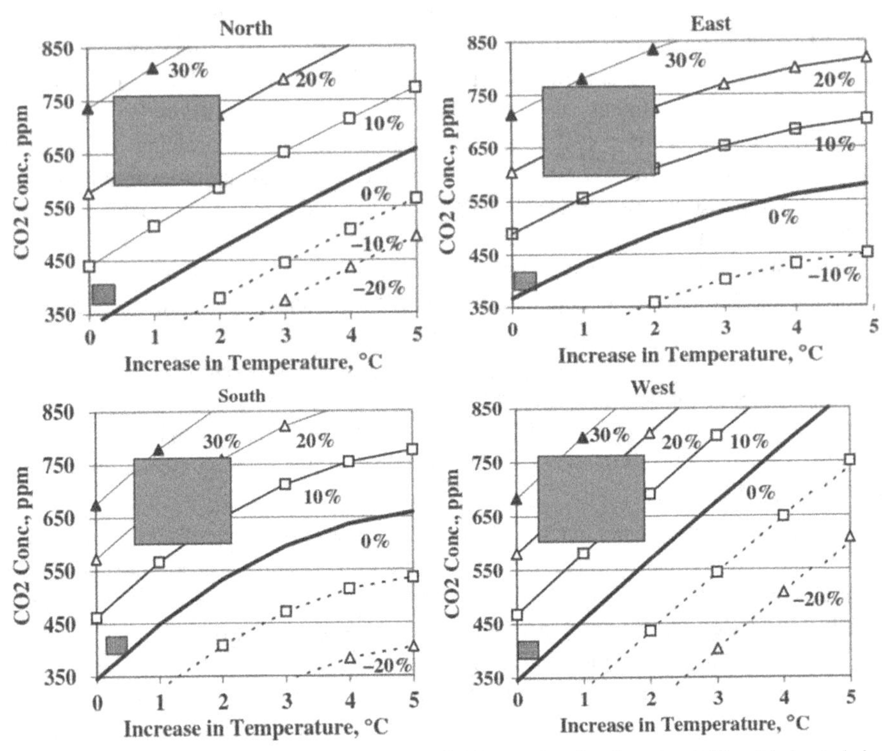

Fig. 8.12. Effect of increase in temperature and CO_2 on simulated grain yields of irrigated rice with improved N management (allowing no N stress) in different regions of India. Lines refer to the equal change in grain yield (per cent change, labelled) at different values of CO_2 and increase in temperature. Large, shaded box refers to bias in impact assessment due to uncertainties in IPCC scenario of 2070 and the small, hatched box refers to the bias due to uncertainties in the scenario of 2010.
(*Source*: Sheehy *et al.*, 2006)

The increase of 1°C temperature without any increase in CO_2 resulted in 5, 8, 5 and 7 per cent decrease in grain yield in north, west, east and southern regions of India respectively. The increase of 2°C temperature resulted in 10–16 per cent reduction in yield in different regions, while a 4°C rise led to 21–30 per cent reduction. Sinha and Swaminathan (1991) reported that a 2°C increase in mean air temperature could decrease rice yield by about 0.75 t/ha in the high yield areas and by about 0.06 t/ha in the low yield coastal regions. Further, a 0.5°C increase in winter temperature would reduce wheat crop duration by seven days and reduce yield by 0.45 t/ha. An increase in winter temperature of 0.5°C would thereby translate into a 10 per cent reduction in wheat production in the high yield States of Punjab, Haryana and Uttar Pradesh. The reduction was lower in eastern India compared to all other regions (Table 8.3). Mean grain yields of control crops in eastern region were 7.9 t/ha as compared to 8.7–9.9 t/ha in other regions. This was because of relatively higher temperatures in east (32.2/25.3°C) both during grain formation and filling phase, accompanied by lower radiation. As a result, these crops had fewer grains and shorter grain filling duration. Although temperatures were high in northern India as well (33.8/25.0°C), the region also had more radiation, which resulted in higher grain yields.

The impact of interactions between carbon dioxide and temperature revealed that at 350 ppm in north India, there was a change of -5, -12, -21, -25 and -31 per cent in grain yield with increase of 1, 2, 3, 4, 5°C temperatures respectively. In the same region, and at the same temperatures but at 550 ppm, these yield changes were 12, 7, 1, -5 and -11per cent, respectively. Similar interaction could be noted for other regions as well. Thus, in eastern and northern regions, the beneficial effect of 450, 550 and 650 ppm CO_2 was nullified by an increase of 1.2–1.7, 3.2–3.5 and 4.8–5.0°C, respectively (Table 8.13). In southern and western regions, positive CO_2 effects were nullified at temperatures lower than these. It can be concluded that in improved management conditions, the regions, such as southern and western parts of India, which currently have relatively lower temperatures are likely to show less increase in rice yields under climate change compared to northern and eastern regions.

Table 8.3. Temperature increases (°C) that will cancel out the positive effect of CO_2 in different regions at two levels of management

Management	CO_2 concentration		
	450 ppm	550 ppm	650 ppm
North India			
Improved	1.7	3.2	>5.0
Current	1.9	2.7	4.8
East India			
Improved	1.2	3.5	>5.0
Current	2.0	4.4	>5.0
West India			
Improved	0.9	1.8	2.8
Current	1.0	2.1	3.4
South India			
Improved	1.0	2.3	4.4
Current	0.9	2.0	3.4

More detailed analysis of rice yields by the International Rice Research Institute forecast 20 per cent reduction in yields over the region per degree celsius of temperature rise. Rice becomes sterile if exposed to temperatures above 35°C for more than one hour during flowering and consequently produces no grain. The rising temperatures will adversely affect the world's food production and India would be the hardest hit, according to the analysis by the Universal Ecological Fund (FEU-US). The crop yield in India, the second largest world producer of rice and wheat, would fall up to 30 per cent by the end of this decade.

Enhanced carbon dioxide concentration, increases the carbon dioxide assimilation and they're partitioning within the source leaf and transport to the sink in mungbean and wheat. Carbon dioxide elevation partially compensates for the negative effect of moisture stress in *Brassica* plants and may possibly help to grow in the drier habitat than they are currently grown. *Brassica spp.* responded differently to elevated carbon dioxide levels. Lal *et al.*, (1999) projected 50 per cent increased yield for soybean for a doubling of CO_2 in Central India. However, a 3°C rise in surface air temperature almost cancels out the positive effects of doubling of carbon dioxide concentration. A decline in daily rainfall amount by 10 per cent restricts the grain yield to about 32 per cent.

Hundal and Kaur (1996) examined the climate change impact on productivity of wheat, rice, maize and groundnut crop in Punjab. If all other climate variables were to remain constant, temperature increase of 1, 2 and 3°C from present day condition, would reduce the grain yield of wheat by 8.1, 18.7 and 25.7 per cent, rice by 5.4, 7.4 and 25.1 per cent, maize by 10.4, 14.6 and 21.4 per cent and seed yield in groundnut by 8.7, 23.2 and 36.2 per cent, respectively. Mandal (1998), Chatterjee (1998) and Sahoo (1999) calibrated and validated the CERES-maize, CERES-sorghum and WOFOST models for the Indian environment and subsequently used them to study the impact of climate change (CO_2 levels: 350 and 700 ppm and temperature rise from 1 to 4°C with 1°C increment) on phenology, growth and yield of different cultivars. Chatterjee (1998) observed that an increase in temperature consistently decreased maize and sorghum yields from the present-day conditions. Increase in temperature by 1 and 2°C, the sorghum potential yields found decreased on an average by 7–12 per cent. An increase in 50 ppm CO_2 increases yields by only 0.5 per cent. The beneficial effect of 700 ppm CO_2 was nullified by an increase of only 0.9°C in temperature.

Mandal (1998) also observed that an increase in temperature up to 2°C did not influence potential and irrigated yields of chickpea as well as above ground biomass significantly. Pre-anthesis and total crop duration got reduced with the temperature rise. Nitrogen uptake and total water use (as evapo-transpiration) were not significantly different up to 2°C rise. The elevated CO_2 increased grain yield under potential, irrigated and rainfed conditions. There was a linear increase in grain yield, as the CO_2 concentration increased from 350–700 ppm. Potential grain yield of pigeon pea decreased over the control when the temperature was increased by 1°C (using WOFOST).

Sahoo (1999) carried out simulation studies of maize for climate change under irrigated and rainfed conditions. Rise in temperature decreased the yield under both the conditions. At CO_2 level of 350 ppm, grain yield decreased continuously with temperature rise till 4°°C. This was possibly due to reduction in days to 50 per cent silking and physiological maturity. At CO_2 level of 700 ppm, grain yield increased by about 9 per cent. The temperature rise effect in reduction of yield was noted in several maize cultivars. Effect of elevated carbon dioxide concentration on growth and yield of maize was established, but less pronounced when compared with crops, like wheat, chickpea and mustard crops. The beneficial effect of 700 ppm CO_2 was nullified by an increase of only 0.6◦C in temperature. Further increase in temperature always resulted in lower yields than control.

The major climate change drivers that could adversely impact agriculture sector in Tamil Nadu State of India are continuous increase in ambient temperature, increase in frequency and intensity of extreme hydro meteorological events such as droughts, cyclones and high intensity rainfall leading to floods. Out of the 5.75 M/ha of cultivable area, around 3.3 M.ha falls under dryland agriculture and this is highly vulnerable to changing climate. There is also shift in monsoon onset over different parts of Tamil Nadu (Fig. 8.13) that impacts the length of the growing period especially in the dry land regions.

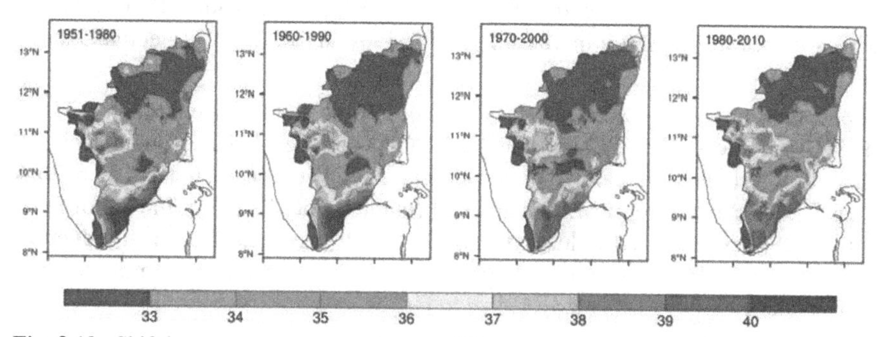

Fig. 8.13. Shift in monsoon onset weeks over different moving periods of Tamil Nadu (*Source*: Dheebakaran *et al.*, 2016)

In Tamil Nadu increase in temperature leads to decrease in yields of majority of the crops. For example, increase in temperature, causes spikelet sterility in rice affecting its productivity. Crop water requirement needs to be increased, and more frequent irrigation is required. The surface water resources are likely to be depleted, creating pressure on ground water. With climate change, the intensity of cyclones is expected to increase, which will have implications on agriculture in the coastal zones. Likely increase in heavy precipitation events would result in flash floods, leading to deterioration in soil health due to heavy loss of topsoil due to erosion, and decline in soil organic matter content which has a direct impact on agriculture productivity.

As per the Vulnerability Atlas of National Initiative for Climate Resilient Agriculture (NICRA) the climate change vulnerability of the people of some of the districts like Ramanathapuram, Perambalur, Ariyalur, Namakkal, Dharmapuri, Krishnagiri, Tiruvannamalai, Villupuram, Tiruvarur, Pudukkottai, Sivagangai, Theni and Karur has to be largely attributed to the decrease in rainfall during the months of June and July. Number of dry spells of 14 days is also showing an increasing trend in the State. Tamil Nadu experiences recurrent droughts and floods, and long-term analysis of yearly climate data, indicates that the probability of occurrence of droughts in Tamil Nadu is every 2.5 years (National Rain-fed Authority of India, 2012). A Study conducted by the Agro Climate Research Centre, Tamil Nadu Agricultural University, Coimbatore on the frequency of drought occurrence at the district level over Tamil Nadu inferred that the districts such as Virudhunagar, Sivagangai, Trichy, Karur, Krishnagiri, Namakkal, Dharmapuri and Kancheepuram were highly prone to drought.

Rice is cultivated in more than 3.1 million ha in Cauvery Delta Zone(CDZ) that contributes to 40 per cent of the food grain production of Tamil Nadu. Due to the region's fertility and contribution of food grains to the State, Cauvery Delta region has been known as the "Granary of the South India". The yields of ADT 43 rice over CDZ for A1B (SRES) scenario simulated by DSSAT without considering the CO_2 fertilization effect showed a reduction of 356 kg ha^{-1} decade^{-1} for PRECIS output, whereas the decline was 217 kg ha^{-1} decade^{-1} for RegCM3 output. However, when CO_2 fertilization effect was considered in DSSAT, the PRECISE output showed decreasing trend at the rate of 135 kg ha^{-1} decade^{-1}(Figure. 8.14), whereas RegCM3 projected increased yield (24 kg ha^{-1} decade^{-1}) (Geethalakshmi *et al.,* 2011).

Geethalakshmi *et al.,* (2011) also tried to simulate the yield of rice through DSSAT and APSIM. The study revealed that the DSSAT predicted reduction in yield in Thanjavur for all the timescales. The reduction ranged between 13.1–18.7, 15.6–26.4 and 16.7–33.9 per cent for near, mid and end century under RCP 4.5. In case of RCP 8.5 also, DSSAT predicted reduction in yield for all the timescales. The reduction ranged between 14.2–17.9, 6.6–17.1 and 8.6–39.2 per cent for near, mid and end century under RCP 8.5 (Table 8.4).

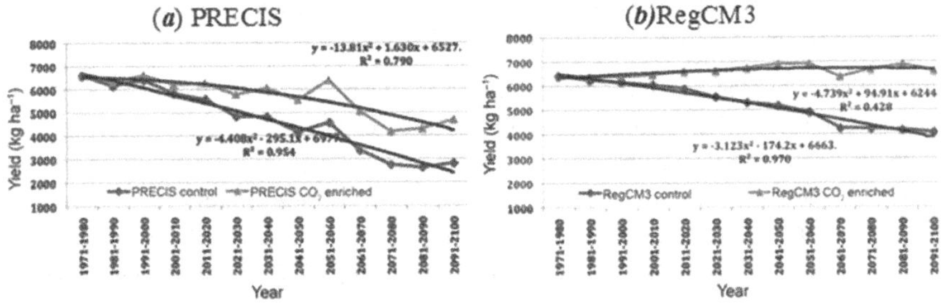

Fig. 8.14. Response of rice yield to climate change. Rice yield predicted using (*a*) PRECIS and (*b*)RegCM3
(*Source*: Geethalakshmi *et al.,* 2011)

Table 8.4 Yield response of Rice to climate change over Thanjavur through DSSAT.

	CCSM4	GFDL-ESM2M	HadGEM2-ES	MIROC5	MPI-ESM-MR
Near	−14.3	−13.1	−18.3	−18.7	−16.2
Mid	−15.9	−15.6	−26.4	−16.3	−16.5
End	−17	−16.7	−33.9	−18	−17.4

APSIM predicted reduction in yield in Thanjavur for all the timescales. The reduction ranged between 3.4–8.0, 9.4–18.2 and 11.9–22.4 per cent for near, mid and end century under RCP 4.5. In case of RCP 8.5, APSIM predicted reduction in yield for all the timescales. The reduction in yield ranged between 7.1 to 9.8, 13.5–23.0 and 20.5–27.6 per cent for near, mid and end century under RCP 4.5 (Table 8.5).

Table 8.5. Yield response of Rice to climate change over Thanjavur through APSIM.

	CCSM4	GFDL-ESM2M	HadGEM2-ES	MIROC5	MPI-ESM-MR
Near	−4.9	−3.4	−7.4	−8	−6.5
Mid	−9.8	−9.4	−18.2	−12	−11.1
End	−12.3	−11.9	−22.4	−14.6	−14.3

8.5.3. Impact of Climate Change on Rice Over Tamil Nadu

Per cent Relative Difference from control years revealed a yield reduction of 12.24 per cent for near future, 26.67 per cent reduction in mid-century and a 44 per cent reduction during late century for PRECIS control. The CO_2 enrichment offsets the yield reduction to some extent and yield reduction in CO_2 enrichment for near future, mid-century and late century are 6.21, 9.93 and 21.95 per cent, respectively (Fig. 8.15)

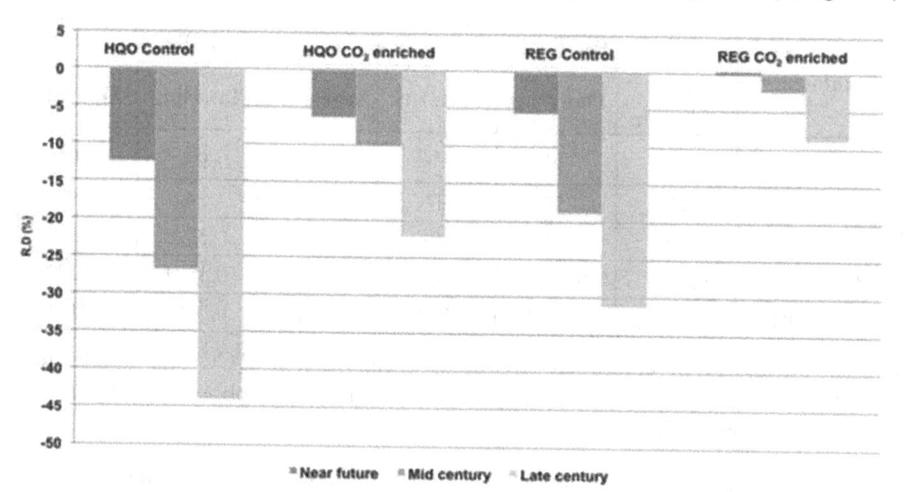

Fig. 8.15. Rice per cent yield different from the base
(*Source*: Rajalakshmi, 2013)

In REG, Per cent Relative Difference from control years showed a yield reduction of 5.42 per cent for near future, 18.64 per cent reduction in mid-century and 31.03 per cent reduction during late century. CO_2 enrichment offset the yield reduction to some extent. Yield reduction in CO_2 enrichment for mid-century and late century are 2.29 and 8.76, respectively while near future had an increment of 0.42 per cent. Yield decreased consistently for near future, mid- century and late century. Yield reduction was sustained in CO2 enrichment in both PRECIS (HQO) and REG (Rajalakshmi *et al.*, 2013).

Impact of elevated temperature on rice productivity

Rice productivity was negatively impacted for elevated temperatures. The yield reduction ranged from 4-6 per cent, 12-15 per cent, 22-25 per cent, 37-40 per cent and 53–56 per cent for 1°C, 2°C, 3°C, 4°C and 5°C respectively (Bhuvaneswari *et al.*, 2014).

8.5.4 Elevated Temperature and CO_2 on C_3 (Rice) and C_4 (Maize) Plants in Tamil Nadu

a. *Impact of* ambient and controlled environment on phenology, growth and yield of C_3 (rice) and C_4 (maize) crops

Phenology: C_3 (rice) and C_4 (maize) crops grown under controlled environment condition in TCC (T_2: ambient temperature + 4°C) and SPAR (T_3: ambient temperature + 4°C with 550 ppm CO_2) attained panicle initiation, flowering and maturity earlier than the crop grown under ambient condition (Table 8.6).

Table 8.6. Duration of different phenophases under open and controlled conditions (in days)

Treatments	RICE (C_3)			MAIZE(C_4)		
	Active tillering	Flowering	Maturity	Vegetative	Tasseling	Maturity
T_1	23	70	114	50	65	110
T_2	16	55	97	38	54	99
T_3	20	59	101	42	57	102
Sed	0.1217	0.3732	0.6826	0.2592	0.3504	0.6247
CD(p = 0.01)	0.2557	0.7841	0.7841	0.5446	0.7361	0.3124

Growth attributes: In controlled environment condition in TCC and SPAR, less number of tillers in rice and number of leaves in maize was observed compared to ambient condition (Table 8.7).

Table 8.7. Effect of ambient and controlled conditions on number of tillers in rice and number of leaves in maize

Treatments	No. of tillers in Rice (C_3)			No. of leaves in Maize (C_4)		
	Active tillering	Flowering	Maturity	Vegetative	Tasseling	Maturity
T_1	32	35	33	9	12	14
T_2	27 (–15.62%)	29 (–17.14%)	28 (–15.2%)	10 (11.11%)	13 (8.3%)	14 (0%)
T_3	31 (–3.13%)	33 (–5.71%)	32 (–3.03%)	11 (22.11%)	13 (8.3%)	15 (7.14%)
Sed	0.182	0.199	0.191	0.063	0.078	0.089
CD (0.01 %)	0.5329	0.5728	0.5502	0.1803	0.2243	0.2549

Dry matter production and grain yield: The treatments T_2 and T_3 had recorded lower rice grain and straw yields due to lesser dry matter production, lesser number of productive tillers and lesser number of filled grains. Maize grain and stover yield also got reduced under controlled environment conditions (T_2 and T_3) due to lesser number of leaves and lower filled kernals. However, under CO_2 enrichment, rice has responded more positively and compensated yield loss by 10 per cent caused by the elevated temperature (ambient temperature + 4°C) while it was only 3 per cent in maize crop (Table 8.8).

Table 8.8. Effect of ambient and controlled conditions on yield of rice and maize crop

Treatments	**RICE (C_3)**				**MAIZE(C_4)**			
	Grain yield (g per plant)	**Straw yield (g per plant)**	**Dry matter produc-tion (g per plant)**	**Grain conversion percent-age**	**Grain yield (g per plant)**	**Straw yield (g per plant)**	**Dry matter produc-tion (g per plant)**	**Grain conversion percent-age**
T_1	38.15	51.12	93.00	41.02	107.25	152.34	270.59	39.64
T_2	21.00 (–45%)	39.06 (–23%)	65.13 (–30.3%)	32.31 (–21.0%)	80.88 (–24%)	133.66 (–12%)	223.55 (–17.4%)	36.18 (–21.0%)
T_3	24.68 (–35%)	47.63 (–15%)	71.06 (–23.7%)	33.78 (–15.3%)	84.59 (–21%)	138.31 (–9.2%)	232.14 (–14.2%)	36.54 (–15.3%)
SEd	0.1713	0.2844	0.4643		0.5507	0.8609	1.47	
CD (p=0.01)	0.4931	0.8187	1.3363		1.5855	2.4785	4.2319	

8.5.5. Impact on physiology of C_3 (rice) and C_4 (maize) crops

Photosynthetic rate: As far as the physiological response of C_3 (rice) and C_4 (maize) crops are concerned, the photosynthetic rate was significantly low when the plants were kept under warmer environment than under ambient condition and the rate of decrease was higher in rice compared to maize (Fig. 8.17). However, the CO_2 enrichment compensated this reduced photosynthetic rate well in rice compared to maize crop.

Transpiration rate and stomatal conductance: The transpiration rate and stomatal conductance were significantly increased in both rice and maize crops under elevated temperature condition compared to ambient condition and the rate of increase was higher in rice than maize (Figs. 8.16, 8.17, 8.18), yet, CO_2 enrichment compensated these ill effects of elevated temperature to certain extent in both rice and maize but the compensation effect was remarkably higher in rice than maize.

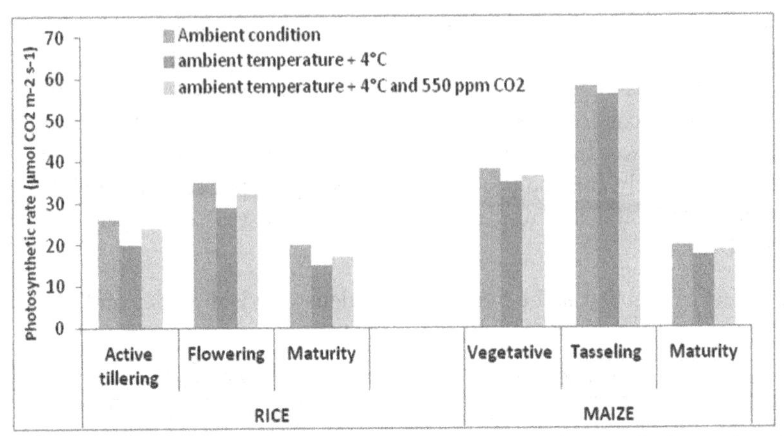

Fig. 8.16. Effect of ambient and controlled conditions on photosynthetic rate (µmol CO_2 m^{-2} s^{-1}) rice and maize crop
(*Source*: Arunkumar *et al.*, 2017)

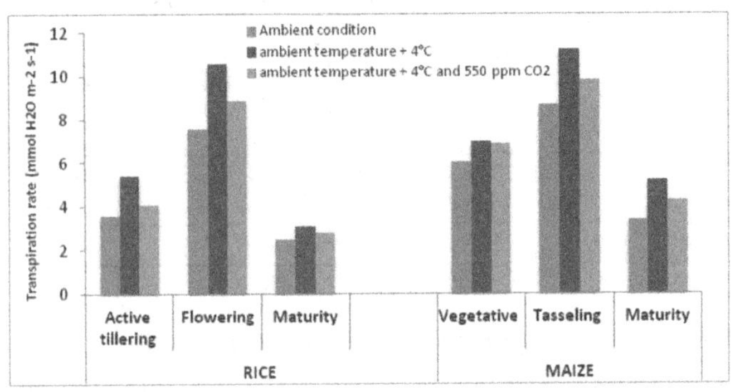

Fig. 8.17. Effect of ambient and controlled conditions on transpiration rate (mmol H_2O m^{-2} s^{-1}) rice and maize crop
(*Source*: Arunkumar *et al.*, 2017)

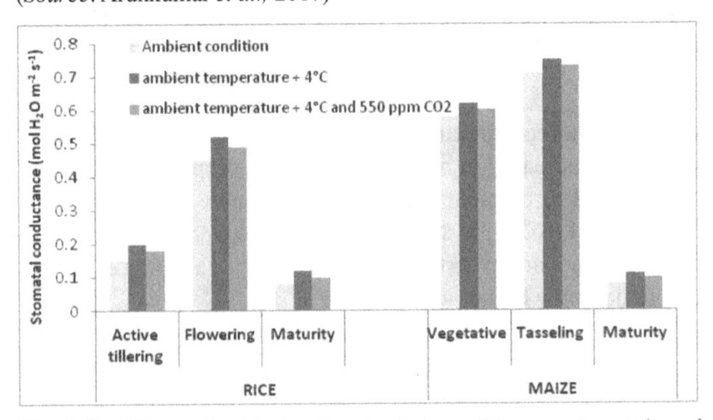

Fig. 8.18. Effect of ambient and controlled conditions on stomatal conductance (mol H_2O m^{-2} s^{-1}) rice and maize crop
(*Source*: Arunkumar *et al.*, 2017)

Increase in temperature with or without CO_2 enrichment impacted the phenology and productivity of C_3 and C_4 plants with different magnitude. Temperature increase alone has negatively impacted the C_3 crop (rice) more compared to C_4 crop (maize). However, CO_2 enrichment has compensated the negative impact of elevated temperature to certain extent. Rice has responded more positively to CO_2 enrichment compared to maize under elevated temperature (Arunkumar *et al.,* 2017).

8.5.6 Climate Change Implication on Water

Climate change will shrink the resources of freshwater. Water scarcity is expected to become an ever-increasing problem in the future, for various reasons. First, the distribution of precipitation in space and time is very uneven, leading to tremendous temporal variability in water resources worldwide (Oki *et al.,* 2006). Second, the rate of evaporation varies a great deal, depending on temperature and relative humidity, which impacts the amount of water available to replenish groundwater supplies. The combination of shorter duration but more intense rainfall (meaning more runoff and less infiltration) combined with increased evapotranspiration (the sum of evaporation and plant transpiration from the earth's land surface to atmosphere) and increased irrigation is expected to lead to groundwater depletion (Konikow and Kendy, 2005).

Projections of changes in total annual precipitation indicate that increases are likely in the tropics and at high latitudes, while decreases are likely in the sub-tropics, especially along its poleward edge. In general, there has been a decrease in precipitation between 10°S and 30°N since the 1980's (IPCC 2007). With the population of these sub-tropical regions increasing, water resources are likely to become more stressed in these areas, especially as climate change intensifies. While some areas will likely experience a decrease in precipitation, others (such as the tropics and high latitudes) are expected to see increasing amounts of precipitation. In addition, warming accelerates the rate of surface drying, leaving less water moving in near-surface layers of soil. Less soil moisture leads to reduced downward movement of water and so less replenishment of groundwater supplies (Nearing *et al.,* 2005). In locations where both precipitation and soil moisture decrease, land surface drying is magnified, and areas are left increasingly susceptible to reduced water supplies.

In water stressed regions, variability of precipitation patterns is likely to further reduce groundwater recharge ability. Water availability is likely to be further exacerbated by poor management, elevated water tables, overuse from increasing populations, and an increase in water demand primarily from increased agricultural production (IPCC 2007). A recent global analysis of variations indicated that the area of land characterized as very dry has more than doubled since the 1970's, while the area of land characterized as very wet has slightly declined during the same time period.

Water supplies can also be affected by warmer winter temperatures that cause a decrease in the volume of snowpack. The result is diminished water resources during the summer months. This water supply is particularly important at the mid-latitudes and in mountainous regions that depend upon glacial runoff to replenish river systems and groundwater supplies. Consequently, these areas will become increasingly susceptible to water shortages with time, because increased temperatures will initially result in a rapid rise in glacial meltwater during the summer months, followed by a decrease in melt as the size of glaciers continue to shrink. This reduction in glacial runoff water is projected to affect approximately one-sixth of the world's population (IPCC, 2007).

A reduction of glacial runoff has already been observed in the Andes, whereby the usual trend of glacial replenishment during winter months has been insufficient. This is due to increased temperatures, which have caused the glaciers to retreat. It is likely that Andean communities such as El Alto in Bolivia have already observed a reduction in glacial runoff due to the scattered distribution of smaller sized glaciers, which further reduces the potential for runoff. In these areas, approximately one-third of the drinking water is dependent upon these supplies, and the recurrent trend of increased melt with diminished replenishment provides a dismal projection for water reserves if this same pattern continues (Goudie, 2006).

8.5.7. Soil and Fertilizer

The most important process affected by climate change is the accelerated decomposition of organic matter, which releases the nutrients in short run, but may reduce the fertility in the long run. Soil temperature influences the rates at which organic matter decompose, nutrients are released and taken up and plant metabolic processes proceed. Chemical reactions, that affect soil minerals and organic matter are strongly influenced by higher soil and water temperature. Soil productivity and nutrient cycling are, therefore influenced by the amount and activity of soil microorganisms. Soil microorganisms fulfill two major functions, *i.e.,* they act as agents of nutrient element transportations well as store carbon and mineral nutrients (mainly N, P and S) in their own living biomass, acting as a liable reservoir for plant available nutrients with a fast turnover. The doubling of CO_2 increases plant biomass production, soil water use efficiency by the plants, and C/N ratios of plants. The changes in the C/N ratios of plant residues returned to the soil, have impact on soil microbial processes and affect the production of trace gases NOx and N_2O.

The interaction of nitrogen, irrigation and seasonal climatic variability, particularly at low input of irrigation, has several implications. Under adequate moisture supply situation, like for Punjab and Haryana, the yield benefits are obtained up to higher nitrogen application, whereas in the regions of limited to moderate water supply situations, the increasing trends in yield are noted up to relatively lower values of nitrogen. At low levels of water availability, it is difficult to decide optimal levels of N fertilizer for maximizing yield returns in view of uncertainty of N response, which is strongly related to a good post monsoon rainfall received during crop growing period (Kalra and Aggarwal, 1996).

Fertilizer use efficiency in India is generally very low (30–50 per cent). Increasing temperature in future is likely to further reduce fertilizer use efficiency. This will lead to increased fertilizer requirements for meeting future food grain production demands which are higher due to income and population growth. At the same time, greater fertilizer use leads to higher emissions of greenhouse gases.

8.5.8 Climate Change Impact on Pest

Pest and diseases cause a significant loss to world food production under different climatic conditions. Development and distribution of pest and diseases are governed by temperature patterns, rainfall or humidity and seasonal length to a great extent. Especially, winter temperatures are important for the survival of pest and studies have shown that increase in temperature accelerates the development of pests in general. Pest-crop interaction will be also directly affected by the rising CO_2 levels through the alteration of host plant attributes, such as C/N ratios and secondary plant nutrient chemistry. In terms of crop production, these fluctuations must be taken into the account while planning agricultural operations.

Incidence of pests and diseases is most severe in tropical regions due to favourable climate/weather conditions, multiple cropping and availability of alternate pests throughout the year. Therefore, in the south Asia, pests and diseases deleteriously affecting the crop yields are prevalent. Climate tors are the causative agents in determining the population fluctuations of pests. Climate change influences plant disease establishment, progression and severity. In fact, a clear understanding of population dynamics, as influenced by abiotic and biotic parameters of environment is of much help in pest forecasting and to formulate control measures.

The global warming may affect growth and development of all organisms including insect-pests themselves. Among all the abiotic factors, temperature is the most important one affecting insect distribution and abundance in time and space, since these are cold-blooded animals. The insects cannot regulate their body temperature and thereby, ambient temperature influences their survival, growth, development and reproduction. With the increase in temperature, the rate of development of insects may also increase, if temperature still lies within the optimal range for the pests. As a consequence, they could complete a greater number of generations for inflicting more loss to our crops. Crop-pest interaction needs to be evaluated in relation to climate change in order to assess the crop losses.

Crop pest interactions will change significantly with climate change leading to impact on pest distribution and crop losses. Diseases and insect populations are strongly dependent upon temperature and humidity. Any increase in these parameters, depending upon their base value, can significantly alter their population, which ultimately results in yield loss. With small changes the virulence of different pest changes. For example, at 16°C, the length of latent period is small for yellow rust. Once the temperature goes beyond 18°C, this latent period increases but that of yellow and stem rusts decreases (Nagarajan and Joshi, 1978).

The swarms of locust produced in the Middle East usually fly eastward into Pakistan and India during summer season and they lay eggs during monsoon period. The swarms as a result of this breeding, return during autumn to the area of winter rainfall, flying to all parts of India and influencing *Kharif* crops (Rao and Rao, 1996). Changes in rainfall, temperature and wind speed may influence the migratory behaviour of locust. At our current rate of greenhouse emissions, several of the main pests that target corn will increase in number and expand their range by the end of 21st century.

8.5.9 Effect on Insecticide Use Efficiency

Warmer temperature requires more number of insecticide applications (*i.e.,* three, more than normal) for controlling corn pests. Entomologists predict more generation of insects in warm climate that necessitates more number of insecticide applications. It will increase cost of protection and environmental pollution. Synthetic pyrethroids and naturalite Spinosad will be less effective in higher temperature. Therefore, it is advisable for the farmers not to use insecticides with similar mode of action frequently to avoid development of resistance in case of more number of applications. Cultural management practices *e.g.* early planting may not be helpful because of early emergence of pests due to warmness.

8.5.10 Effect on Natural Pest Control

Global warming is expected to make regional climates more varied and unpredictable which could affect relationship between insects and their natural enemies. In years of most variable rainfall, the caterpillars have significantly less number of parasitoids. This could be because the parasitoids use cues *e.g.* changing in local climate to determine the best time for laying eggs. Unpredictable rains might disrupt the parasitoids ability to track their caterpillar hosts. The wasps use start of the rain as cues to hatch out of their cocoons and look for a caterpillar to lay their eggs. If the rains are late, they emerge late and may not find larval stage of host resulting in reduced natural pest control. Due to changes in climate, the frequency of occurrence of droughts, heat waves, windstorms and floods etc. will increase disrupting the natural ecosystems.

8.5.11. Impact of Climate Change on Disease

Any direct yield gains caused by increased CO_2 or climate change could be offset partly or entirely by losses caused by phytophagous insects, plant pathogens and weeds. It is, therefore, important to consider these biotic constraints on crop yields under climate change.

i. **Impacts on Plant Path systems:** Climate change has the potential to modify host physiology and resistance and to alter stages and rates of development of the pathogen. The most likely impacts would be shifts in the geographical distribution of host and pathogen, changes in the physiology of host-pathogen interactions and changes in crop loss. Another important impact may be through changes in the efficacy of control strategies.

ii. **Geographical Distribution of Host and Pathogen:** New disease complexes may arise and some diseases may cease to be economically important if warming causes a pole ward shift of agro climatic zones and host plants migrate into new regions. Pathogens would follow the migrating hosts and may infect remnant vegetation of natural plant communities not previously exposed to the often more aggressive strains from agricultural crops. The mechanism of pathogen dispersal, suitability of the environment for dispersal, survival between seasons, and any change in host physiology and ecology in the new environment will largely determine how quickly pathogens become established in a new region.

Changes may occur in the type, amount and relative importance of pathogens and affect the spectrum of diseases affecting a particular crop. This would be more pronounced for pathogens with alternate hosts. Plants growing in marginal climate could experience chronic stress that would predispose them to insect and disease outbreaks. Warming and other changes could also make plants more vulnerable to damage from pathogens that are currently not important because of unfavorable climate.

iii. **Physiology of Host-Pathogen Interactions**: Elevated CO_2: Increases in leaf area and duration, leaf thickness, branching, tillering, stem and root length and dry weight are well known effects of increased CO_2 on many plants. Scientists have suggested that elevated CO_2 would increase canopy size and density, resulting in a greater biomass of high nutritional quality. When combined with increased canopy humidity, this is likely to promote foliar diseases such as rusts, powdery mildews, leaf spots and blights. The decomposition of plant litter is an important factor in nutrient cycling and in the saprophytic survival of many pathogens. Increased C:N ratio of litter is a consequence of plant growth under elevated CO_2. Scientists have indicated that decomposition of high-CO_2 litter occurs at a slower rate. Increased plant biomass, slower decomposition of litter and higher winter temperature could increase pathogen survival on over wintering crop residues and increase the amount of initial inoculum available to infect subsequent crops. Host pathogen interactions in selected fungal patho systems, two important trends have emerged on the effects of elevated CO_2. First, the initial establishment of the pathogen may be delayed because of modifications in pathogen aggressiveness and/or host susceptibility. The second important finding has been an increase in the fecundity of pathogens under elevated CO_2.

iv. **Elevated Temperature*:*** Increases in temperature can modify host physiology and resistance. Considerable information is available on heat-induced susceptibility and temperature-sensitive genes. In contrast, signification of cell walls increased in forage species at higher temperatures to enhance resistance to fungal pathogens. Impacts would, therefore, depend on the nature of the host-pathogen interactions and the mechanism of resistance. Agricultural crops and plants in natural communities may harbor pathogens as symptom less carriers, and disease may develop if plants are stressed in a warmer climate. Host stress is an especially important factor in decline of various forest species.

8.5.12 Crop-weed Competition

Crop-weed competitions will be affected depending on their photosynthetic pathway. The C_3 crop growth would be favoured over C_4 weeds affecting the need for weed control. The accompanied temperature increase may further alter the competition depending upon the threshold ambient temperatures. The increased temperature would offset the benefits of elevated CO_2 and allow sleeper weeds to become invasive. The increment in temperature leads to expansion of weeds into higher latitudes or higher altitudes. The temperature increase also helps very aggressive weeds that are currently found in the lower latitudes to expand to higher latitudes also. For example, Itch grass, a profusely tillering, robust grass weed could invade the central Midwest and California with a 3°C warming trend (Patterson, 1995).

8.6 Adaptation to Climate Change in Agriculture

Agriculture is one of the most climate-sensitive sectors. Climate change is expected to have significant impacts on crop yields through changes in temperature and water availability. Adaptive capacity of the agricultural system should be enhanced using scientific information, model outputs and indigenous traditional knowledge of farmers. It is a proven fact that, farm-level adaptation practices are helpful in coping with climate variability, but their effectiveness can be enhanced by institutional and policy support from government.

Farm level adaptation measures include changing cropping calendars and patterns (for example, from a rice-rice to rice-maize cropping pattern to optimize the use of available water for crop growth); intercropping; mixed cropping; low external input sustainable agriculture; diversified farming; variations in crop rotation patterns; changing sowing date; improved farm management; and use of drought-resistant and heat-resilient crop varieties.

Agro-technological strategies such as natural rainfall management, cropping pattern adjustment, and improved access to detailed information on soil fertility and taxonomy were chosen primarily because they are low-cost options.

8.6.1 Soil Management

The poor soil management could speed up climate change because soils contain around twice the amount of carbon in the atmosphere and three times the amount found in vegetation, so soil is both "a source and a sink of greenhouse gases" so the soil management practices for maintaining the soil fertility and erosion control are the most important adaptation practices followed in agriculture.

Crop rotation with legume and green manure crop

Crop rotation practices refer to cultivating a variety of crops continuously in a fixed rotation to preserve the fertility of the soil. Through crop rotation practices, legumes help to build-up nitrogen fixation, control of off-site pollution and increasing the biodiversity at very low cost (Mishra *et al.*, 2009).

Vermicompost

Vermicompost has been shown to be richer in many nutrients than compost produced by other composting methods. Improvement of soil aeration, enrichment of soil with micro-organisms, improvement of water holding capacity, enhances germination and plant growth, and crop yield (Nagavallemma *et al.*, 2004) would occur with vermicompost application. Vermicompost are environment friendly and no imported inputs is required and this technology is labour extensive and highly profitable

Soil health card

The soil health card evaluates the health or quality of a soil as a function of its characteristics, water, plant and other biological properties. The card is a tool to help the farmer to monitor and improve soil health and give an indication on how much fertilizers need to be applied for the crop that will be grown in the ensuing season. Over use of chemical fertilizer could be avoided and at the same time any secondary or micro nutrient deficiency could also be rectified to maintain the soil health.

Soil Test Crop Response (STCR) based nutrient recommendation

It is a kind of fertilizer prescriptions developed for different soils, different varieties of crops and for different yield targets (Sharma *et al.,* 2016).

Integrated nutrient management

Integrated Nutrient Management refers to the maintenance of soil fertility and of plant nutrient supply at an optimum level for sustaining the desired productivity through optimization of the benefits from all possible sources of organic, inorganic and biological components in an integrated manner (Sharma *et al.,* 2016).

Nutri-seed Pack technology

Nutri-seed pack contains seed at top, enriched manure in the middle and encapsulated fertilizer at bottom. Nutri-seed pack gives support for each plant in the root zone in terms of optimum nutrient supply, biological activity and consequently enables the fullest utilization of nutrients by plants. There is no wastage of fertilizer nutrients with Nutri-seed Packs (Hota and Arulmozhiselvan, 2016)

Organic farming

It is a method of farming system which primarily aimed at cultivating the land and raising crops in such a way, as to keep the soil alive and in good health by use of organic wastes (crop, animal and farm wastes, aquatic wastes) and other biological materials along with beneficial microbes (biofertilizers) to release nutrients to crops for increased sustainable production in an ecofriendly pollution free environment (Yadhav, 2010).

Soil related constraints and their management measures

Saline soils could be reclaimed through leaching of salts and application of farm yard manure at 5 t/ha. Reclamation of sodic soils could be done through leaching of soluble salts, application of gypsum, *In situ* incorporation of green manure at 5 t/ha. Reclamation of acid soils can be done by application of lime. Iron and aluminum toxicity can be managed by application of lime with recommended dose of NPK and also organic manure will suppress the toxicity. Fluffy paddy soils can be reclaimed by passing of 400 kg stone roller along with addition of lime @ 2t/ha once in three years. Sandy soils can be managed by compacting the soil with 400 kg stone roller once in three years and this could reduce the percolation losses and addition of tank silt for coastal sandy soils is recommended for enhancing their productivity. Hard pan soils can be reclaimed by chiseling the soils and application of FYM or composted coir pith at 12.5 t/ha.

Zero tillage

Reduced water use, C sequestration, similar or higher yield, increased income, reduced fuel consumption and reduced GHG emission are the benefits from zero tillage.

8.6.2 Water Management

Water scarcity is a looming threat to agriculture and it is further aggravated by climate change. Cultivating less water intensive crops; increasing water use efficiency by practicing improved crop cultivation systems like SRI and increasing water productivity of crops are essentials for fulfilling the motto "More crops per drop of water". Micro-level water conservation measures like mulching, zero-tillage, maintaining recharge ponds and other rainwater harvesting structures are to be carried out at farm scale.

Micro irrigation

Drip Irrigation: It refers to application of water in small quantity at the rate of mostly less than 12 lph as drops to the zone of the plants. Merits of drip irrigation includes:

1. Increased water use efficiency,
2. Better crop yield,
3. Uniform and better quality of the produce,
4. Efficient and economic use or fertiliser through fertigation,
5. Less weed growth,
6. Minimum damage to the soil structure,
7. Avoidance of leaf burn due to saline soil,
8. Usage in undulating areas and slow permeable soil,
9. Low energy requirement *i.e.,* labour saving and
10. High uniformity suitable for automization (Sankaranarayanan *et al.,* 2011).

Sprinkler irrigation: water is sprayed into the air and allowed to fall on the ground surface somewhat resembling rainfall. Advantages of sprinkler irrigation includes:

1. No conveyance loss,
2. Suitable to all types of soil,
3. Suitable for irrigating crops where the plant population per unit area is very high,
4. Water saving,
5. Higher water application efficiency, and
6. Increase in yield (Shiva Shankar *et al.,* 2015).

Special irrigation techniques

Special irrigation practices include paired row technique, alternate furrow system, ridges and furrows with coir pith mulch and rain-fed irrigation were followed in India according to crop's nature. These techniques are highly helpful in saving irrigation water ranging from 15 to 30 per cent (Mandal *et al.,* 2019).

Laser-aided land levelling

Reduced water use, reduced fuel consumption, reduced GHG emissions, increased area for cultivation, and increased productivity are some of the merits to be obtained.

Direct drill seeding of rice
Less requirement of water, time saving, better postharvest condition of field, deeper root growth, and more tolerance to water and heat stress, reduced methane emission are the benefits to be accrued from this technique.

8.6.3 Fertilizer Management

GHGs emissions are often directly related to nutrients added to the soil in the form of mineral fertilizers and animal manure. By helping to maximize crop-nitrogen uptake, improved nutrient management has a significant and cost-effective role to play in mitigating GHG emissions from agriculture. Effective nutrient management can also help reduce methane (CH_4) emissions from rice production and increase carbon sequestration in agricultural soils.

Leaf colour chart (LCC) for N management and nitrification inhibitors
Reduces fertilizer N requirement, reduced N loss and environmental pollution and reduced nitrous oxide emission are some of the benefits anticipated.

8.6.4 Crop Management

To manage climate variability and its change in crop production systems, it requires strategies such as: (1) changes in planting and harvesting times (2) diversification and intensification of food and plantation crops (3) diversified farming, intercropping, crop rotation (4) changes in management and farming techniques (4) growing short duration varieties (5) revisiting to the traditional varieties and (6) alternate cropping system

Crop Diversification
Efficient use of water, increased income, increased nutritional security, conserve soil fertility and reduced risk are some benefits

System of Rice Intensification (SRI) in rice
SRI is a system of production with four main components, viz., soil fertility management, planting method, weed control and water (irrigation) management. SRI requires the root zone to be kept moist, not submerged. Water applications can be intermittent, leaving plant roots with sufficiency, rather than surfeit of water. Rice grown under SRI has larger root system, profuse and strong tillers with big panicles and well-filled spikelets with higher grain weight. The rice plants develop about 30–80 tillers and the yields are reported to be higher (Haldar *et al.,* 2012).

SSI in sugarcane
The major principles that govern SSI are raising nursery in portrays using single budded chips, transplanting young seedlings (25–35 days old), maintaining wider spacing (5 × 2 feet) in the main field and providing sufficient moisture through efficient water management technologies *viz*., drip fertigation (sub or sub surface). Benefits of SSI are improved water use efficiency, optimum use of fertilizers, favouring balanced availability of nutrients, better aeration and more penetration of sunlight and that

favours higher sugar content, reduced cost of cultivation and increased returns through intercropping (Chogatapur *et al.*, 2017).

Plastic mulching for crop production

Mulching is the process or practice of covering the soil/ground to make more favourable conditions for plant growth, development and efficient crop production. The suppression of evaporation also has a supplementary effect (Lalitha *et al.*, 2010).

Integrated Farming Systems

This approach introduces optimal utilization of resources like farm wastes that are better recycled for maximum production in the cropping system/pattern. A judicious mix of agricultural allied enterprises like dairy, poultry, piggery, fishery, and sericulture suited to the given agro-climatic condition and socio-economic status of the farmers would bring prosperity in the farming (Kumar *et al.*, 2018).

Dry Farming Practices

A set of practices are being recommended in dry farming for increasing crop production which includes bunding across the slope & levelling done before onset of monsoon, deep summer ploughing, application of organic manures like FYM/compost/green manuring, basal application of fertilizers at a depth of 7.5–10 cm in the soil and seeds sown in the same furrows about 3 cm above the feriliser band. Suitable crops/varieties according to their suitability to a particular region/micro climate, soaking seeds in plain water helps to get higher germination, intercropping for moisture harvesting, adopting integrated weed control measures, mulching with tree leaves, uprooted weeds are used to check the evaporation of water, water harvesting between the rows by growing some pulse crops and collection of run-off water in nearby ponds and used for lifesaving irrigation (Singh *et al.*, 2016).

Protected cultivation

Greenhouse technology is the technique of providing favourable environment condition to the plants. It is rather used to protect the plants from the adverse climatic conditions such as wind, cold, precipitation, excessive radiation, extreme temperature, insects and diseases. It improves yield by 10–12 times higher than conventional cultivation. Ideally suited for vegetables and flower crops, helpful for year-round production of floricultural crops and off-season production of vegetable crops (Sabir *et al.*, 2010).

Plant boosters and growth regulators

There are number of crop boosters and plant growth regulators developed for variety of crops and utilized for enhancing the crop yield, which includes application of coconut tonic for decreasing button shedding, increasing number and size of nuts, increasing nut yield up to 20 per cent, and imparting resistance to pests, diseases and environmental stresses in coconut. Pulse wonder decreases flower shedding, increases yield up to 20 per cent and increases drought tolerance in legume crops. Groundnut rich increases flower retention, improves pod filling, increases pod yield up to 15 per cent and improves drought tolerance in groundnut. Cotton plus reduces flower and square shedding, improves boll bursting, increases seed cotton yield up to

18 per cent and increases drought tolerance in cotton crop. Sugarcane booster enhances cane growth and weight, improves internodal length, improves cane yield up to 20 per cent, improves sugar content, and increases drought tolerance in sugarcane. Pink Pigmented Facultative Methylotrops (PPFM) are aerobic, gram-negative bacteria and its application fasten seed germination and seedling growth, accelerating vegetative growth, increase leaf area index and chlorophyll content, earliness in flowering, fruit set and maturation, improves fruit quality, color and seed weight, yield increase by 10 per cent and mitigate drought.

Crop Improvement Activities

Heat and drought tolerant varieties should be developed to cope up with increased temperature and change in precipitation trends. Also change in pest dynamics and increased salinity affected land area necessitates development of pest resistant and saline tolerant varieties.

Forewarning system

Future climate projections indicate increase in intensity and frequency of extreme weather events. Especially occurrence of prolonged dry period or heavy rainfall spell coinciding with the critical stages of crop growth and development may lead to significant reduction in crop yields and extensive crop losses. So, more thrust should be given in developing timely and effective forewarning system.

Weather based response farming

Response farming is a method of using weather prediction (seasonal and sub seasonal) for farming decisions in order to minimize risk and to optimize crop production. Goal of this practice is to design sustainable cropping systems for low resource farmers in marginal rainfall zones, characterized by great seasonal rainfall variability, uncertainty and recurrent drought. This saves the input and increases the farm income (Dakshina Murthy *et al.,* 2017).

Choice of crop based on seasonal rainfall prediction

Choice of crop mainly depends on the availability of water for the entire crop growth. Hence, seasonal rainfall prediction information is used for selecting the crop as well as variety. If the rainfall prediction is above normal for the season, high water requiring crops are selected and vice-versa (Dakshina Murthy *et al.,* 2017).

Land allocation based on seasonal rainfall prediction

Similar to crop choice, land allocation is also done based on the expected seasonal rainfall. If above average rainfall is expected in the season, more area is allocated to high water requiring crops and less area to low water requiring crops. Opposite condition is maintained if the rainfall prediction for the season is lesser than average.

Right time of planting (Best sowing window) based on medium range weather forecast

Field preparation for rainfed crops is weather- dependent. In any dryland areas the amount of rainfall is very meagre and farmers should take advantage of even the first drop of rain that falls on the ground in the growing season for the crop production. Minimal tillage is the current agronomic mantra for conserving moisture, retaining nutrients and keeping weeds out. The prediction of the exact time of occurrence of rainfall in a particular location helps to initiate field preparation and time of sowing. Knowing the occurrence of sowing rain, the farmers can take up pre-monsoon sowing which will help in avoiding the terminal season drought (Kaur *et al.,* 2012).

Weather based tactical farming decisions

Weather and climate are the most pervasive factors of crop environment. It has practical utility in timing of agriculture operations so as to make the best use of favourable weather conditions and make adjustments for adverse weather. Tactical farming decisions such as intercultural operations (weeding, earthing up, irrigation operation, fertilizer topdressing, etc.), plant protection measures and the time of harvest could be decided based on the expected weather in the coming 5–7 days. This would increase the input use efficiency, labour use efficiency and help in reducing the cost of cultivation by minimizing the number of sprays of plant protection chemicals. Agrometeorological information can also be used in land planning, risk analysis of climatic hazards, production and harvest forecasts and linking similar crop environments for crop adaptability and productivity (Kaur *et al.,* 2012).

Managing extreme weather events

Extreme weather events like high temperature, excessive rainfall, drought, hailstorm and foggy and cloudy weather are prevailing in different parts of the country. Therefore, to sustain the productivity, developing the adoptive responsive varieties, managing the risk by providing weather linked value-added advisory services and crop insurance to the farmers are required. Action also needed to stabilise the income of farmers to ensure their continuance in farming and encouraging farmers to adopt innovative and modern agricultural practices and encouraging to ensure flow of credit to the agriculture sector (Verma *et al.,* 2013).

Outreach programs

Creating awareness among the farmers about dwindling of agricultural resources in the wake of climate crisis and importance of resource–conservation, through out-reach programs. Policy-level measures have been devised to provide an adequate incentive framework to promote private adaptation.

8.7. Mitigation and Resilience

8.7.1 Mitigation

Mitigating climate change is about reducing the release of greenhouse gas or fossil fuels emissions until their total eradication that are warming our planet. Mitigation strategies include retrofitting buildings to make them more energy efficient; adopting renewable energy sources like solar, wind and small hydro; helping cities develop more sustainable transport such as bus rapid transit, electric vehicles, and biofuels; and promoting more sustainable uses of land and forests.

They also include the improvement of the sumps to increase the absorption capacity of these gases. Likewise, programs such as carbon taxes and incentives for voluntary GHG reduction and clean energy substitution are considered.

Removing CO_2 and carbon storage from the atmosphere

The carbon market approach is world-wide concept and it tries to solve the issue by using market forces to make carbon use less affordable but vulnerable host communities that are the intended beneficiaries have been found to receive little to no benefit. Developed Nations that have often prioritized growth of their own gross national product over implementing changes that would address climate change concerns by taxing carbon, which might damage GDP. In addition, the pace of change necessary to implement a carbon market approach is too slow to be effective at most international and national policy levels.

Alternatively, a study by Mathur, *et al.,* (2014) proposes a multi-level approach that focuses on addressing some primary issues concerning climate mitigation at local and international levels. The approach includes:

1. Developing the capacity for a carbon market approach
2. Focusing on power dynamics within local and regional government
3. Managing businesses with regard to carbon practices
4. Special attention given to developing countries

Carbon capture and storage (CCS) uses established technologies to capture, transport and store carbon dioxide emissions from large point sources, such as power stations. It also has an important role to play to ensure manufacturing industries, such as steel and cement and can continue to operate without the associated emissions. The process is a low carbon technology which captures CO_2 from the burning of coal and gas for power generation and from the manufacturing of steel, cement and other industrial facilities. The carbon dioxide is then transported by either pipeline or ship for safe and permanent underground storage, preventing it from entering the atmosphere and contributing to anthropogenic climate change.

In several countries, the emissions of power stations have been reduced by a great expansion of renewable wind and solar energy, but, alarmingly, nearly 40 per cent of the world's electricity is still generated from coal and only 10 per cent is from wind and solar and 16 per cent from hydro power. The rate of removal of power stations emissions is far too slow to avoid the tipping points. The CCS is still in its infancy. The UK Government for example plans a first "Carbon capture, usage and storage" project by the mid-2020's, when climate campaigners say we need to be carbon neutral by 2030.

Planting trees

Currently, earth's forests and soil absorb about 30 per cent of atmospheric carbon emissions, partially through forest productivity and restoration. While deforestation has occurred throughout human history, the practice has increased dramatically in the past 50 years. However, planting 950 million hectares of new forests could help to limit the increase in global average temperature to 1.5°C above pre-industrial levels by 2050.

Tree planting could be beneficial but tree growth takes time and the covering of more of the earth with forests will affect the earth's surface albedo (reflectivity) and evapotranspiration. A darker surface will absorb more heat, while evapotranspiration will have a cooling effect. While the uncertain balance between these effects is being investigated, the prime task for forest management must be to stop further deforestation and restore depleted and damaged forests to their original cover of 50 years ago. The use of tree planting as the input for carbon offset schemes, such as those operated by some airlines, has to be examined carefully to be sure that all the benefits are claimed.

Mitigation by technologies: the need for changes in human behavior

The Costs of Carbon Capture and storage (CCS) and tree planting on a large scale show that we must mitigate climate change for reducing greenhouse gas emissions as quickly as possible in as many ways as possible. Here is a summary of some things that can be done, and are being done already in some places and societies around the world.

Individuals: consume less meat; buy fewer clothes; walk preferably, use public transport, switch off lights and appliances that are not being used; travel less and avoid air travel where-ever possible.

Households: in modern cities, ensure buildings are designed appropriately for the climate, to be cool in hot climates and well-insulated in cold climates; use cooking and heating appliances sensibly. Help poorer households to cook with electricity rather than with wood or charcoal. Grow food locally where possible; consider the food-miles of purchases made in shops.

Communities: support local food production; share community transport; sustain local markets and shops.

Municipalities: develop local climate change mitigation plans; support solar and wind power installations on all public buildings, including schools and hospitals; support local public transport; raise parking charges and related fees to discourage private vehicle use; develop pedestrian precincts in large cities, with bans on vehicles with high emissions. Encourage municipal tree-planting and urban agriculture schemes.

National Governments: Subsidize and encourage renewable energy schemes at all scales form the household to major power plants; close down coal-fired power stations quickly. Support public transport that serves people well at the frequencies and times that encourage high use. Ensure that all government operations and building become carbon-neutral rapidly.

Business corporations: make sure that the whole business is working to become carbon neutral in the next few years. Ensure that all buildings are energy efficient and well insulated; that supply chains use the most sustainable form of transport; that land around premises has tree-planting or is used for food production; that all forms of energy saving, renewable electricity generation, water and materials recycling are in place and used to the fullest extent.

8.7.2 Resilience

The Intergovernmental Panel on Climate Change (IPCC) defines resilience as "the ability of a social or ecological system to absorb disturbances while retaining the same basic structure and ways of functioning, the capacity of self-organization, and the capacity to adapt from stress and change.

Climate resilience is the ability to anticipate, prepare for, and respond to hazardous events, trends, or disturbances related to climate. Improving climate resilience involves assessing how climate change will create new, or alter current, climate-related risks, and taking steps to better cope with these risks.

Climate resilience can be generally defined as the capacity for a socio-ecological system to: (1) absorb stresses and maintain function in the face of external stresses imposed upon it by climate change and, (2) adapt, reorganize, and evolve into more desirable configurations that improve the sustainability of the system, leaving it better prepared for future climate change impacts.

Currently, climate resilience efforts encompass social, economic, technological, and political strategies that are being implemented at all scales of society. From local community action to global treaties, addressing climate resilience is becoming a priority, although it could be argued that a significant amount of the theory has yet to be translated into practice. Despite this, there is a robust and ever-growing movement fuelled by local and national bodies alike geared towards building and improving climate resilience.

Historical overview of climate resilience

Theoretical basis for many of the ideas central to climate resilience have actually existed since the 1960's. Originally an idea defined for strictly ecological systems, resilience was initially outlined by Holling (1973) as the capacity for ecological systems and relationships within those systems to persist and absorb changes to state variables, driving variables, and parameters.

By the mid-1970's, resilience began gaining momentum as an idea in anthropology, culture theory, and other social sciences. Even more compelling is the fact that there was significant work in these relatively non-traditional fields that helped facilitate the evolution of the resilience perspective as a whole. Part of the reason resilience began moving away from an equilibrium-centric view and towards a more flexible, malleable description of social-ecological systems was due to work such as that of Andrew Vayda and Bonnie McCay in the field of social anthropology, where more modern versions of resilience were deployed to challenge traditional ideals of cultural dynamics.

As the issues of global warming and climate change have gained traction and become more prominent since the early 1990's, the question of climate resilience has also emerged. Considering the global implications of the impacts induced by climate change, climate resilience has become a critical concept that scientific institutions, policymakers, governments, and international organizations have begun to rally around as a framework for designing the solutions that will be needed to address the effects of global warming (Stern and Stern, 2007).

Theoretical foundations for building climate resilience

As the threat of environmental disturbances due to climate change becomes more and more relevant, so does the need for strategies to build a more resilient society.

1. Urban resilience

Systems concern both physical infrastructure in the city and ecosystems within or surrounding the urban centre; while working to provide essential services like food production, flood control, or runoff management. For example, city electricity, a necessity of urban life, depends on the performance of generators, grids, and distant reservoirs. The failure of these core systems jeopardizes human well-being in these urban areas, with that being said, it is crucial to maintain them in the face of impending environmental disturbances. Resilient systems work to "ensure that functionality is retained and can be re-instated through system linkages".

Efforts to improve the resiliency of housing and workplace buildings involves not only fortifying these buildings through use of updated materials and foundation, but also establishing better standards that ensure safer and health conditions for occupants. Better housing standards are in the course of being established through calls for sufficient space, natural lighting, provision for heating or cooling, insulation, and ventilation. Another major issue faced more commonly by communities in the Third World are highly disorganized and inconsistently enforced housing rights systems. In countries such as Kenya and Nicaragua, local militias or corrupted government bodies that have reserved the right to seizure of any housing properties as needed: the end result is the degradation of any ability for citizens to develop climate resilient housing–without property rights for their own homes, the people are powerless to make changes to their housing situation without facing potentially harmful consequences.

In more urban areas, construction of a "green belt" on the peripheries of cities has become increasingly common. Green belts are being used as means of improving climate resilience–in addition to provide natural air filtering, these belts of trees have proven to be a healthier and sustainable means of mitigating the damages created by heavy winds and storms.

2. Business Resilience Solutions

Business initiatives to build resilience include developing disaster recovery plans, adding onsite energy resources like combined heat and power systems or rooftop solar, and identifying backup supply and distribution chains. Small businesses may deploy different strategies like installing green roofs for water retention and communicating preparedness information to employees.

3. Financing Resilience

Governments and businesses are obtaining capital to invest in resilience projects through innovative finance mechanisms like green bonds and climate funds. The States that participate in emissions-trading systems also allocate proceeds to resilience projects. In addition, many Federal and State insurance offices and private insurers offer lower rates for taking steps to reduce climate risks, providing additional savings later.

4. Building resilience through forestry

There are lot of ways to resilience the climate from changing through forest ecosystem *viz*., maximize the resilience of the forest ecosystem, use forest and trees to reduce the CO_2 level through photosynthesis, build resilience landscape, adopt forest policies and build institutions conductive to resilience.

Barriers to resilient investments

There are a number of reasons why some organizations and individuals may not invest in climate resilience. Among those are the inability to understand how to assess and value climate risk and the lack of knowledge regarding ways in which it can be managed. Further, not understanding when to mitigate the risk, when to transfer the risk, and when to accept the risk, can be other challenges. Market conditions, cash flow, outstanding debt, contractual obligations, and leadership can also create barriers to investment. Other additional barriers are regulatory and governance policies (beyond those faced by investor-owned utilities). Even political perspectives about climate change in the United States can limit investments in resilience.

8.8 Protected Agriculture

Protected agriculture (PA) refers to the use of technology to modify the natural environment (temperature, rainfall, humidity, wind, etc.) that surrounds a crop to harvest higher yields, of better quality, during an extended season. The PA often employs cutting-edge technologies to enhance the productivity of commercial crops, as well as the quality and safety of agri food products for greater profitability. Some examples of PA are floating covers, low tunnels, walking tunnels, and greenhouses.

(*Source*: https://krishijagran.com/authors/abha-toppo/)

Fig. 8.19. Cultivation of high-value crops under protected agriculture.

Protected agriculture–the cultivation of high-value vegetables and other horticultural crops in greenhouses–allows farmers to grow cash crops on small plots in marginal, water-deficient areas where traditional cropping is not viable (Fig. 8.19).

This technology, originally developed and tested by ICARDA in the Arabian Peninsula, has been adapted and successfully used in smallholder farming systems in Afghanistan, Pakistan, and Yemen.

Importance of Protected Cultivation

Although agriculture has been the backbone of India's economy since ages, yet our experience during the last 50 years indicates a relationship between the agricultural practices and its growth *vis-à-vis* economic wellbeing. The wide variations in the climatic conditions across the diverse topography through the length and breadth of the country allow a large number of cropping patterns. India also experiences climatic extremes such as floods, droughts and other climatic abnormalities that cause crop losses regularly or damages resulting in economic losses to the farmers. Simultaneously, the demand for quality agricultural produce has increased over the last decade. This provides better opportunities for the Indian farmers to adopt protected cultivation technologies as per region and suitability of the crops.

Greenhouses are being commercially used for production of exotic (non-native) and off-season vegetables, export-quality cut flowers and also for raising quality seedlings. Economic returns from the high value agricultural produce can be increased substantially when grown under greenhouse conditions. For the crops under protected environment, the use of chemical pesticides and insecticides can be kept minimal to avoid their residues on the crop produce. Greenhouses are mostly used as rain shelters, particularly in high rainfall areas of India such as Northeastern States and coastal regions.

Objectives of Protected Cultivation

- Protection of plants from abiotic stress (physical or by non-living organism) such as temperature, excess/deficit water, hot and cold waves, and biotic factors such as pest and disease incidences, etc.
- Efficient water use with minimum weed infestation.
- Enhancing productivity per unit area.
- Minimizing the use of pesticides in crop production.
- Promotion of high value, quality horticultural produces.
- Propagation of planting material to improve germination percentage; healthy, uniform, disease free planting material and better hardening.
- Year-round and off-season production of flower, vegetable or fruit crops.
- Production of disease-free and genetically better transplants.

Advantages

- Large increases in yield, produce quality, and revenue
- High water productivity, saving significant amounts of water
- Significant reduction in pesticide use for lower production costs and healthier produce

- Year-round production, allowing farmers to take advantage of market seasonality and higher prices.
- Small-scale entrepreneurs fabricate low-cost greenhouses locally. Training on greenhouse fabrication, installation, and maintenance is provided to farmers, technicians, and entrepreneurs, and integrated crop management methods have been developed to maximize the benefits of greenhouse cultivation.

Limitations of Protected Cultivation

- High cost of initial infrastructure (capital cost).
- Non-availability of skilled human power and their replacement locally.
- Lack of technical knowledge of growing crops under protected structures.
- All the operations are very intensive and require constant effort.
- Requires close supervision and monitoring.
- A few pests and soil-borne pathogens are difficult to manage.
- Repair and maintenance are major hurdles.
- Requires assured marketing, since the investment of resources like time, effort and finances, is expected to be very high.

Protective structures

Following are the various protective structures, which are used to increase the crop production and quality of the produce.

1. Net houses - these are used to reduce adverse effect of scorching sun and rains in vegetables, ornamentals and herbs. Depending upon the cladding (covering) material used, the net houses may be classified as insect-proof net houses and shade net houses. An insect-proof net house can be fabricated as a temporary or permanent structure in different designs. It can be in a walk-in tunnel design and shape, with double door facility at one end of the structure. It is covered with UV-stabilised insect-proof net of 40–50 mesh for effective control of pests and diseases. The minimum size of insect-proof net house is 100 sq m (Fig. 8.20).

Fig. 8.20 Insect proof net house

2. Plastic low tunnels–these are used to raise early nurseries of vegetables and flowering annuals (Fig. 8.21).

Fig. 8.21 Plastic low tunnels

3. Green houses–these are framed structure covered with a transparent material in which crops could be grown under controlled environment. The environmental conditions refer to light, temperature, air composition and nature of root medium. Where, Off-season cultivation of crops is possible round the year and crop cultivation is possible under harsh environmental conditions (Fig. 8.22).

Fig. 8.22. Greenhouse

4. Glasshouses–glass is used as a Glazing material in the green house. Glasshouses are fitted with the help of wooden or metal frame.
5. Walk-in tunnel - It is a temporary structure made by using GI pipes or bamboo, and is covered with different cladding material depending upon the season in which the cultivation is proposed. Walk-in tunnels are used for off-season cultivation of vegetables and flower seedlings. They give an advantage of better prices of the off-season produce, giving more profit per unit area. Optimum size of the walking tunnel is 60–75 sq m, with 2–2.5 m width and up to 30 m length

with a 2–2.5 m central height. Overall, the height is enough for the worker to walk comfortably during operation (Fig. 8.23).

Fig. 8.23. Walk in tunnels

6. Shade Net House - It is primarily constructed to protect plants from highly intense solar radiation. The structure is made of wood, stone, bamboo or GI pipes. When wood or bamboo is used, the poles are treated with turpentine and tar on one side before inserting them in the ground. Cladding material used on the top and sides of the structure is generally a shade net (Fig. 8.24).

Fig. 8.24. Shade Net house

7. Mist Chamber - The main purpose of such a structure is to create high humidity and droplet-free presence of water for propagating delicate soft wood cuttings, vegetable crops, root plants and shrubs, etc. Cuttings are misted intermittently in place of continuous water application or drenching. The intermittent water misting is done using a high-pressure pump, pipeline system and a timer switch. The mist nozzles are connected to the main pipelines for misting the plant material growing inside the growth chambers or structures. A mist chamber of 15–25 sq m is sufficient for a nursery. The frequency of misting depends upon ambient temperature and type of plant material being propagated (Fig. 8.25).

Fig. 8.25. Mist Chamber
(*Source*: http://ncert.nic.in/vocational/pdf/kepc102.pdf)

Major Components of a Greenhouse

A greenhouse is constructed with different material and their components. The major components used in greenhouse construction and their features and functions are described (Fig. 8.26).

1. **Cladding Material**: Polythenes or other transparent material used for walls and roof of a greenhouse for protection as well as transparency, which simulates climatic conditions inside the greenhouse is called cladding material. The material could be made of polycarbonate, glass or poly sheets. The polycarbonate and glasshouses are temporary structures and mostly used for research or academic purposes. The polythene sheet as a cladding material is most commonly used and these films are normally UV-stabilised, 200 microns thick and fixed with aluminium profiles using zigzag springs. It is important to select a proper film for the polyhouse, which has direct relation with the quality of the crop as well as the quantity of the produce. Polythene should be properly UV-stabilised and a minimum life span of at least three years. With 1 kg poly-film, a maximum area of 5.4 sq m can be accommodated.
2. **Polyhouse Film**
   i. Compulsory properties: UV stabilisation, diffusion/clear (light transmission)
   ii. Optional properties: UV blocking/antivirus, sulphur resistant, thermic, anti-drip, anti-mist, anti-dust, three-layer/five-layer films

   Crop-wise Recommendations

   iii. Dutch roses: Cladding—200 micron thick, UV stabilised, anti-dust, anti-sulphur, with cooling effect, light diffusion
   iv. Gerbera, Bell pepper, Anthurium and Orchids: Cladding—200 micron thick, UV-stabilised, anti-dust, with cooling effect, light diffusion
   v. Carnation: Cladding—200 micron thick, UV stabilised, anti-dust, with cooling effect for IR protection polythene at high altitudes

3. **Gutter:** It is used for collecting rainwater from the roof of the greenhouses and are placed at an elevated level (at least 4–4.5 m from ground level) between two spans. Gutters are made of galvanised sheet of 2 mm thickness in trapezoidal shape (preferably of single length without joint). It should be leak-proof. Minimum of 1 per cent slope is required for the gutter. Gutter orientation is in North–South direction in multi-span greenhouse and may change according to the direction of the wind.
4. **Foundation Pipe:** It connects the structure and the ground.
5. **Micro Irrigation System:** It is the best way for watering plants in a polyhouse as per the daily needs and the stage of the crop. Besides this, care should be taken that water does not trickle directly on the leaves or the flower, which may lead to disease and scorching of leaves or flowers.
6. **Fertigation Equipment:** For providing fertilisers to the plants as per their daily needs, water-soluble or liquid fertilisers are injected in the irrigation mainlines feeding the greenhouse crops. Fertiliser dosers and tanks are used for injecting soluble fertilisers. They can also be connected to automatic mixing and dispensing unit. The fertilisers are dissolved in different tanks as per compatibility and are mixed in discrete proportions for supply to the plants through drip irrigation systems.
7. **Spraying System:** This system is used for spraying required chemicals on the crop to control pests and diseases, if any. The spraying machines are normally portable but may be equipped with high pressure motorised piston pumps and nozzles.
8. **Exhaust Fan and Cooling Pads:** For removing hot air from the greenhouses in forced ventilated greenhouses, cooling pads are used for cooling the air entering into the greenhouses. These systems are operated as and when the climatic parameters like temperature, humidity, etc., inside the greenhouse need manipulation as per crop growth requirement.
9. **Shading Nets:** These are used for controlling light intensity falling on the crops inside the greenhouse. Various shading nets with shading capacities like 35 per cent, 50 per cent, 75 per cent are used for different crops and seasons.
10. **Sensors and Controllers:** They are used for controlling climatic parameters automatically inside hi-tech greenhouses. These systems are generally used for very high-value crops and sensitive activities like soil-less cultivation, tissue culture plant and hardening activities.
11. **Tubular Structural Members, Foundation and Labelling:** These are the galvanised iron tubular/square pipe and angles. These items are used to erect a stable frame to support the cladding material and other systems in the greenhouse. These items include horizontal and vertical structure members in any polyhouse.

- Purlin: It is a member that connects cladding supporting bars to the columns.
- Ridge: It is the highest horizontal section on top of the roof.
- Girder: It is a horizontal structural member connecting columns on gutter height.

- Bracings: These support the structures against wind.
- Arches: These support covering or cladding materials.

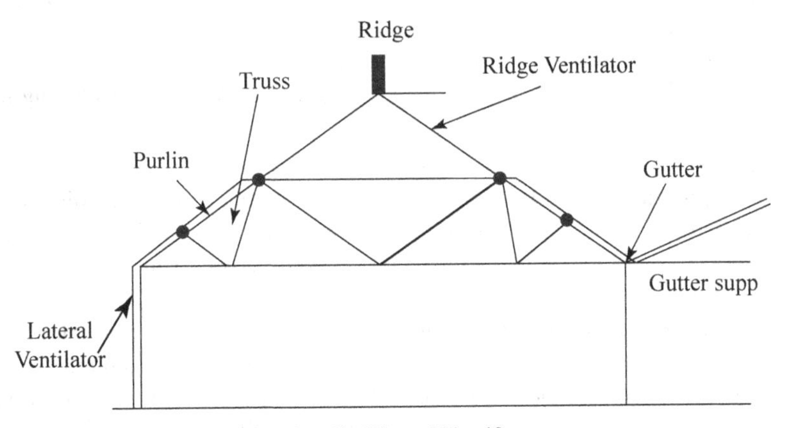

(*Source*: http://ncert.nic.in/vocational/pdf/kepc102.pdf)
Fig. 8.26. Components of Greenhouse

Crops Grown under Protected Cultivation

- Flower crops viz., Chrysanthemum, Carnation, Gerbera, Rose, Lilium, Orchid, Gladiolus, etc.
- Vegetable crops viz., Tomato, Coloured Capsicum (Yellow and Red Bell Peppers), Cucumber, Broccoli, Red Cabbage, Leafy vegetables, Radish, etc.
- Fruit crop–Strawberry
- Seedling and Nurseries - Vegetables, Flowers, Tissue Culture, Clonal for Forestry, Fruit Grafting (like Lemon, Citrus, Mango, Pomegranate, Guava, Litchi, etc.)

Site Selection for Protected Cultivation

While protected cultivation practices such as drip irrigation, raised bed farming, mulching can be practised on any site, even where cultivation is still being done. The criteria for site selection in case of protected cultivation structures like shade net houses and greenhouses are as follows:

1. **Exposure to ample sunlight:** The site should not be near tall trees, buildings or by the leeward side of hills.
2. **Appropriate distance from a low-lying area:** The site should not be in an area prone to waterlogging.
3. **Levelled ground surface:** A slope of 0–2 per cent is recommended. Levelling is required to be done in case the slope is beyond the recommended range. For steep terrains, it is recommended to build several separate greenhouses with axes parallel to contour lines.
4. **pH and electrical conductivity of soil:** It should have a pH of 6.0–6.5 and electrical conductivity should be less than 0.5 dS/m.

5. **Availability of continuous source of good quality water in sufficient volume:** The approximate water requirement is 1–2 L/m^2/day, which can be adjusted based on the season and the stage of cultivation.
6. **pH and electrical conductivity of water:** The pH and electrical conductivity of irrigation water should be in the range 6.5–7.0 and less than 0.7 dS/m respectively.
7. **Continuous supply of electricity:** This is particularly necessary during the day time.
8. **Good transportation facilities:** This is important to enable the transportation of greenhouse produce to nearby markets in time.
9. **Availability of sufficient land for future expansion:** A gap of 10–15 m should be maintained between two greenhouses, considering the possibility of expansion in future.
10. **Easy availability of labourers in surrounding area:** This should also be kept into consideration. Usually, four labourers are required for flower cultivation in a one-acre greenhouse.
11. **Good communication facilities:** These should be available at the site.
12. **Plantation of windbreaks:** The plants that breaks the flow of the wind from a particular direction. These plants are tall and have strong root base. These include poplar, silver oak, casuarina, etc., which are planted on the western side about 20 m away from the greenhouse because west winds are the strongest.
13. Awareness of relevant occupational safety and health standards.

Polyhouse Length and Width, Orientation

i. Polyhouse length is the dimension of the polyhouse in the direction of gable. (Length is side along the gable or side along the truss lines)
ii. Polyhouse width is the dimension of the polyhouse along the gutter.

Orientation of polyhouse for single-span structures should be East–West. For multi-span structures, the orientation should be North–South. The distance of trees adjacent to the greenhouse should be about 2.5 times the height of the greenhouse, to avoid shade.

Summarising the chapter content, climate change, the outcome of the "Global Warming" has now started showing its impacts worldwide. Global climate change has considerable implications in Indian agriculture and hence our food security and farmers livelihood are at stake. Agriculture sector is the most sensitive sector to the climate change because the climate of a region/country determines the nature and characteristics of vegetation and crops. It is estimated that India needs 320 MT of food grains by the year 2025. For a country like India, sustainable agricultural development is essential not only to meet the food demands, but also for poverty reduction through economic growth by creating employment opportunities in non-agricultural rural sectors.

We need to urgently take steps to increase our adaptive capacity. Right kind of technologies and policies are required to strengthen the capacity of communities to cope effectively with both climatic variability and changes. This would require increased support to adaptation research, developing regionally differentiated contingency plans for temperature and rainfall related risks, enhanced research on seasonal weather forecasts and their application for reducing production risks, and evolving new land use systems, including heat and drought tolerant varieties, adapted to climatic variability and changes and yet meeting food demand. Mechanisms for integrated management of rainwater, surface and ground water need to be developed. Once a crop is planted farmers need insurance cover to manage risks associated with extremes of temperature and precipitation events. Weather derivatives should be provided to increasing number of farmers at an early date. Crop simulation technique offers an opportunity to link the climate change with the other socio-economic and bio-physical aspects. These models can effectively work out the impact and also suggest suitable mitigation options to sustain the agricultural productivity. The crop-pest-weather interaction studies, conducted in the past, need to be thoroughly investigated for developing a sub-routine to link with the crop growth models to give the realistic estimates.

This adaptation and mitigation potential is nowhere more pronounced than in developing countries like India where agricultural productivity remains low; poverty, vulnerability and food insecurity remain high; and the direct effects of climate change are expected to be especially harsh. Coping with the impact of climate change on agriculture will require careful management of resources like soil, water and biodiversity. To cope with the impact of climate change on agriculture and food production, India will need to act at the global, regional, national and local levels.

9

Livestock Climatology

Livestock population meet the protein requirement of humans in addition to meeting from plants. The food chain starts from fodder crops to animals and from animals to humans. In human physiology, digestion of plant protein is much easier than animal protein, but however the digestion depends on climate (temperate climate, sub-tropical climate, tropical climate *etc*.,) wherein human kind live. Similar to crops, there is strong interaction between livestock and meteorological weather elements (temperature, relative humidity, wind *etc*.). But detailed study has not been published elaborately in scientific journals in the past. The weather elements have greater influence on livestock productivity in terms of milk, meat, skin, wool etc., pest and disease infestation and their housing needs. In this chapter how weather elements influence shed management, pest and disease problem and animals' productivity and are discussed. In majority of the Veterinary and Animal Science University in India, they have one division on animal climatology, wherein both research and extension aspects are being attended. However, in depth research is the need of the present decade, since little works have been done in the past on animal climatology.

Irrespective of weather elements, the heat stress due to the combination of temperature and relative humidity plays important role in affecting the productivity of animals. There is a range of thermal conditions within which animals are able to maintain a relatively stable body temperature by means of behavioral and physiological means. Heat stress results from the animal's inability to dissipate sufficient heat to maintain homeothermy (Sejian *et al.*, 2017). High ambient temperature, relative humidity and radiant energy compromise the ability of the animals to dissipate heat. As a result, there is an increase in body temperature, which in turn initiates compensatory and adaptive mechanisms to re-establish homeothermy and homeostasis. Homeostasis refers to the tendency to maintain a balanced or constant internal state that is optimal for functioning. Animals have a specific balanced or normal body temperature. When there is a problem with the internal functioning of animal body, this temperature may increase, signaling imbalance. As a result, body attempts to solve the problem and restore homeostasis for normal body temperature. These readjustments generally referred as adaptations and this is essential for survival of the animals.

9.1 Productivity

9.1.1 Cattle

In a study by Singh *et al.*, (2019), Tharparkar and Karan Fries cows were exposed to 25 °C, 35 °C and 42 °C with relative humidity of 50 + 5 per cent, corresponding to Temperature Humidity Index (THI) value of 72.2, 85.3 and 91.0 respectively for three hours continuously at different climate chamber consecutively for three days.. The results based on Respiration Rate per minute (RR/m), Pulse Rate/minute (PR/min), Rectum Temperature (RT, °F), surface body temperature, skin blood flow and milk yield indicated that the more sensitivity of Karan Fries' cows as compared to Tharparkar cows at higher THI. Therefore, the Karan Fries cow needs protection in terms of management from extreme heat for sustainable productivity. Further it is inferred that based on breed characteristics, the productivity varies with environment. Heat stress depressed the milk producing capability of the cattle and also reduced the per centage of fat in milk as per the report of Krishna Rao (2013). Reduced milk production was due to heat stress. Reproduction in cattle also get affected with heat stress. Experimental studies indicated that milk yield in cross bred cows in India is negatively correlated with THI as reported by many authors including Shinde *et al.*, (1990).

Both Krishna Rao (2013) and Sejian *et al.,* (2017), reported the effect of Thermal Humidly Index (THI) on cattle performance and those are given in the Table below

S. No.	THI	Stress level	Animal response and Respiration rate/m
1.	< 72	None	Normal
2.	72–79	Mild	Seek shade and Respiration rate is 90–110/minute
3.	80–89	Moderate	Increase in saliva production, Respiration rate is 110–130/minute
4.	90–98	Severe	Very uncomfortable, Respiration rate is >130/minute
5.	> 98	Danger	Potential death

The productivity of lactating animals during thermal stress may be sustained up to some extent by modifying the microclimate of animals (Maurya *et al.*, 2013).

Reproductive pattern of goats in Kerala was reviewed with reference to seasonality of male goat reproduction (Ibraheem Kutty, 2013). Impact of climatic variables was correlated with semen parameters (body weight of buck, scrotal circumstance, testicular length and diameter). Most of the reproductive parameters that varied significantly between seasons and showed an inverse relationship with humidity. Thus, high humidity together with prevailing moderately high ambient temperature could be major climate factor affecting reproductive process across seasons. Hence it is concluded that humidity is the major single factor regulating seasonal variations in reproductive processes of goats in humid tropical climate of Kerala State. Similarly, in cows and buffaloes for effective fertilization through artificial insemination, season and time of insemination are very important.

Prasad *et al.* (2017), reported that when constant exposure of cattle to high temperature causes a rise in rectal temperature, decline in food intake, increase in water intake, decrease in milk production, changes in milk composition, reduction in growth and even loss in body weight. This is the reason for deterioration in the performance of temperate cattle when introduced in tropical countries like India.

Sejian *et al*., (2017) did find the reasons for low productive performance of cattle under heat stress. Those were;

- Due to dehydration and low feed intake, the growth and weight of the animals got affected greatly.
- The milk production also got down and this is true for cold stress also. Reduced milk production is due to reduced feed intake, mammary cell proliferation and less dry matter in take. Sheep and goat also felt the same result with heat and cold stress.
- Wool production also got affected because of reduced fiber length.
- Due to negative correlation between heat stress and egg production, layers would produce less.
- Reproductive performance got affected due to imbalance in hormone action and decreased embryo development
- Reduced the length and intensity of estrus with more incidence of anestrus
- Decline in conception rate and change in sexual behavior

9.1.2 Poultry

The hot, humid climate has a negative impact on the performance and well-being of broilers. Particularly during summer, the temperature and humidity remains much above the thermally comfortable zone of broilers. The temperature-humidity index can be used to evaluate the impact of heat stress on the performance of broiler.

When bird's body temperature is in thermo-neutral zone, the energy from the feed can be directed to immune system development apart from the growth and reproduction. During the heat stress, bird's body makes several physiological changes to maintain body temperature causing reduced immune response (Daghir, 2009). When broilers were subjected to chronic heat stress, it was reported from the study made earlier that the body weight had significantly reduced to the tune of (-) 32.6 per cent.

Vimal Antony (2017) reported that the body temperature of birds(poultry) has a diurnal variation. The body temperature is found increased during day time and lesser during night time. This is mainly due to prevailing ambient temperature and feed intake. With increase in heat stress the birds drink more water and keep panting to increase the heat loss. The production also gets down followed by decline in body weight. The egg shell quality also do get affected drastically on the negative side. There is decrease in breast muscle development, while it is opposite for thigh muscles.

9.2 Pest and Disease Impact

9.2.1 Cattle

Among the weather factors, temperature both minimum and maximum, rainfall, relative humidity, wind speed, dew, fog and mist would influence the organisms and those cause pest and disease problem in livestock. The age of the animals, stress intensity of the animals, nature of stress, nutrient deficiency in the metabolic process of the animals and immunity level of the animals count more for manifesting the severity of different diseases and this again depend on climate and environment. The heat stress as caused by temperature is very important, as high temperature would induce stress and this leads to reduction of immune status of the animals. When immune status is reduced the animal becomes susceptible to diseases (Balasubramanian and Rajendran, 2017).

It was established that high humidity, low temperature along with dampness as a result of continuous rain would affect animals' health. The diseases like foot and mouth, anthrax, black quarter and *Hemorrhagic septicemia* are associated with weather parameters. Sivakumar *et al.*, (2012) studied the relationship between seasonal influence and infectious disease of bovine in Tamil Nadu. The study revealed that the incidence of anthrax was found highest during pre-monsoon season (June-Aug). Foot and mouth and Hemorrhagic septicemia diseases were more during monsoon season (Sept-Nov). The incidence of black quarter was highest during summer season (March-may).

The hot and humid weather condition is found to aggravate the infestation of cattle ticks namely, *Boophilus microplus*, *Haemaphysalis bispinosa* and *Hyalomma anatolicum*.

9.2.2 Poultry

When bird's body temperature is in thermo-neutral zone, the energy from the feed can be directed to immune system development apart from the growth and reproduction. During the heat stress, the bird's body makes several physiological changes to maintain body temperature causing reduced immune response (Daghir, 2009).

9.3 Shed Requirement

Under Indian condition, cow is kept under pucca shed, while buffaloes are not kept like cows because of variation in skin thickness and colour. Similarly, goats like shade management very well as compared sheep. As per the report of Krishna Rao (2013) provide the sheds to the cows to minimize heat gain by blocking the solar radiation and maximize the heat loss by using sprinkler and fans to provide evaporative cooling. Suraj and Sivakumar (2013) studied the benefits of existing housing structures of animals at different agro-climatic zones of Tamil Nadu. Based on dairy housing comfort index parameter, they found that tile roofed shelter was best for North eastern zone and North western zone, while cement sheet roofed shelter was best for Western zone and Hilly zone among the existing housing structure studied by using correlation studies.

Monisha (2017) concluded from her thesis experiment that the weather that prevails inside the poultry shed had much greater impact on the feed intake, water intake, and this in turn on body weight and livability and finally on the broiler production. Among the environmental factor temperature plays vital role than others. Zhou and Yamamoto (1997) found that the broiler body temperature did increase to 3°C and skin temperature to 6°C, when exposed to heat stress(30°C) for three hours.

9.4 Micrometeorology / Microclimatology

The meteorological factors that affect domestic animals and their productions are air temperature, relative humidity, pressure, wind, solar radiation and rain (John Abraham, 2017). Micro climate modification in animal and poultry sheds can be done through blowing of cool air through air pipes with holes, providing air conditioner, dehumidifier, room heater, humidifier and fan. Environmentally controlled shed also can be constructed to modify the comfort weather requirement of both animals and poultry. This area needs in-depth research to make the animals comfort and productivity.

9.5 Physical Stress

Thermal Humidity Index (THI) as induced by the combination of temperature and relative humidity does affect animal health and its productivity. A value of <72 is favorable for cow management. When the level gets increased from 72, the animals show symptom of discomfort and dies at a THI value of > 98.

10

Astro-Meteorology

Astro-meteorology is considered as one of the branches of Astronomy along with Astrophysics, Astrobiology and Astrometry. Astronomy is the science that deals with the material universe beyond the earth's atmosphere. It studies the nature and constitution of Celestial (heavenly) bodies and the laws which govern them including the effect they produce upon one another. It seeks to unravel the probable past history of heavenly bodies and their development. There exists a confusion between the branches of Astronomy viz., Astrophysics and Astro-meteorology with the term Astrology. Astrophysics uses principles and laws of physics to explain how Celestial bodies form and function or in other words it will help to ascertain the nature of the astronomical objects, rather than their positions or motions in space. However, the Astrology and Astro-meteorology are the subjects deal in future prediction and often considered as pseudo-science which uses position and movement of stars and planets for that purpose. Astrology links these to predict the human behaviour, while that of Astro-meteorology links the same to predict the future weather.

From ancient days, these predictions are based on one's wisdom and knowledge in astronomy and often these astrologers are astronomers who keenly watch the stars and planets movement and guide their rulers and farmers with their guesstimates. Till date it is being lore, mainly passed to followers of this group and the methodology is not documented clearly and properly. There are many incidences of falsified predictions of astrology as reported by many reviewers elsewhere and we are not discussing the same here, owing to our interest in weather prediction *i.e.,* astro-meteorology only. However, we have to recognize the works of the ancient astrologers who helped in developing the astronomical science.

Basically, there are two approaches in astrology *viz*., geocentric and heliocentric as used by the practitioner of astrology worldwide. As the name implies the Geocentric astrology is based on the calculation of planetary position as seen by the observer from the earth and thus earth is centre, while Heliocentric astrology bases its interpretation upon position within solar system with reference to the sun as centre. Though in true sense, the sun is the centre, the effect of motion as manifest on the earth is the basis of most astrological interpretation. Hence, vast majority of astrologers employs the geocentric calculations of the planetary positions for their predictions.

10.1 History of Ancient Forecasting

From time immemorial, people in all walks of life like hunters, farmers, warriors, shepherds and sailors know the importance of weather in order to carry out their respective works smoothly. There are weather wizards in those times who had traditional knowledge and provided future predictions of weather to the needy society. There are records which indicated that the Babylonians as early as 650 BCE were able to predict the weather from cloud patterns as well as using astrology. Aristotle's *Meteorologica* is the oldest comprehensive treatise on the subject of meteorology written during 340 BCE in which he described weather patterns in three books, while the fourth is about the chemistry. Later, Theophrastus, named as his successor by Aristotle, compiled a book on weather forecasting, called the Book of Signs. The notable scientist, Hippocrates, known as "the Father of Medicine" was also very much involved with the weather. The opening of his work begins with the advice that those who wish to investigate medicine must first begin with an understanding of seasons and weather. Chinese weather prediction lore extends at least as far back as 300 BCE, which was also around the same time ancient Indian astronomers developed weather-prediction methods. In 904 CE, Ibn Wahshiyya's Nabatean Agriculture, translated into Arabic from an earlier Aramaic work, discussed the weather forecasting of atmospheric changes and signs from the planetary astral alterations; signs of rain based on observation of the lunar phases; and weather forecasts based on the movement of winds.

The earliest reference to rain gauges and regional distribution of rainfall is contained in the Arthasastra, a treatise on statecraft authored by Kautilya in the 4th century BCE. He introduced first scientific measurements of rainfall and its application to revenue and relief work. Kautilya also emphasized observing and understanding the weather for better management of agricultural crops. The classical work Brihat Samhita written in sanskirit by Varahamihira around 500 CE provides clear evidence that a deep knowledge of atmospheric process existed in those times. It was understood that rains come from the sun (Adidyat Jayate Vrishti) and that good rainfall in the rainy season was the key to bountiful agriculture and food for the people (Attri *et al.*, 2011). Varahamihira also postulated different principles for the prediction of rainfall and weather. In the Indian sub-continent, the ancient astronomers and astrologers made a systematic study of meteorology and this indigenous knowledge is unique to India. Even today, it is common that village astrologers (pandits) able to rightly predict weather with high percentage accuracy (Varshneya, 2008). Though meteorology was considered the oldest of all sciences, evidence indicates that it, as a "science" is comparatively young. However, as a branch of knowledge, it dates back to the origin of human civilization.

Bruce Scofield (1987, 2010) described in his paper about Jhon Goad's *Astro-meteorologica*, which is a comprehensive work of 500 pages written on 1686. He is of the opinion that Goad is one of the few astrologers of that time who attempted to apply the newly emerging experimental method to his subject. A few lines from Scofield's work is reproduced below to understand better about Goad's book. "During the mid-17th century John Goad applied the astrological scientific methodology of the time, mostly data collection and correlation, to meteorological astrology and produced a major work on the subject. In it he analysed, the geocentric angular separations between the planets, Sun and Moon, and correlated them with weather records he made over several decades. Goad's *Astro-Meteorologica* was probably the greatest single effort to make a science out of astrology in the 17th century. Unfortunately, his research was conducted at a time before instrumentation and statistical methodologies for processing data were readily available". His paper was written to examine *Astro-meteorologica* for its content and to introduce a generally unrecognized reason for the decline of astrology in the 17th century.

10.2 Role of Panchangs in Astro-Meteorology

The traditional ancient Indian astrological almanac known as Panchaangam, prepared for public use from Vedaanga Jyotisham period (1400 BCE - 1300 BCE) (Sivaprakaasam and Kanakasabai, 2009) stands out as the best exemplification for ancient traditional texts that employ theoretical methods. The book published yearly gives information on daily basis and extensively used by the astrologers for making astrological calculations and the farmers to start the farming activity based on the prediction of rainfall (Bharadwaaj Dinesh 2004). This practice of referring to zodiacal signs continues to this day even in the United States, which publish the popular Farmer's Almanac for the past consecutive 170 years. Based on folklore, astrology, rituals and ancient literature, 30 odd panchangs are in force across the country, which are the closest equivalents to the Farmer's Almanac. As there are many panchangs in India, Prime Minister Pandit Jawaharlal Nehru during 1952 formed a committee to reform this and to develop a unified National Calendar on the basis of the most accurate modern astronomical data for the interest of National integrity. The committee recommended preparation of the Indian Ephemeris and Nautical Almanac incorporating therein along with the usual astronomical data calculated with most modern astronomical formulae, the National Calendar of India (using Saka Era) with timings of thithis, nakshatras, yoga etc., and also festival dates. The committee also recommended the preparation of the Rashtriya Panchang with solar calendar system for civil purposes and luni-solar calendar system for religious purposes. The Government of India decided that the work should be continued by a special unit attached to a scientific department and hence it was administered by India Meteorological Department from the Council of Scientific and Industrial Research on 1st December, 1955. The unit, which was functioning as the office of the Calendar Reform Committee at the then Institute of Nuclear Physics, Calcutta, was brought under the control of the Regional Meteorological Centre, Calcutta as one of its sections named 'Nautical Almanac Unit'. The unit undertook the preparation of 'The Indian Ephemeris and Nautical Almanac' for 1958, which was the first issue published in March 1957. Simultaneously the first issue of Rashtriya Panchang was started from 1879 Saka Era (1957–1958).

The Council of Meteorology and Atmospheric Sciences (CMAS) of the Government of India during 1979 approved the formation of "Positional Astronomy Centre" as a separate Directorate in the place of Nautical Almanac Unit directly under the control of the Director General of Meteorology, New Delhi. Though the implementation of the approval has been made with effect from the 1st December 1979, the formal inauguration of the Positional Astronomy Centre (PAC) took place on the 26th April 1980. The PAC is now responsible for bringing publications mainly the Indian Astronomical Ephemeris, Tables of Sunrise and Sunset and Moon rise and Moon set and Rashtriya Panchang in 14 languages. The Indian Astronomical

Ephemeris contains about 500 pages of astronomical data on the position of the Sun, Moon, Planets and Bright Stars and a section on the National Calendar of India (Saka Calendar) with the timings of Tithis, Nakshatras etc. and the festival dates. Rashtriya Panchang is a popular publication brought out with the aim of providing a standard unified calendar for the whole country and promoting a scientific basis for Panchang calculation. The Centre meets the data requirements of a large number of users. Lunar data for prediction of tides are computed for the Survey of India and data of the Sun's daily path are useful to civil engineers, architects and telecommunications engineers. They provide data for determining the dates of festivals of all communities in India in advance for holiday declaration by the Government, tourist promotion abroad and for use of Panchang Makers.

Another important milestone in the history of Indian Astronomy is the establishment of the Indian Institute of Astrophysics, which is a premier institute devoted to research in astronomy, astrophysics and related physics. It traces its origins back to an observatory set up in 1786 at Madras, which from the year 1792, began to formally function at its Nungambakkam premises as the Madras Observatory. In 1899, the observatory moved to Kodaikanal. In the year 1971, the Kodaikanal Observatory became an autonomous society, the Indian Institute of Astrophysics. The headquarters were shifted to Bengaluru into its present campus in Koramangala in 1975. Today, funded by the Department of Science and Technology, the Institute ranks as a premier institution devoted to research and education of astronomy and physics in the country. The main observing facilities of the Institute are located at Kodaikanal, Kavalur, Gauribidanur and Hanle in India. Though the Institute has no direct relevance to the Astro-meteorological work but their contribution to the Astronomy in turn helped the astrologers to develop their prognostics.

10.3 Astro-Meteorology in 20th Century

The development in the science of Agricultural Meteorology and setting up of separate Department of Agrometeorology in India paved way for research in Astro-meteorology. The research being done either to verify statistically the findings of ancient workers like Varahmihir or trying to correlate the planetary movement and zodiac signs to the weather events. Besides researchers, there are independent astrologers across the country also doing prediction, mainly seasonal rainfall for farmers of the country. As pointed out by Varshneya (2008), there are good agreements of observed values with that of prediction by the local Astrologers but there is no systematic study by the astrologers that is acceptable by the scientists of meteorology. Thus, the work of these astrologers is still considered as myth because most of them keep their astrological calculation for prognostics as a secret.

The Tamil Nadu Agricultural University, Coimbatore, Gujarat Agricultural University, Anand, Mahatma Phule Krishi Vidyapeeth (MPKV), Pune are some of the institutes having Agricultural Meteorology wings involved in astro-meteorological research work. Besides agricultural universities, some of the conventional universities have shown interest in astro-meteorological research and published prediction results. Some of the verification of ancient work and forecast produced by research workers are discussed here.

At Agro Climate Research Centre of Tamil Nadu Agricultural University, work has been done to verify the saint Idaikkadar Siddhar's rainfall prediction written in sixty verses for each of the sixty Tamil years. The total annual rainfall of Tamilnadu for the last 180 years was collected from Indian Institute of Tropical Meteorology and verified with the verses of the saint and found that the accuracy of the prediction is around 50 per cent. Iyengar (2004) simulated the rainfall for the years 1901 to 2002 as described by Varaha Mihir in his book Brahat Samhita and compared with that of India Meteorological Department's observations. The study done in the western part of Madhya Pradesh, including Ujjain (Varaha Mihir's birth place) was validated and he found that the coefficient of variability is around 37 per cent which is close to the present-day variability. A report accessed from Shodhganga, INFLIBNET Centre web pages in which verification of almanac (Panchangam) prediction of annual and seasonal rainfall was discussed in detail. The authors used Andhra Patrika Panchangam which was prepared as per the 'Indian Ephemeris and Nautical Almanac' of the Positional Astronomy Centre, Kolkatta. The India Meteorological Department rainfall observations were verified with Panchangam predictions for the period from 1992-'93 to 2002-2003 and they found on an average 57 per cent accuracy and concluded that Panchangam can be conferred with the status of a full-fledged 'Scientific Weather Prediction Model'.

Another study in Tamilnadu (Sivaprakasam & Kanakasabai, 2009) used *Asal 28 No Manonmani Vilasa Suddha Vakiya Panchangam,* a 150 yrs. old almanac which is being prepared for the King of Tanjore in Tamil Nadu. The almanac used theory suggested by Parashara having a particular planet as a ruler, another planet as a minister, and a particular cloud defining the amount of rainfall for every year. They verified 60 years of almanac prediction to that of actual observation over Tamil Nadu for the period from 1936–1995. Though they have experienced difficulty in quantifying the rainfall from the said almanac, it was observed that out of 60 years 32 years rainfall occurred above the mean. As a final observation they indicated that though the traditional methods of forecasting rainfall may be riddled with inaccuracies, but cannot be ignored altogether.

The basis by which the rainfall prediction is done in the almanac is not clearly known and is also not mentioned in the published almanac. These almanacs are written in regional languages spoken by individual State and thus everyone considers the prognosis given in the almanac is hold good for the entire State. If rain is forecasted for a particular day which means that entire State will receive the rainfall but in real sense this may not happen and there are regional differences with in a State which is not taken by the almanac. State-wide rain occurrence is also a rare phenomenon in meteorological sense. The forecast inaccuracies in respect of almanac may be attributed to the fact that it provides a generic forecast and not the area-wise. We need to recollect even Varahamihir's prognosis written in Brihat Samhita is being test verified for the location (Ujjain) where he lived because every astrological prediction is based on position of planets and zodiac signs in respect of that particular location.

Hence, there is possibilities for improved accuracies, if the prognosis is done for specific locations.

Building on the ideas of Kepler and Goad, Bruce Scofield (2010) in his Ph.D. dissertation made correlation between geocentric Sun-Saturn alignments and cold temperatures, using modern daily temperature data from New England, Central England, Prague and other locations. A correlation was shown in daily temperature records, between cooling trends in specific regions and the geocentric alignments of the Sun and the planet Saturn. His hypothesis was supported by a number of tests that showed lower temperatures on days when Sun-Saturn alignments occur, especially when near the equinoxes. It was further explained that the astronomy of this positioning suggests that tidal forces on the atmosphere may be part of a mechanism that would explain the apparent effect.

While doing validation of astro-meteorological rainfall forecast for Gujarat Varshneya *et al.,* (2008) briefly described the influence of different planets on the weather. They indicated that Sun, Mars and Pluto are responsible for higher temperature; Moon, Venus and Neptune are responsible for rain; Mercury causes wind; Jupiter causes temperature increase and Saturn causes cold. Conjunction and inter-junction of these planets have variation in different types of weather. Using astro-meteorological technique, they predicted rainfall for 2005–2007. The results revealed that the accuracy of forecast varied between 42–73 per cent during 2005, 42–74 per cent during 2006 and for 2007, the accuracy of forecast varied between 59–73 per cent for various regions studied with average accuracy of 66 per cent for Gujarat State. In continuation Vaidya *et al.,* (2019) made predictions for 19 districts of Gujarat and prepared Monsoon Research Almanac-2018 for Gujarat. For different districts, the accuracy skill differed which ranged from 48–77 per cent.

Using Astro-meteorology as an option for improving the accuracy of Numerical Weather Prediction (NWP) model, Dheebakaran *et al.*, (2017) opined that in general, the planets viz., Venus, Neptune and Saturn are considered as positive for rainfall, the Mars and Uranus are considered as negative for rainfall, whereas the Mercury and Jupiter have varied influence depending on its position to the earth. They tried prediction of daily rainfall employing the NWP model (WRF ingesting GFS forecast outputs), probability-based forecast and astro-meteorological forecast. They found that the astro-meteorological forccast developed using planetary azimuth from respective observed location has highest accuracy (0.74–0.87) among the three techniques, while their integration resulted in better accuracy of 0.75 to 0.88. The occurrence of cyclone was investigated by Sankar *et al.,* (2019) by relating the azimuth of planets viz., Saturn, Uranus, Moon, Mercury, and Venus. These planets were found to record 59–75 per cent of cyclonic events in the azimuth range of 61–120 and 241–300 degrees which happened during the years 1990–2016. Similar studies using azimuth of planets correlating with wind speed in 31 districts of Tamil Nadu were carried out by Rathika *et al.* (2019) and they also found that the azimuth range of 61–120 and 241–300 degrees in the above study also responsible for the wind speed events by different planet.

The radiation from the celestial bodies of the solar system and beyond said to influence the earth's atmosphere process as being influenced by the sun. This being the basis for inclusion of these celestial bodies and the zodiac signs for prognosis of weather variables. These kind of relationships has not been studied fully and there are lot of scope for this subject. Hence Astro-meteorology as a science need to be given impetus in order to provide the stake-holders better prediction. The numerical weather prediction coupled with astro-meteorological prediction will definitely pave way for better weather prediction for the future especially on seasonal scale.

References

Adams, R.M., Rosenzweig, C., Peart, R.M., Ritchie, J.T., McCarl, B.M., Glyer, J.D., Curry, R.B., Jones, J.W., Boote, K.J and L.H. Allen Jr. 1990. Global climate change and U.S. agriculture: an interdisciplinary assessment. Nature, 345:219–224.

Adger, W.N., Arnell, N.W and E.L. Tompkins. 2005. Successful adaptation to climate change across scales. Global Environmental Change 15, 77–86.

Aggarwal, P.K and Mall, R.K. 2002. Climate change and rice yields in diverse agro environments of India. II. Effect of uncertainties in scenarios and crop models on impact assessment. Climate change, 52: 331-343.

Aggarwal, P.K, Kalra, N. 1994. Analyzing the limitations set by climatic factors, genotype, water, and nitrogen availability on productivity of wheat. II. Climatically potential yields and optimal management strategies. Field Crops Res 38:93–103

Aggarwal, P.K. and Sinha, S.K. 1993. Effect of probable increase of carbon dioxide and temperature on productivity of wheat in India. Jour. Agric. Meteorol, 48: 811-814.

Aggarwal, P.K., Bandyopadhyay, S.K., Pathak, H., Kalra, N., Chander, S and S.K. Sujith. 2000. Analyses of yield trends of the rice–wheat system in north-western India. Outlook Agric, 29 (4):259–268.

Ajith, K., Geethalakshmi, V., Ragunath, K.P. Pazhanivelan, S and S. Panneerselvam. 2017. Rice Acreage Estimation in Thanjavur, Tamil Nadu Using Lands at 8 OLIIMAGES and GIS Techniques. International Journal of Current Microbiology and Applied Sciences. 6(7):2327-2335.

Akponikpè, P., Irénikatché, B., Gérard, B., Michels, K and C. Bielders. 2010. Use of the APSIM Model in Long Term Simulation to Support Decision Making Regarding Nitrogen Management for Pearl Millet in the Sahel. European Journal of Agronomy, 32 (2): 144–54. doi.org/10.1016/j.eja.2009.09.005.

Alazzy, A.A., Lu, H., Chen, R., Ali, A.B., Zhu, Y and L. Su. 2017. Evaluation of satellite precipitation products and their potential influence on hydrological modelling over the Ganzi River Basin of the Tibetan Plateau. Advances in Meteorology,1-23.

Allen, L.H. Jr., Boote, K.J., Jones, J.W., Jones, P.H., Valle, R.R., Acock, B., Rogers, H.H. and R.C. Dahlman. 1987. Response of vegetation to rising carbon dioxide: Photosynthesis, biomass, and seed yield of soybean. Global Biogeochemical Cycles **1**: 1-14.

Andarzian, B., Bakshshandeh, A.M., Bannayan, M and Y. Emam. 2007. Modeling and Simulation Growth, Development and Yield of Wheat. Shahid Chamran University of Ahvaz, Ahvaz, Iran.

Anil Kumar., Chander Shekhar., Anurag and Raj singh.2020. Paper presented in National Seminar on Agrometeorological Interventions for enhancing farmers income (Agmet-2020) 20-22 Jan. 2020. Kerala Agricultural University, Thrissur, Kerala. Abstract Compendium. Theme: 0.11.15. P 155.

Annamalai, H., Hamilton, K.P and K.R. Sperber. 2007. South Asian Summer Monsoon and its relationship with ENSO in the IPCC AR4 Simulations: J. Climate., 20:1071-1092.

Antle, J.M., Basso, B.O., Conant, R.T., Godfray, C., Jones, J.W., Herrero, M., Howitt, R.E., Keating, B.A., Munoz-Carpena, R., Rosenzweig, C., Tittonell, P and T.R. Wheeler. 2017. Towards a new generation of agricultural system models, data, and knowledge products: model design, improvement and implementation. Agricultural Systems, 155, 255–268.

Apan, A., Held, A., Phinn, S and J. Markley. 2004. Detecting sugarcane 'orange rust' disease using EO-1 Hyperion hyperspectral imagery. International Journal of Remote Sensing. 25:489-498.

Aplin, P., Atkinson, P.M and P.J. Curran. 1998. Identifying missing field boundaries to increase the accuracy of per-field classification of fine spatial resolution satellite sensor imagery. 27th International Symposium on Remote Sensing of Environment, Tromso, Norway, 399-402

Arrhenius, S. 1896. On the influence of carbonic acid in the air upon the temperature of the ground. The London, Edinburgh, and Dublin Philosophical Magazine and Journal of Science, 41(251), 237-276.

Arunkumar, P., Geethalakshmi, V., Maragatham, N., Panneerselvam, S., Bhuvaneswari, K and V. Kumar. 2017. Impact of elevated CO_2 and temperature on rice (C3) and maize (C4). Editorial Board, 329.

Asseng, S., Ewert, F., Rosenzweig, C., Jones, J.W., Hatfield, J.L., Ruane, A.C., Boote, K.J., Thorburn, P.J., Rotter, R.P., Cammarano, D., Brisson, N., Basso, B., Martre, P., Aggarwal, P.K., Angulo, C., Bertuzzi, P., Beirnath, C., Challinor, A.J., Doltra, J., Gayler, S., Goldberg, R., Grant, R., Heng, L., Hooker, J., Hunt, L.A., Ingwersen, J., Izaurralde, R.C., Kersebaum, K.C., Müller, C., Naresh Kumar, S., Nendel, C., O'Leary, G., Olesen, J.E., Osborne, T.M., Palosuo, T., Priesack, E., Ripoche, D., Semenov, M.A., Shcherbak, I., Steduto, P., Stöckle, C., Stratonovitch, P., Streck, T., Supit, I., Tao, F., Travasso, M., Waha, K., Wallach, D., White, J.W., Williams, J.R and J. Wolf. 2013. Uncertainty in simulating wheat yields under climate change. Nature Climate Change, 3:827–832. doi.org/10.1038/nclimate1916

Atkinson, P.M and P. Lewis. 2000. Geostatistical classification for remote sensing: an introduction. Computers and Geosciences, 26, 361-371.

Attri, S.D., L.S. Rathore, M.V.K. Sivakumar, S.K. Dash. 2011. Challenges and Opportunities in Agrometeorology Springer Science & Business Media. 600 pages.

Baker, J.E., Allen, L.H., Boote, K.E and N.B. Pickering. 2000. Direct effects of atmospheric carbon dioxide concentration on whole canopy dark respiration of rice. Global Change Biol. 6:275-286.

Bala, 2018. Crop Insurance - Remote Sensing. https://skymapglobal.com/crop-insurance-remote-sensing/.

Balasubramanian, T. N and A. Rajendran. 2017. Effect of climate on animal diseases. Livestock Meteorology (edts. G.S.L.H.V. Prasad Rao., Grish Varma, G and V. Beena). New India Publishing Agency, New Delhi. PP. 251-260

Balasubramanian, T. N and N. Gopalasamy. 2002. A Tamil text book on agricultural meteorology. Palani Paramount Publishers, Palani. P.179: pp.103-148.

Balasubramanian, T.N., Jagannathan, R., Maragatham, N., Sathyamoorthi, K and R. Nagarajan. 2014. Generation of weather windows to develop agro-advisories for Tamil Nadu under automated weather forecast system. Journal of Agrometeorology, 16(1):60-68.

Balasubramanian, T.N., Jagannathan, R., Maragatham, N., Sathyamoorthi, K., Nagarajan, R., Malliga Vanangamudi., Sathyamoorthy, N.K., Poongulali, S., Sakthivel, P., Sataraj, S., Arun Prakash, P., Ramesh Kumar, P and J. Abdul Hameed. 2016.Designing agro-met advisories for selected weather windows under automated weather based advisory system in Tamil Nadu- a case study. Journal of Agrometeorlogy, 18(1):33-40.

Balasubramanian, T.N.., Arivudai Nambi, A and R. Isaac Manuel. 2009. Practical Hand manual on village Level Mini Agricultural Meteorology Observatory and Processing Data for Farm Decision making. M.S. Swaminathan Research Foundation, Chennai. P.33.

Balasubramanian, T.N.2008. Concept paper on Farmer and weather. Abstract of International symposium on Agrometeorology and Food security. CRIDA, Hyderabad. Feb.18-21. S10-01, p221.

Balwinder-Singh, Humphreys, E., Yadav, E and D. S. Gaydon. 2015. Options for increasing the productivity of the rice-wheat system of north-west india while reducing groundwater depletion. part 1. rice variety duration, sowing date and inclusion of mungbean. Field Crops Research, 173: 68–80. https://doi.org/10.1016/j.fcr.2014.11.018.

Bannayan, M., Crout, N.M.J., Eyshi Rezaei, E and G. Hoogenboom. 2013. Application of the CERES-Wheat model for within-season prediction of wheat yield in United Kingdom. Agronomy Journal, 95:114–125.

Baranowski, P., Krzyszczak, J., Slawinski, C., Hoffmann, H., Kozyra, J., Nieróbca, A., Siwek, K and A. Gluza. 2015. Multifractal analysis of meteoro logical time series to assess climate impacts. Climate Research, 65: 39–52

Barbosa, P.M., Casterad, M.A and J. Herrero. 1996. Performance of several Landsat 5 Thematic Mapper (TM) image classification methods for crop extent estimates in an irrigation district. International Journal of Remote Sensing, 17, 3665-3674.

Barrs, H. D and S.A. Prathapar. 1994. An inexpensive and effective basis for monitoring rice areas using GIS and remote sensing. Australian Journal of Experimental Agriculture, 34, 1079-1083.

Barrs, H.D and P.E. Weatherly. 1962. Physiological indices for high yield potential in wheat. Indian Journal of Plant Physiology,25: 352-357

Bassu, S., Asseng, A., Motzo, R and F. Giunta. 2009. Optimizing sowing date of durum wheat in a variable Mediterranean environment. Field Crops Research, 111:109–118.

Bassu, S., Brisson, N., Durand, J.L., Boote, K.J., Lizaso, J., Jones, J.W., Rosenzweig, C., Ruane, A.C., Adam, M., Baron, C., Basso, B., Biernath, C., Boogaard, H., Conijn, S., Corbeels, M., Deryng, D., De Sanctis, G., Gayler, S., Grassini, P., Hatfield, J., Hoek, S., Izaurralde, C., Jongschaap, R., Kemanian, A., Kersebaum, K., Müller, C., Nendel, C., Priesack, E., Virginia, M., Nin, P., Sau, F., Shcherbak, I., Tao, F., Teixeira, E., Makowski, D., Timlin, D and K. Waha. 2014. How do various maize crop models vary in their responses to climate change factors. Global Change Biology, 20 (7):2301–2320.

Batchelor, W.D., Jones, J.W and K.J. Boote. 1993. Extending the use of crop models to study pest damage. Transactions - American Society of Agricultural Engineers, 36 (2): 551–558.

Bates, P.D and A.P.J. De Roo. 2000. A simple raster based model for flood inundationsimulation. J.Hydrol. 2000, 236, 54–77

Bauriegel, E., Brabandt, H., Garber, U and W.B. Herppich. 2014. Chlorophyll fluorescence imaging to facilitate breeding of Bremia lactucae -resistant lettuce cultivars. Computers and Electronics in Agriculture, 105:74-82.

Bauriegel, E., Giebel, A., Geyer, M., Schmidt, U and W.B. Herppich. 2011. Early detection of Fusarium infection in wheat using hyper-spectral imaging. Computers and Electronics in Agriculture, 75:304-312.

Berdugo, C., Zito, R., Paulus, S and A.K. Mahlein. 2014. Fusion of sensor data for the detection and differentiation of plant diseases in cucumber. Plant Pathology, 63:1344-1356.

Bergstrasser, S., Fanourakis, D., Schmittgen, S., Cendrero-Mateo, M. P., Jansen, M., Scharr, H and U. Rascher. 2015. HyperART: non-invasive quantification of leaf traits using hyperspectral absorption-reflectance-transmittance imaging. Plant Methods, 11:1.

Bharadwaaj Dinesh M. 2004. Panchangam - The Indian Almanac, About Astrology, www.explocity.com/Channels/Astrology/Panchangam.asp.

Bhuvaneswari, K., Geethalakshmi, V., Lakshmanan, A., Anbhazhagan, R and D. N. U. Sekhar. 2014. Climate change impact assessment and developing adaptation strategies for rice crop in western zone of Tamil Nadu. Journal of Agrometeorology, 16(1), 38.

Bock, C.H., Parker, P.E., Cook, A.Z and T.R. Gottwald. 2008. Visual rating and the use of image analysis for assessing different symptoms of citrus canker on grapefruit leaves. Plant Diseases, 92:530-541.

Böhler, J.E., Schaepman, M.E and M. Kneubühler. 2020. Crop Separability from Individual and Combined Airborne Imaging Spectroscopy and UAV Multispectral Data. Remote Sensing. 12, 1256; doi:10.3390/rs12081256.

Boote, K.J and M. Tollenaar. 1994. Modeling genetic yield potential. p. 533-565. In K.J. Boote et al., (ed.) Physiology and determination of crop yield. ASA-CSSA-SSSA, Madison, WI.

Boote, K.J., Allen Jr., L.H., Vara Prasad, P.V and J.W. Jones. 2010. Testing effects of climate change in crop models. In: Hillel, D., Rosenzweig, C. (Eds.), Handbook of Climate Change and Agroecosystems. Imperial College Press, London UK.

Boote, K.J., Jones, J.W and N. B. Pickering. 1996. Potential uses and limitations of crop models. Agronomy Journal, 88 (5):704–716.

Boote, K.J., Jones, J.W., Hoogenboom, G and N.B. Pickering. 1998. The CROPGRO model for grain legumes. In: Tsuji, G.Y., Hoogenboom, G., Thornton, P.K. (Eds.), Understanding Options for Agricultural Production. Kluwer Academic Publishers, Dordrecht, pp. 99–128.

Boote, K.J., Jones, J.W., Mishoe, J.W and R.D. Berger. 1983. Coupling pests to crop growth simulators to predict yield reductions. Phytopathology, 73:1581–1587.

Bouma, J and J.W. Jones. 2001. An international collaborative network for agricultural systems applications (ICASA). Agricultural Systems, 70 (2):355–368.

Bouman, B.A.M., Kropff, M.J., Tuong, T.P., Woperies, M.C.S and H.H. van Laar. 2001. ORYZA 2000: Modeling Lowland Rice. International Rice Research Institute and Wageningen University and Research Center, p. 235.

Bouman, B.A.M., van Keulen, H., van Laar, H.H and R. Rabbinge. 1996. The 'school of de wit' crop growth simulation models: a pedigree and historical overview. Agricultural Systems, 52 (2–3):171–198.

Brabandt, H., Bauriegel, E., Garber, U and W. B. Herppich. 2014. ΦPSII and NPQ to evaluate Bremia lactucae -infection in susceptible and resistant lettuce cultivars. Scientia Horticulturae (Amsterdam), 180:123-129.

Bravo, C., Moshou, D., West, J., McCartney, A and H. Ramon. 2003. Early disease detection in wheat fields using spectral reflectance. Biosystems Engineering, 84:137-145.

Bruce Scofield. 1987. John Goad's Astro-Meteorologica, NCGR Journal, Winter 1986-1987, pp. 58-61

Bruce Scofield. 2010. A history and test of planetory waether forecasting. Ph.D. dissertation submitted to the Department of Geosciences, Graduate School of the University of Massachusetts

Buis, A. 2019. The Atmosphere: Getting a Handle on Carbon Dioxide. NASA. Global Climate Change. Retrieved May, 1, 2020.

Burling, K., Hunsche, M and G. Noga. 2011. Use of blue-green and chlorophyll fluorescence measurements for differentiation between nitrogen deficiency and pathogen infection in wheat. Journal of Plant Physiology, 168:1641-1648.

Camargo, A and J.S. Smith. 2009. Image pattern classification for the identification of disease causing agents in plants. Computers and Electronics in Agriculture, 66:121-125.

Campbell, J.B. 1981. Spatial correlation effects upon accuracy of supervised classification of land cover. Photogrammetric Engineering and Remote Sensing, 47, 355-363

Cassman, K.G., Grassini, P and J. van Wart. 2010. Crop yield potential, yield trends, and global food security in a changing climate. In: Rosenzweig, C., Hillel, D. (Eds.), Handbook of Climate Change and Agroecosystems. Imperial College Press, London, pp. 37–51.

Chaerle, L., Hagenbeek, D., De Bruyne, E and D. Van der Straeten. 2007. Chlorophyll fluorescence imaging for disease-resistance screening of sugar beet. Plant Cell Tissue and Organ Culture, 91:97-106.

Chaerle, L., Hagenbeek, D., De Bruyne, E., Valcke, R and D. Van Der Straeten. 2004. Thermal and chlorophyll-fluorescence imaging distinguish plant- pathogen interactions at an early stage. Plant Cell Physiol. 45:887-896.

Chatterjee, A. 1998. Simulating the impact of increase in carbon dioxide and temperature on growth and yield of maize and sorghum. M.Sc Thesis, Division of Environmental Sciences, IARI, New Delhi.

Chatterjee, S. Huang, J and A.E. Hartemink. 2020. Establishing an Empirical Model for Surface Soil Moisture Retrieval at the U.S. Climate Reference Network Using Sentinel-1 Backscatter and Ancillary Data. Remote Sensing. 12, 1242; doi:10.3390/rs12081242.

Chauhan Kalyan Singh Kiransingh and M.M. Lunagaria,2020. Spatial and temporal variation in microclimate over capsicum under open ventilated condition. Paper presented in National Seminar on Agrometeorological Interventions for enhancing farmers income (Agmet-2020) 20-22 Jan. 2020. Kerala Agricultural University, Thrissur, Kerala. Abstract Compendium. Theme: 0.11.5. P 147.

Chogatapur. S.V., Chandranath, H.T and R.B. Khandagave. 2017. Sustainable Sugarcane Initiative (SSI), An Approach to Enhance Sugarcane Production: A Review. Int. J. Pure App. Biosci. 5 (6): 241-246.

Coleman, D.C., Swift, D.M and J.E. Mitchell. 2004. From the frontier to the biosphere: a brief history of the USIBP grasslands biome program and its impact on scientific research in North America. Rangelands, 26 (4):8–15.

Conway, G.R. 1987. The properties of agroecosystems. Agricultural Systems, 24 (2): 95–117.

Cook, S.E., Corner, R.J., Groves, P.R and G.J. Grealish. 1996. Use of airborne gamma radiometric data for soil mapping. Australian Journal of Soil Research, 34, 183-194.

Cure, J.D. and B. Acock. 1986. Crop responses to carbon dioxide doubling: A literature survey. Agric. For. Meteorol. 38: 127-145.

Curry, R.B., Peart, R.M., Jones, J.W., Boote, K.J and L.H. Allen Jr. 1990. Response of crop yield to predicted changes in climate and atmospheric CO2 using simulation. Transactions - American Society of Agricultural Engineers, 33: 1383–1390.

Daghir, N.J. 2009. Nutritional strategies to reduce heat stress in broilers. Lohmann Information. 44(1):6-15

Dakshina Murthy., Rao, K.P.C., Anthony Whitbread and V. NageswaraRao. 2017. Seasonal climate forecast (SCF) based crop management options for managing climate variability and change in semi-arid tropics. The Third International Conference on Bio-resource and Stress Management.

Das, H.P. 2000. Monitoring the incidence of large scale droughts in India. 181-195. In Drought A Global Assessment (Donald A. Wilhite (Ed.)), Vol. 1, Routeldge London and New York.

De Fraiture, C., Smakhtin, V., Bossio, D., McCornick, P., Hoanh, C.T., Noble, A., Molden, D., Gichuki, F., Giordano, M., Finlayson, M and H. Turral. 2007. Facing climate change by securing water for food, livelihoods and ecosystems. Journal of Semi-Arid Tropical Agricultural Research, 4 (1):12.

De Wit, C.T and F.W.T. Penning de Vries. 1982. La synthese et la simulations des systems des production primaire. In: Penning de Vries, F.W.T., Djitèye, M.A. (Eds.), La productivité des paturage sahéliens. Pudoc, Wageningen, pp. 23–27.

Deering, D.W. 1978. Rangeland reflectance characteristics measured by aircraft and spacecraft sensors. Ph.D. Dissertation, Texas A & M University, College Station, TX, 338 pp.

Delalieux, S., van Aardt, J., Keulemans, W and P. Coppin. 2007. Detection of biotic stress (Venturia inaequalis) in apple trees using hyperspectral data: non-parametric statistical approaches and physiological implications. European Journal of Agronomy, 27:130-143.

Dempster, J.P. 1983. The natural control of populations of butterflies and moths. Biological Reviews, 58:461–481

Dent, J.B and M.J. Blackie. 1979. Systems Simulation in Agriculture. Applied Science Publishers, London, p. 180.

Department of Science and Technologies,1999. Guide for Agro-Meteorological Services (eds. S.V. Singh, L.S. Rathore and H.K.N. Trivedi). New Delhi: National Centre for Medium Range Weather Forecasting. P 201.

Dheebakaran, G. A., Jagannathan, R., Paulpandi, V. K and S. Kokilavani 2015. Impact of climate change on monsoon onset over Tamil Nadu. International Journal of Current Research, 7(12), 23818-23820.

Dheebakaran, GA., S.Arulprasad, S.Kokilavani and S.Panneerselvam. 2017. Astrometeorology: An option to improve the accuracy of numerical daily rainfall forecast of Tamil Nadu. Journal of Agrometeorology 19:205-211.

Dillehay, B.L., Calvin, D.D., Roth, G.W., Hyde, J.A., Kuldau, G.A., Kratochvil, R.J., Russo, J.M and D.G., Voight. 2005. Verification of a European corn borer (Lepidoptera: Crambidae) loss equation in the major corn production region of the Northeastern United States. Journal of Economic Entomology, 98:103–112.

Donatelli, M., Magarey, R. D., Bregaglio, S., Willocquet, L., Whish, J. P.M and S. Savary. 2017. Modelling the impacts of pests and diseases on agricultural systems. Agricultural Systems, 155: 213–24. doi.org/10.1016/j.agsy.2017.01.019.

Dong, Q., Chen, X., Chen, J., Zhang, C., Liu, L., Cao, X., Zang, Y., Zhu, × and X. Cui. 2020. Mapping Winter Wheat in North China Using Sentinel 2A/B Data: A Method Based on Phenology-Time Weighted Dynamic Time Warping. Remote Sensing. 12, 1274; doi:10.3390/rs12081274.

Duncan, W.G. 1972. SIMCOT: a simulation of cotton growth and yield. In: Murphy, C.M. (Ed.), Proceedings of a Workshop for Modeling Tree Growth. Duke University, Durham, North Carolina, pp. 115–118.

Duncan, W.G., Loomis, R.S., Williams, W.A and R. Hanau. 1967. A model for simulating photosynthesis in plant communities. Hilgardia, 38 (4): 181–205.

Dwivedi, R.S. 1992. Monitoring and the study of the effects of image scale on delineation of salt-affected soils in the Indo-Gangetic plains. International Journal of Remote Sensing, 13, 1527-1536.

El-Asmar, H.M., Hereher, M.E and S.B. El Kafrawy. 2013. Surface area change detection of the Burullus Lagoon, North of the Nile delta, Egypt, using water indices: a remote sensing approach. The Egyptian Journal of Remote Sensing and Space Science, 16, 119–123.

Elliott, J., Deryng, D., Müller, C., Frieler, K., Konzmann, M., Gerten, D., Glotter, M., Flörke, M., Wada, Y., Best, N., Eisner, S., Fekete, B., Folberth, C., Foster, I., Gosling, S., Haddeland, I., Khabarov, N., Ludwig, F., Masaki, Y., Olin, S., Rosenzweig, C., Ruane, A., Satoh, Y., Schmid, E., Stacke, R., Tang, Q and D. Wisser. 2014a. Constraints and potentials of future irrigation water availability on agricultural production under climate change. Proceedings of the National Academy of Sciences of the United States of America, 111 (9):3239–3244.

Elliott, J., Kelly, D., Chryssanthacopoulos, J., Glotter, M., Jhunjhnuwala, K., Best, N., Wilde, M., and I. Foster. 2014b. The parallel system for integrating impact models and sectors (pSIMS). Environmental Modelling & Software, 62:509–516.

Ennaffah, B., Bouslim, F., Benkirane, R., Ouazzani Touhami, A and A. Douira. 1997. Helminthosporium spiciferum, foliar parasite of rice in Morocco. Agronomie, 17, 299-300.

Esker, P.D., Savary, S and N. McRoberts. 2012. Crop loss analysis and global food supply: focusing now on required harvests. CAB Reviews: perspectives in Agriculture, Veterinary Science, Nutrition and Natural Resources 052. CAB Reviews, 2012, pp. 1–147.

Evans, L.T. 1993. Crop Evolution, Adaptation and Yield. Cambridge University Press, Cambridge.

FAO, 1996. Guidelines: Agro-ecological zoning. Food and Agricultural Organisation, Soils Bulletin, Rome, Italy.

FAO, 2013. Greenhouse gas emissions from ruminant supply chains. A Global Life Cycle Assessment. Food and Agriculture Organisation of the United Nations, Rome, Italy.

FAO, World Agriculture Towards 2030/2050 (FAO, Rome, Italy, 2003).

FAO, World Agriculture Towards 2030/2050 (FAO, Rome, Italy, 2006).

FAO. 2010. Global action on climate change in agriculture: Linkages to food security, markets and trade policies in developing countries. Trade and Markets Division, Food and Agriculture Organization of the United Nations.

FAO. 2012. World Agriculture Towards 2030/2050: The 2012 revision. Agricultural Development Economic Division.

Farquhar, G.D., von Caemmerer, S and J.A. Berry. 1980. A biochemical model of photosynthetic CO_2 assimilation in leaves of C_3 species. Planta, 149:78–90.

Fischer, G., Shah, M and van H. Valthuizen. 2002. Climate change and agricultural vulnerability. International institute for applied systems analysis. Laxenburg, Austria.

Fischer, G.K.F., Rosenzweig, C and M.L. Parry. 1995. Impacts of potential climate change on global and regional food production and vulnerability. In: Downing, T.E. (Ed.), Climate Change and World Food Security. Springer-Verlag, New York.

Flohn, H. 1977. Climate and energy: A scenario to a 21st century problem. Climatic Change, 1(1), 5-20.

Fox Strand, J. 2000. Some agrometeorological aspects of pest and disease management for the 21st century. Agricultural and Forest Meteorology, 103, 73-82.

Fritz, S., See, L., Justice, C., Becker-Reshef, I., Bydekerke, L., Cumani, R and P. Defourney. 2013. The need for improvement maps of global cropland. Eos Transcations, Am. Geophys. Union, 94 (3): 31. doi.org/10.1002/2013EO030005.

García-Vila, M and E. Fereres. 2012. Combining the Simulation Crop Model AquaCrop with an Economic Model for the Optimization of Irrigation Management at Farm Level. European Journal of Agronomy 36 (1): 21–31. doi.org/10.1016/j.eja.2011.08.003

Geethalakshmi, V., Lakshmanan, A., Rajalakshmi, D., Jagannathan, R., Gummidi Sridhar, A., Ramaraj, P., Bhuvaneswari, K., Gurusamy, L and R. Anbhazhagan. 2011. Climate change impact assessment and adaptation strategies to sustain rice production in Cauvery Basin of Tamil Nadu. Current science, 101 (3): 56-65. http://www.ias.ac.in/currsci/10aug2011/342.pdf.

Geethalakshmi, V., Ramesh, T., Palamuthirsolai, A and A. Lakshmanan. 2011. Agronomic evaluation on rice cultivation systems for water and grain productivity. Archives of Agronomy and Soil Science, 57(2): 159-166.

Geethalakshmi, V., Selvaraju, R., Vasanthi, C and T.N. Balasubramanian. 2000. Guidelines for Agromet Observatory. Department of Agricultural Meteorology, Soil and Crop Management Studies, Tamil Nadu Agricultural University, Coimbatore.P.22(For Official Circulation).

Godfray, H.C.J., Beddington, J.R., Crute, I.R., Haddad, L., Lawrence, D., Muir, J.F., Pretty, J., Robinson, S., Thomas, S.M and C. Toulmin. 2010. Food Security: The Challenge of Feeding 9 Billion People. Science, 327:812-818.

Gomez, S. 2014. Infection and spread of Peronospora sparsa on Rosa sp. (Berk.) - a microscopic and a thermographic approach. Dissertation, University of Bonn, Germany.

Goswami, B.N. 1998. Interannual variations of Indian summer monsoon in a GCM: External conditions versus internal feedback. Journal of Climate, 11, 507–522.

Goudie, Andrew. 2006. Global Warming and Fluvial Geomorphology. Geomorphology. (79). 3-4. 384-394.

Gowtham, R., Panneerselvam, S and V. Geethalakshmi. 2017. Rainfed maize productivity under changing climate in Tamil Nadu. Editorial Board, 212.

Gutierrez, A.P., Mills, N.J., Schreiber, S.J and C.K., Ellis. 1994. A physiologically based tritrophic perspective on bottom-up-top-down regulation of populations. Ecology, 75: 2227–2242.

Haldar, S., Honnaiah and G. Govindaraj. 2012. System of Rice Intensification (SRI) method of rice cultivation in West Bengal (India): An Economic analysis. International Association of Agricultural Economists (IAAE) Triennial Conference, Foz do Iguaçu, Brazil, 18-24 August 2012.

Harrison, B.A., Jupp, D. L. B., Ibrahim, A. A. and J.F. Angus. 1984. The use of Landsat data for monitoring growth of irrigated crops. Third Australasian Remote Sensing Conference, Queensland, 36-43.

Hassell, M.P. 1986. Parasitoids and population regulation. In: Waage, J.K and D. Greathead (Eds.), Insect Parasitoids — 13th Symposium of the Royal Entomolgical Society of London. Blackwell Scientific Publications, Oxford, pp. 201–224.

Havlik, P., Valin, H., Herrero, M., Obersteiner, M., Schmid, E., Rufino, M.C., Mosnier, A., Thornton, P.K., Bottcher, H., Conant, R.T., Frank, S., Fritz, S., Fuss, S., Kraxner, F and A. Notenbaert. 2014. Climate changemitigation through livestock systemtransitions. Proceedings of the National Academy of Sciences of the United States of America, 111: 3709–3714.

Heady, E.O and J.L. Dillon. 1964. Agricultural Production Functions. Ames Iowa State University Press.

Heady, E.O. 1957. An econometric investigation of agricultural production functions. Econometrica, 25(2):249–268.

Heng, L.K., Asseng, S., Mejahed, K and M. Rusan. 2007. Optimizing wheat productivity in two rain-fed environments of the West Asia-North Africa region using a simulation model. European journal of Agronomy, 26:121–129.

Herrero, M., Havlík, P., Valin, H., Notenbaert, A., Rufino, M.C., Thornton, P.K., Blümmel, M., Weiss, F., Grace, D and M. Obersteiner. 2013. Biomass use, production, feed efficiencies, and greenhouse gas emissions from global livestock systems. Proceedings of the National Academy of Sciences of the United States of America, 110 (52):20888–20893.

Higgins.G.M and A.H. Hassam.1981. The FAO agro-ecological zone approach to determination of land potential. *Pedologie* xxxi 2: 146-148.

Hillnhutter, C., Mahlein, A.K., Sikora, R. A and E.C. Oerke. 2011. Remote sensing to detect plant stress induced by Heterodera schachtii and Rhizoctonia solani in sugar beet fields. Field Crops Research, 122:70-77.

Hoffmann, H., Zhao, G., van Bussel, L.G.J., Enders, A., Specka, X., Sosa, C., Yeluripati, J., Tao, F., Constantin, J., Raynal, H., Teixeira, E., Grosz, B., Doro, L., Zhao, Z., Wang, E., Nendel, C., Kersebaum, K., Haas, E., Kiese, R and F. Ewert. 2015. Variability of effects of spatial climate data aggregation on regional yield simulation by crop models. Climate Research, 65: 53–69.

Holling, CS 1973. Resilience and Stability of Ecological Systems (PDF). Annual Review of Ecology and Systematics. 4: 1–23. doi:10.1146/annurev.es.04.110173.000245.

Hong, J.H., Su, Z.L.T and E.H.C. Lu. 2020. Spatial Perspectives toward the Recommendation of Remote Sensing Images Using the INDEX Indicator, Based on Principal Component Analysis. 2020. Remote Sensing. 12, 1277; doi:10.3390/rs12081277.

Hoogenboom, G., Jones, J.W., Wilkens, P.W., Porter, C.H., Boote, K.J., Hunt, L.A., Singh, U., Lizaso, J.L., White, J.W., Uryasev, O., Royce, F.S., Ogoshi, R., Gijsman, A.J., Tsuji, G.Y and J. Koo. 2012. Decision Support System for Agrotechnology Transfer (DSSAT) Version 4.5 [CD-ROM]. University of Hawaii, Honolulu, Hawaii.

Hoogenboom, G., Jones, J.W., Wilkens, P.W., Porter, C.H., Boote, K.J and L.A. Hunt. 2010. Decision Support System for Agrotechnology Transfer (DSSAT) Version 4.5. University of Hawai, Honolulu (CD ROM).

Hota, S and K. Arulmozhiselvan. 2016. Effect of nutriseed pack placement on growth, yield and nutrient uptake of tomato (*Lycopersiconesculentum*) under drip irrigation. Journal of Applied and Natural Science 8(4): 2128-2132.

Huang, W., Lamb, D.W., Niu, Z., Zhang, Y., Liu, L and J. Wang. 2007. Identification of yellow rust in wheat using in-situ spectral reflectance measurements and airborne hyperspectral imaging. Precision Agriculture, 8: 187-197.

Hume, I.H., Beecher, H.G and B.W. Dunn. 1999. EM to estimate groundwater recharge from rice growing. Rural Industries Research and Development Corporation RIRDC RIRDC-98, pp. 37.

Humphreys, E., Kukal, S.S., Christen, E.W., Hira, G.S., Balwinder-Singh, Yadav, S and R.K. Sharma. 2010. Halting the groundwater decline in north-west india-which crop technologies will be winners? Advances in Agronomy., 109:155–217.

Hundal, S.S and P. Kaur. 1996. Application of CERES-Wheat model to yield prediction in the irrigated Plains of the Indian Punjab. J Agric Sci (Cambridge) 129:13–18.

Hydrologic Engineering Center (HEC). 2002. River Analysis System: Hydraulic Reference Manual; US Army Corps of Engineers Hydrologic Engineering Center: Davis, CA, USA, 2002

Ibraheem Kutty, 2013. Role of climate in reproductive pattern of small ruminants in the humid tropics. P.194-202. IN. Fundamentals of Livestock Meteorology. Vol. II (edts.G.S.L.H.V. Prasada Rao and G. Girish Varma). Centre for Animal Adaptation to Environment and Climate Change Studies, KVASU. Mannuthy, Thrissur, Kerala, India,295p.

IBSNAT, 1984. Experimental design and data collection procedures for IBSNAT. The Minimum Data Set for Systems Analysis and Crop Simulation, first ed. IBSNAT Technical Report no 1. http://pdf.usaid.gov/pdf_docs/PNABK919.pdf

India Meteorological Department,1987. Instruction to Observers at the Surface Observatories. Part.1. New Delhi. The Director General of Meteorology, New Delhi.

Inoue, Y., Moran, M.S and T. Horie. 1998. Analysis of spectral measurements in paddy field for predicting rice growth and yield based on a simple crop simulation model. Plant Production Science, 1, 269-279.

IPCC, 1990. In: Houghton, J.T., Jenkins, G.J and J.J. Ephraums (Eds.), Climate change: the IPCC scientific assessment. Cambridge University Press, Cambridge, Great Britain, New York, NY, USA, p. 410.

IPCC. 2001. Climate change impacts, adaptation and vulnerability. IPCC Report of the working group II. Cambridge, UK.

IPCC. 2007. Summary for Policy Makers. In:ClimateChange 2007: The Physical Science Basis. Contributing of Working Group It other Fourth assessment Report of the Intergovermental Panel on climate Change. Solomon S., Qin D., Manning M., Chen Z., Marquis M., Averyt K.B., TignorM., and Miller H.L. (eds). CambridgeUniversity Press, UK.

Isard, S.A., Russo, J.M., Magarey, R.D., Golod, J and J.R. VanKirk. 2015. Integrated pest information platform for extension and education (iPiPE): progress through sharing. Integrated Pest Management Reviews., 6:15.

Islam, Z and A.N.M.R. Karim. 1997. Whiteheads associated with stem borer infestation in modern rice varieties: an attempt to resolve the dilemma of yield losses. Crop Protection, 16, 303-311.

Iwata. F., 1984. Heat unit concept of crop maturity. In physiological aspects of dry land farming. Gupta.U.S. (edt.) Oxford and IBH. New Delhi. Pp 351-370

Iyengar, R.N. 2004. Description of rainfall variability in Brihat Samhita of Varaha-mihira Current Science, VOL. 87, NO. 4, 25.

Janssen, S., Porter, C.H., Moore, A.D., Athanasiadis, I.N., Foster, I., Jones, J.W and J.M. Antle. 2017. Towards a new generation of agricultural system models, data, and knowledge products: building an open web-based approach to agricultural data, system modelling and decision support. Agricultural Systems, 155:200–212.

Jeevananda Reddy, S.1983. Agroclimatic classification of the SAT 1. Method for the computation of classificatory variables. Agricultural Meteorology .30: 185-200. Elsevier Science Publisher, The Netherlands.

Jensen, J.R. 1986. Introductory digital image processing: A remote sensing perspective, Prentice-Hall, Englewood Cliffs.

Jeyaseelan, A.T and K. Chandrasekar. 2002. Satellite based identification for updation of Drought prone area in India. ISPRS-TC-VII, International Symposium on Resource and Environmental Monitoring, Hyderabad.

John Abraham. 2017. Micro climate and Animal husbandry. Pp. 55-76. IN Livestock Meteorology (edts. G.S.L.H.V. Prasada Rao., Girish Varma and V. Beena.,) New India Publishing Agency, New Delhi.P.522.

Jones, J.W. 1993. Decision support systems for agricultural development. In: Penning de Vries, F., Teng, P. and K. Metselaar (Eds.), Systems Approaches for Agricultural Development. Kluwer Academic Press, Boston, pp. 459–471.

Jones, J.W., Antle, J.M., Basso, B., Boote, K.J. Conant, R.T., Foster, I., Godfray, H.C.J., Herrero, M., Howitt, R.E., Janssen, S., Keating, B.A., Carpena, R.M., Porter, C.H., C. Rosenzweig and T.R.Wheeler. 2016. Brief history of agricultural systems modelling. Agricultural Systems, 155: 240–254.

Jones, J.W., Hoogenboom, G., Porter, C.H., Boote, K.J., Batchelor, W.D., Hunt, L.A., Wilkens, P.W., Singh, U., Gijsman, A.J and J.T. Ritchie. 2003. The DSSAT cropping system model. European Journal of Agronomy, 8:235–265.

Joseph, P.V., Sooraj, K.P and C.K. Rajan. 2003. Conditions leading to monsoon over Kerala and associated Hadley cell. Mausam, 54(1): 155-16

Joy Deep Mukherjee., Anu Radha Sharma., V.K. Sehgal., Das, D. K and P. Krishnan. 2020. Seasonal variation of energy fluxes over Maize -Wheat cropping system in North-west semi-arid region of India. Paper presented in National Seminar on Agrometeorological Interventions for enhancing farmers income (Agmet-2020) 20-22 Jan. 2020. Kerala Agricultural University, Thrissur, Kerala. Abstract Compendium. Theme: 0.11.1. P 145.

Kaladevi, R.1998. Studies on the planting dates, methods, levels and times of phosphorus and potassium application on the growth and yield of Thaladi/*rabi season* rice (Oct-March). M.Sc. (Ag) thesis submitted to TNAU, Coimbatore.

Kalra, N and P.K. Aggarwal. 1996. Evaluating the growth response for wheat under varying inputs and changing climate options using wheat growth simulator–WTGROWS. In: Abrol YP, Gadgil S, Pant GB (eds), Climate Variability and Agriculture, Narosa Publishing House, New Delhi, pp 320–338.

Kalsi, S.R and D.K. Mishra. 1983. On some meteorological aspects of snowfall over Himalayas and observed by satellite. Proc. of the First National Symposium on Seasonal Snow Cover, 28-30 April, 1983, New Delhi, p125-132.

Kalsi, S.R., Rao, A.V.R.K., Misra, D.K., Jain R.K and V.R. Rao. 1996. Structural variability of tropical disturbances in the Bay of Bengal. In Advances in Tropical Meteorology, Ed. by R.K. Datta. Concept Publishing Company, New Delhi. 449-458.

Kanter, D. R., Musumba, M., Wood, S. L., Palm, C., Antle, J., Balvanera, P., & Thornton, P. 2018. Evaluating agricultural trade-offs in the age of sustainable development. Agricultural Systems, 163, 73-88.

Kanwar, J. 1972. Cropping patterns, scope and concept. In: Proceedings of symp. on cropping pattern in India, ICAR, New Delhi. Pp 11-32.

Karimi, M.M and K.H.M. Siddique. 1992. The relationship between grain and yield of spring wheat and rainfall distribution in Western Australia using regression models. Iran Agricultural Research, 11, 1-23.

Kaur, P., Sandhu, S.S., Singh, H., Gill, K.K., Bal, S.K., Bala, A and A. Singh. 2012. Weather based decisions for wheat cultivation in Punjab. In: All India Coordinated Research Project on Agrometeorology, Department of Agricultural Meteorology, Punjab Agricultural University, Ludhiana.

Keating, B.A., Carberry, P.S., Hammer, G.L., Probert, M.E., Robertson, M.J., Holzworth, D., Huth, N.I., Hargreaves, J.N.G., Meinke, H., Hochman, Z., McLean, G., Verburg, K., Snow, V., Dimes, J.P., Silburn, M., Wang, E., Brown, S., Bristow, K.L., Asseng, S., Chapman, S., McCown, R.L., Freebairn, D.M and C.J. Smith. 2003. An overview of APSIM, a model designed for farming systems simulation. European Journal of Agronomy, 18:267–288.

Khan, S.A., Kumar, S., Hussain, M.Z., N. Kalra. 2009. Climate change, climate variability and Indian agriculture: Impact vulnerability and adaptation starategies. In: Singh, S.N (eds). Climate change a crops. Springer-Verlag, Berlin, Germany, 19-38.

Kholová, J., Tharanya M., Sivasakthi K., Srikanth M., Rekha B., Graeme L. H., McLean, G., Deshpande, S.P., Hash, C.T., Craufurd, P.Q and V. Vadez. 2014. Modelling the Effect of Plant Water Use Traits on Yield and Stay-Green Expression in Sorghum. Functional Plant Biology, 41(11): 1019. doi.org/10.1071/FP13355.

Kimball, B.A. 1983. Carbon dioxide and agricultural yield: An assemblage and analysis of 430 prior observations. Agron. J. **75**: 779-788.

Kimball, B.A., 2010. Lessons from FACE: CO_2 effects and interactions with water, nitrogen, and temperature. In: Hillel, D., Rosenzweig, C. (Eds.), Handbook of Climate Change and Agroecosystems: Impacts, Adaptation, and Mitigation. Imperial College Press, London UK, pp. 87–107.

Kiniry. J.R., Blanchet.R., Williams. J.R., Texoer. V., Jones. C.A and M. Cabelguenne. 1992. Sunflower simulation using the DPIC and almanac modules. Field Crop Research. 30: 403-423.

Konanz, S., Kocsa´nyi, L and C. Buschmann. 2014. Advanced multi-color fluorescence imaging system for detection of biotic and abiotic stresses in leaves. Agriculture, 4:79-95

Konikow, Leonard and Eloise Kendy. 2005. Groundwater Depletion: A Global Problem. Hydrogeology (13). 317-320.

Koning, N and M.K. van Ittersum. 2009. Will the world have enough to eat? Current Opinion in Environmental Sustainability, 1:77–82.

Krishna Rao. 2013. Role of weather in Animal husbandry and livestock production and management. P. 99-142 IN. Fundamentals of Livestock Meteorology. Vol. I (edts.G.S.L.H.V.

Prasada Rao and G. Girish Varma). Centre for Animal Adaptation to Environment and Climate Change Studies, KVASU.Mannuthy, Thrissur, Kerala, India,312p.

Krishnakumar, S.K., Patwardhan, A., Kulkarni, K., Kamala, K., Koteswara Rao and R. Jones. 2010. Simulated projections for summer monsoon climate over India by a high-resolution regional climate model (PRECIS). Curr. Sci. 2011, 101, 312–325.

Krupnik, T.S., Schulthess, U., Ahmed, Z.U., and McDonald, A.J., 2017. Sustainable crop intensification through surface water irrigation in Bangladesh. Land Use Policy, 60 (2017), pp. 206-222.

Kumar, P., Sharma, L.K., Pandey, P.C., Sinha, S and M.S. Nathawat. 2013. Geospatial strategy for tropical forest-wildlife reserve biomass estimation. IEEE Journal of Selected Topics in Applied Earth Observations and Remote Sensing, 6 (2), 917–923.

Kumar, S., Bhatt, B.P., Dey, A., Shivani, U., Kumar, M.D., Idris, J.S., Mishra and S. Kumar. 2018. Integrated farming system in India: Current status, scope and future prospects in changing agricultural scenario. Indian Journal of Agricultural Sciences 88 (11): 1661–75.

Kurosu, T., Fujita, M and K. Chiba. 1997. The identification of rice fields using multi-temporal ERS-1 C band SAR data. International Journal of Remote Sensing, 18, 2953-2965.

Kushka, M., Wahabzada, M., Leucker, M., Dehne, H.W., Kersting, K., Oerke, E.C., Steiner, U and A.K. Mahlein. 2015. Hyperspectral phenotyping on microscopic scale–towards automated characterization of plant-pathogen interactions. Plant Methods, 11:28.

Lal, M., Cubasch, U., Voss, R and J. Waszkewitz. 1995. Effect of transient increases in greenhouse gases and sulphate aerosols on monsoon climate. Current Sci., 69(9): 752 - 763.

Lal, M., Singh, K.K., Srinivasan, G., Rathore, L.S and D. Naidu. 1999. Growth and yield response of soybean in Madhya Pradesh, India to climate variability and change. Agric Forest Meterol. 93:53–70.

Le Hegarat-Mascle, S., Quesney, A and D. Vidal-Madjar. 2000. Land cover discrimination from multitemporal ERS images and multispectral Landsat images: a study case in an agricultural area in France. International Journal of Remote Sensing, 21, 435-456.

Lenka, D.1998. Climate, weather and crops in India. Kalyani Publishers, New Delhi.P490: 287-374.

Li, T., Hasegawa, T., Yin, X., Zhu, Y., Boote, K., Adam, M., Bregaglio, S., Buis, S., Confalonieri, R., Fumoto, T., Gaydon, D., Manuel Marcaida, I.I.I., Nakagawa, H., Oriol, P., Ruane, A.C., Ruget, F., Singh, B., Singh, U., Tang, L., Tao, F.,Wilkens, P., Yoshida, H., Zhang, Z and B. Bouman. 2015. Uncertainties in predicting rice yield by current crop models under a wide range of climatic conditions. Global Change Biology, 21 (3):1328–1341.

Linker, R and I. Kisekka. 2017. Model-based deficit irrigation of maize in Kansas. American Society of Agricultural and Biological Engineers, 60(6): 2011-2022.

Liu, W., Yang, H., Liu, J., Azevedo, L.B., Wang, X., Xu, Z., Abbaspour, K.C and R. Schulin. 2016. Global Assessment of Nitrogen Losses and Trade-Offs with Yields from Major Crop Cultivations. Science of the Total Environment, 572: 526–37. doi.org/10.1016/j.scitotenv.2016.08.093.

Long, S.P., Ainsworth, E.A., Leakey, A.D.B., Nösberger, J and D.R. Ort. 2006. Food for thought: lower-than-expected crop yield stimulation with rising CO_2 concentrations. Science, 312:1918–1921.

Loomis, R.S., Rabbinge, R and E. Ng. 1979. Explanatory models in crop physiology. Annual Review of Plant Physiology, 30:339–367.

Lunt, T., Jones, A. W., Mulhern, W. S., Lezaks, D. P., & Jahn, M. M. (2016). Vulnerabilities to agricultural production shocks: An extreme, plausible scenario for assessment of risk for the insurance sector. Climate Risk Management, 13, 1-9.

Maas, S.J. 1988. Use of remotely-sensed information in agricultural crop growth models. Ecological Modelling, 41, 247-268.

Madden, L.V., Hughes, G and F. Van Den Bosh. 2007. The Study of Plant Disease Epidemics. APS Press, Saint Paul, USA.

Magarey, R.D., Sutton, T.B. and C.L. Thayer. 2005. A simple generic infection model for foliar fungal plant pathogens. Phytopathology, 95:92–100

Magarey, R.D., Travis, J.W., Russo, J.M., Seem, R.W and P.A. Magarey. 2002. Decision support systems: quenching the thirst. Plant Disease, 86:4–13.

Mahato, A. 2014. Climate change and its impacts on agriculture. International Journal of Scientific and Research Publications. 4 (4): 1-6.

Mahlein, A.K., Steiner, U., Dehne, H.W and E.C. Oerke. 2010. Spectral signatures of sugar beet leaves for the detection and differentiation of diseases. Precision Agriculture, 11:413-431.

Mananze, S., Pôças, I and M. Cunha. 2020. Mapping and Assessing the Dynamics of Shifting Agricultural Landscapes Using Google Earth Engine Cloud Computing, a Case Study in Mozambique. Remote Sensing. 12, 1279; doi:10.3390/rs12081279.

Mandal, K.G., Thakur, A.K and S. Mohanty. 2019. Paired-row planting and furrow irrigation increased light interception, pod yield and water use efficiency of groundnut in a hot sub-humid climate. Agricultural water management, 213:968-977.

Mandal, N. 1998. Simulating the impact of climatic variability and climate change on growth and yield of chickpea and pigenonpea crops. M.Sc Thesis, Division of Environmental Sciences, IARI, New Delhi.

Mandal, D.K., C.Mandal and S.K.Singh,2016. Delineating agro-ecological regions. e publication; ICAR-NBSS&LUP Technologies.P8. (http://www.nbsslus.in/assets/uploads/clinks/Delineating Agroecological regions.Pdf.

Mathews, R., Stephens, W., Hess, T., Middleton, T., Middleton, T and A. Graves. 2002. Application of crop/soil simulation model in tropical agricultural systems. Advances in Agronomy, 76:31–124.

Mathur, Vivek N., et al., 2014. Experiences of Host Communities with Carbon Market Projects: Towards Multi-Level Climate Justice. Climate Policy 14.1: 42–62. Web.

Matsui, T., Omasa, K and T. Horie. 1997. High temperature-induced spikelet sterility of japonica rice at flowering in relation to air temperature, humidity and wind velocity condition. Japanese Journal of Crop Science 66: 449-455.

Maurya, V.P., Gyanendra Singh., Sejian, V and Mihir Sarkar.2013. Temperature effects on milk composition and production in livestock. 183-192. IN. Fundamentals of Livestock Meteorology. Vol. I (edts.G.S.L.H.V. Prasada Rao and G. Girish Varma). Centre for Animal Adaptation to Environment and Climate Change Studies, KVASU.Mannuthy, Thrissur, Kerala, India,312p.

McCloy, K.R., Smith, F.R and M.R. Robinson. 1987. Monitoring rice areas using Landsat MSS data. International Journal of Remote Sensing, 8, 741-749.

McCown, R.L., Hammer, G.L., Hargreaves, J.N.G., Holzworth, D.P and D.M. Freebairn. 1996. APSIM: a novel software system for model development, model testing and simulation in agricultural systems research. Agricultural Systems, 50:255–271

Medhavy, T.T., Sharma, T., Dubey, R.P., Hooda, R.S., Mothikumar, K.E., Yadav, M., Manchanda, M.L., Ruhal, D.S., Khera, A.P and S.D. Jarwal. 1993. Crop classification accuracy as influenced by training strategy, data transformation and spatial resolution of data. Journal of the Indian Society of Remote Sensing, 21, 21-28.

Mishra S., Chaudhury, S.S., Swain, S and T. Ray. 2009. Multiple cropping system for conservation and sustainable use in Jeypore Tract of Orissa, India.13 (1): 39-51.

Mishra, D.K., Kalsi, S.R and R.K. Jain. 1988. The estimation of heavy rainfall using INSAT-1B. Mausam, 39(4): 19-40.

Mohamed, A. B., Van Duivenbooden, N and S. Abdoussallam. 2002. Impact of climate change on agricultural production in the Sahel–Part 1. Methodological approach and case study for millet in Niger. Climatic Change, 54(3), 327-348.

Monfreda, C., Ramankutty, N and J.A. Foley. 2008. Farming the planet 2: geographic distribution of crop areas, yields, physiological types, and net primary production in the year 2000. Global Biogeochemical Cycles, 22 (1), GB1022. doi.org/10.1029/2007GB002947.

Monisha, M, 2017. Effect on season on broiler production. M.Sc. thesis submitted to TNAU, Coimbatore.3

Moody, A. 1997. Using landscape spatial relationships to improve estimates of land-cover area from coarse resolution remote sensing. Remote Sensing of Environment 64, 202-220.

Moshou, D., Bravo, C., West, J., Wahlen, S., McCartney, A and H. Ramon. 2004. Automatic detection of 'yellow rust' in wheat using reflectance measurements and neural networks. Computers and Electronics in Agriculture, 44:173-188.

Mota Babiloni, A. 2016. Analysis of low global warming potential fluoride working fluids in vapour compression systems. Experimental evaluation of commercial refrigeration alternatives (Doctoral dissertation).

Moya, T.B., Ziska, L.H., Namuco, O.S and D. Olszyk. 1998. Growth dynamics and genotypic variation in tropical, field-grown paddy rice (Oryza sativa L.) in response to increasing carbon dioxide and temperature. Global Change Biol. 4:645-656.

Murdoch, W.W. 1994. Population regulation in theory and practice — the Robert H Macarthur award lecture presented August 1991 in San Antonio, Texas, USA. Ecology, 75, 271–287.

Murthy, C.S., Thiruvengadachari, S., Raju, P.V and S. Jonna. 1996. Improved ground sampling and crop yield estimation using satellite data. International Journal of Remote Sensing, 17, 945-956.

Nagarajan, S and L.M, Joshi. 1978. Epidemiology of brown and yellow rusts of wheat over northern india. II. Associated meteorological conditions. Please disease Reporter, 62: 186-188.

Nagavallemma, K.P., Wani, S.P., Stephane, L., Padmaja, V.V., Vineela, C., BabuRao, M and K.L. Sahrawat. 2004. Vermicomposting: Recycling wastes into valuable organic fertilizer. Global Theme on Agrecosystems Report no. 8. Patancheru 502 324, Andhra Pradesh, India: International Crops Research Institute for the Semi-Arid Tropics. 20 pp.

Nageswara Rao, P.P and V.R. Rao. 1984. An approach for agricultural drought monitoring using NOAA/AVHRR and Landsat imagery. Proc. IGARSS '84 symposium, Strasbourg, 1984. Vol. 1, (ESA SP-215; distributed ESTEC, Noordwijk). 225-229.

Nain, A.S and K. Kersebaum. 2007. Calibration and validation of CERES-Wheat model for simulating water and nutrients in Germany. In: Kersebaum, K.Ch. (Ed.), Modeling Water and Nutrient Dynamics in Soil-crop-systems. Springer, pp. 161–181.

Nanda, T., Sahoob, B., Beriaa, H and C.A. Chatterje. 2016. A wavelet-based non-linear autoregressive with exogenous inputs (WNARX) dynamic neural network model for real-time flood forecasting using satellite-based rainfall products. Journal of Hydrology, 539, 57–73

NASA. 2020. Global Annual Mean Surface Air Temperature Change. Retrieved 23 February 2020. https://data.giss.nasa.gov/gistemp/graphs_v4/.

NATCOM. 2004. India's Initial National Communication to the United Nations Convention on Climate Change, Ministry of Environment and Forests, Government of India.

Nearing, M.A., Jetten, V., Baffaut, C., Cerdan, O., Couturier, A., Hernandez, M., Le Bissonnals, Y., Nichols, M.H., Nunes, J.P., Renschler, C.S., Souchere, V and K. Van Oost. 2005. Modeling Response of Soil Erosion and Runoff to Changes in Precipitation and Cover. Catena (61). 131–154.

Nelson, G.C., Rosegrant, M.W., Koo, J., Robertson, R., Sulser, T., Zhu, T., Ringler, C., Msangi, S., Palazzo, A., Batka, M., Magalhaes, M., Valmonte-Santos, R., Ewing, M and D. Lee. 2009. Climate change. Impact on Agriculture and Costs of Adaptation. International Food Policy Research Institute. International Food Policy Research Institute, Washington, D.C.

Nelson, G.C., Valin, H., Sands, R.D., Havlik, P., Ahammad, H., Deryng, D., Elliott, J., Fujimori, S., Hasegawa, T., Heyhoe, E., Kyle, P., Von Lampe, M., Lotze-Campen, H., d'Croz, D.M., van Meijl, H., van der Mensbrugghe, D., Muller, C., Popp, A., Roberston, R., Robinson, S., Schmid, E., Schmitz, C., Tabeu, A and D. Willenbockel. 2013. Climate change effects on agriculture: economic responses to biophysical shocks. Proceedings of the National Academy of Sciences of the United States of America, 111 (9): 3274–3279. doi.org/10.1073/pnas.1222465110.

Neumann, M., Hallau, L., Klatt, B., Kersting, K and C. Bauckhage. 2014. Erosion band features for cell phone image based plant disease classification. Pages 3315-3320 in: Proceeding of the 22nd International Conference on Pattern Recognition (ICPR), Stockholm, Sweden, 24-28 August 2014.

Newman, J.A., Gibson, D.J., Parsons, A.J and J.H.M. Thornley. 2003. How predictable are aphid population responses to elevated CO_2. Journal of Animal Ecology, 52:556–566.

Nguyen, P., Thorstensen, A., Sorooshian, S., Hsu, K and A. AghaKouchak. 2015. Flood forecasting and inundation mapping using HiResFlood-UCI and near-real-time satellite precipitation data: The 2008 Iowa Flood. Journal of Hydrometeorology, 16, 1171–1183

Oerke, E.C., Frohling, P and U. Steiner. 2011. Thermographic assessment of scab disease on apple leaves. Precision Agriculture, 12:699-715.

Oerke, E.C., Steiner, U., Dehne, H.W and M. Lindenthal. 2006. Thermal imaging of cucumber leaves affected by downy mildew and environmental conditions. Journal of experimental Botany, 57:2121-2132.

Oki, Taikan and Shinjiro Kanae. 2006. Global Hydrological Cycles and World Water Resources. Science (313): 5790. 1068-1072.

Olmstead, S. M and R. N. Stavins 2012. Three key elements of a post-2012 international climate policy architecture. Review of Environmental Economics and Policy, 6(1), 65-85.

Osborne, T.M., Slingo, J.M., Lawrence, D.M and T.R. Wheeler. 2009. Examining the interaction of growing crops with local climate using a coupled crop-climate model. Journal of Climate, 22:1393–1411.

Palosuo, T., Kersebaum, K.C., Angulo, C., Hlavinka, P., Moriondo, M., Olesen, J.E., Patil, R.H., Ruget, F., Rumbaur, C and J. Takac. 2011. Simulation of winter wheat yield and its variability in different climates of Europe: a comparison of eight crop growth models. European Journal of Agronomy, 35:103e114.

Pandey, P.C., Sharma, L.K and M.S. Nathawat. 2012. Geospatial strategy for sustainable management of municipal solid waste for growing urban environment. Environmental Monitoring and Assessment, 184 (4), 2419–2431.

Panigrahy, S., Manjunath, K.R., Chakraborty, M., Kundu, N and J.S. Parihar. 1999. Evaluation of RADARSAT standard beam data for identification of potato and ice crops in India. ISPRS Journal of Photogrammetry and Remote Sensing, 54, 254262.

Parihar, J.S and M.P. Oza. 2006. FASAL: an integrated approach for crop assessment and production forecasting. Proc. of SPIE Vol. 6411641101 Agriculture and Hydrology

Applications of Remote Sensing, edited by R.J. Kuligowski, J.S. Parihar and G. Saito (doi: 10.1117/12.713157).

Parry, M.L., Rosenzweig, C., Iglesias, A., Livermore, M and G. Fischer. 2004. Effects of climate change on global food production under SRES emissions and socio-economic scenarios. Global Environmental Change, 14:53–67.

Passel, J.V., Keersmaecker, W.D and B. Somers. 2020. Monitoring Woody Cover Dynamics in Tropical Dry Forest Ecosystems Using Sentinel-2 Satellite Imagery. Remote Sensing. 12, 1276; doi:10.3390/rs12081276.

Passioura, J.B. 1996. Simulation models: science, snake oil, education, or engineering. Agronomy Journal, 88 (5):690–694.

Pasupalak, S.2008. Impact of climate change over Orissa,147-161. In: Climate Change and Agriculture over India, (edits.) G.S.L.H.V. Prasad Rao., G.G.S.N. Rao., V.U.M. Rao and Y.S. Ramakrishna, KAU and AICRP on Agro -meteorology, Hyderabad. P.256.

Patel, N.R., Mandal, U.K and L.M. Pande. 2000. Agro-ecological zoning system. A Remote Sensing and GIS Perspective. Journal of Agrometeorology, 2 (1) : 1-13.

Patterson, D. T. 1995. Effects of environmental stress on weed/crop interactions. Weed Science, 483-490.

Pecetti, L and P.A. Hollington. 1997. Application of the CERES-Wheat simulation model to durum wheat in two diverse Mediterranean environments. European Journal of Agronomy, 6:125–139.

Penning de Vries, F.W.T., van Laar, H.H and M.J. Kropff (Eds.). 1991. Simulation and systems analysis for rice production (SARP) 1991. PUDOC, Waneningen, Netherlands, p. 369.

Persson, T., Kværnø, S and M. Höglind. 2015. Impact of soil type extrapolation on timothy grass yield under baseline and future climate conditions in southeastern Norway. Climate Research, 65: 71–86

Pimentel, D., Peshin, R. (Eds.), 2014. Use and Benefit of Pesticides in Agricultural Pest Control Integrated Pest Management, Pesticide Problems vol 3. Springer (2014).

Pinter Jr, P. J., Ritchie, J. C., Hatfield, J. L., & Hart, G. F. (2003). The Agricultural Research Service's Remote Sensing Program. Photogrammetric Engineering & Remote Sensing, 69(6), 615-618.

Pirttioja, N, Carter, T.R., Fronzek, S., Bindi, M., Hoffmann, H., Palosuo, T., Ruiz-Ramos, M., Tao, F., Trnka, M., Acutis, M., Asseng, S., Baranowski, P., Basso, B., Bodin, P., Buis, S., Cammarano, D., Deligios, P., Destain, M.F., Dumont, B and R.P. Rötter. 2015. Temperature and precipitation effects on wheat yield across a European transect: a crop model ensemble analysis using impact response surfaces. Climate Research, 65: 87–105

Polder, G., van der Heijden, G.W.A.M., van Doorn, J and T.A.H.M.C. Baltissen. 2014. Automatic detection of tulip breaking virus (TBV) in tulip fields using machine vision. Biosystems Engineering, 117:35-42.

Pooja, K., Kumar, M and J.S. Rawat. 2012. Application of remote sensing and GIS in land use and land cover change detection: a case study of Gagas Watershed, Kumaun Lesser Himalaya, India. Quest. 6 (2), 342–345.

Prasad, A., Kannan, A and V.L. Gleeja. 2017. Thermal indices and dairy cattle management. PP127-143. IN Livestock Meteorology (edits. G.S.L.H.V. Prasada Rao., Girish Varma and V. Beena.,) New India Publishing Agency, New Delhi.P.522.

Prasada Rao, G.S.L.H.V., Girsh Varma, G and V. Beena. (editors) 2017.Livestock Meteorology. New India Publishing Agency. New Delhi. India. 522.

Pratap Narain., Sharma, K.D., Rao, A.S., Singh, D.V., Mathar, B.K and Usha Rani Ahuja.2000. Strategy to combat drought and famine in the Indian arid zone, Technical bulletin 1. Central Arid Zone Research Institute, Jodhpur, p.65.

Puliti, S., Breidenbach, J and R. Astrup. 2020. Estimation of Forest Growing Stock Volume with UAV Laser Scanning Data: Can It Be Done without Field Data?. Remote Sensing. 12, 1245; doi:10.3390/rs12081245

Purdom, J.F.W. 2003. Local severe storms monitoring and prediction using satellite systems. Mausam, 54(1): 141-154.

Quarmby, N.A., Milnes, M., Hindle, T.L and Silleos, N. (1993). The use of multi-temporal NDVI measurements from AVHRR data for crop yield estimation and prediction. International Journal of Remote Sensing, 14, 199-210.

Radha Krishna Murthy, V. 2002. Basic Principles of Agricultural Meteorology. BSP BS Publications. Hyderabad. P 259:234-240

Raes, D., Steduto, P., Hsiao, T.C and E. Fereres. 2009. AquaCrop-The FAO crop model to simulate yield response to water: II. Main algorithms and software description. Agronomy Journal 101 438e447.

Rahman, A., Mojid, M.A and S. Banu. 2018. Climate change impact assessment on three major crops in the north-central region of Bangladesh using DSSAT. International Journal of Agricultural and Biological Engineering, 11(4):135-142.

Rainfall-Runoff-Inundation (RRI) Model. 2017. Available online: http://www.icharm.pwri.go.jp/ research/rri/rri_top.html (access on 26 September 2017).

Rajalakshmi, D., Jagannathan, R and V. Geethalakshmi. 2013. Uncertainty in seasonal climate projection over Tamil Nadu for 21st century. African Journal of Agricultural Research, 8(32), 4334-4344.

Rajesh Kumar., Ram Niwas and Yogesh Kumar. 2020. Study of microclimatic profiles and PAR interception in pigeon pea (*Cajanus cajan*) cultivars under different environments. Paper presented in National Seminar on Agrometeorological Interventions for enhancing farmers income (Agmet-2020) 20-22 Jan. 2020. Kerala Agricultural University, Thrissur, Kerala. Abstract Compendium. Theme: 0.11.4. P 147.

Rajiv Kumar Chaturvedi., Jaideep Joshi., Mathangi Jayaraman., Bala, G and N. H. Ravindranath. 2012. Multi-model climate change projec.tions for India under representative concentration pathways. Current science, vol. 103 (7): 1-12.

Rajput. R.P.1980. Response of soybean crop to climatic and soil environments. PhD thesis. P.G. School. IARI. New Delhi.

Raktim Jyoti Saikia., Neog, P and R.L. Deka.2020. Assessment of radiation use efficiency (RUE) of potato under different microclimate at UBVZ of Assam. Paper presented in National Seminar on Agrometeorological Interventions for enhancing farmers income (Agmet-2020) 20-22 Jan. 2020. Kerala Agricultural University, Thrissur, Kerala. Abstract Compendium. Theme: 0.11.22. P 159.

Raman., C.R.V.1974. Analysis of commencement of monsoon rains over Maharashtra State for agricultural planning.India Meteorological Department. Poona. IndiaP.216.

Ramankutty, N and J. A. Foley. 1999. Estimating historical changes in global land cover: croplands from 1700 to 1992. Global Biogeochemical Cycles, 13, 997–1027.

Ramankutty, N., Evan, A.T., Monfreda, C and J.A. Foley. 2008. Farming the planet: 1. Geographic distribution of global agricultural lands in the year 2000. Global Biogeochemical Cycles, 22: GB1003 (doi: 1010.1029/2007GB002952).

Rao, B., Sankar, T., Dwivedi, R., Thammappa, S., Venkataratnam, L., Sharma, R and S. Das. 1995. Spectral behaviour of salt-affected soils. International Journal of Remote Sensing, 16, 2125–2136.

Rao, B.V.R and M.R. Rao. 1996. Weather effects on pest. In: Abrol YP, Gadgil S, Pant GB (eds), Climate Variability and Agriculture, Narosa Publishing House, New Delhi, pp 281–296.

Rasche, L and R.A.J. Taylor. 2019. EPIC-GILSYM: Modelling Crop-Pest Insect Interactions and Management with a Novel Coupled Crop-Insect Model. Journal of Applied Ecology, 56 (8): 2045–56. doi.org/10.1111/1365-2664.13426.

Rasmussen, M.S. 1992. Assessment of millet yields and production in northern Burkina Faso using integrated NDVI from AVHRR. International Journal of Remote Sensing, 13, 3431-3442

Rasmussen, M.S. 1997. Operational yield forecast using AVHRR NDVI data: reduction of environmental and inter-annual variability. International Journal of Remote Sensing, 18, 1059-1077.

Rathika, K, Ga.Dheebakaran, S.Panneerselvam, Patel Santosh Ganapati and S.Kokilavani. 2019. Studying the Astrometeorological Relationship between Planet's Azimuth and Wind Speed Events, Madras Agric. J., 106:4-6,pages 330-334. (doi:10.29321/MAJ 2019.000268).

Rawat, J.S., Biswas, V and M. Kumar. 2013. Changes in land use/cover using geospatial techniques-A case study of Ramnagar town area, district Nainital, Uttarakhand, India. The Egyptian Journal of Remote Sensing and Space Science, 16, 111–117.

Ray., S.K. 2018. Use of Satellite Remote Sensing for Crop Insurance. Mahalanobis national Crop Forecast Centre, DAC&FW, New Delhi.

Reynolds, M., Martin K., Jose C., Jawoo K., Gideon K., Anabel M.M., Jessica R., Urs S., Balwinder-Singh, Kai K., Henri T and V. Vincent. 2018. Role of Modelling in International Crop Research: Overview and Some Case Studies. Agronomy, 8:291.

Richards, J.A. 1996. Classifier performance and map accuracy. Remote Sensing of Environment, 57, 161-166.

Richards, L.A. 1931. Capillary conduction of liquids through porous mediums. Physics, 1 (5): 318–333.

Riley, J.R. 1989. Remote sensing in entomology. Annual Review of Entomology. 1989. 34:247-71

Riley, J.R., Armes, N.J. Reynolds, D.R and A.D. Smith. 1992. Nocturnal observations on the emergence and flight behaviour of Helicoverpa armigera (Lepidoptera: Noctuidae) in the post-rainy season in central India. Bulletin of Entomological Research. 82. 243 - 256. 10.1017/S0007485300051798.

Ritchie, J and S. Otter. 1985. Description of and performance of CERES-Wheat: a user-oriented wheat yield model. In: Willis, W.O. (Ed.), ARS Wheat Yield Project. Department of Agriculture Agricultural Research Service. ARS-38, Washington, DC, pp. 159–175.

Ritchie, J.T and S. Otter. 1984. Description and performance of CERES-wheat: a user-oriented wheat yield model. In: Wheat Yield Project, A.R.S. (Ed.), ARS-38. National Technical Information Service, Springfield, Missouri, pp. 159–175.

Ritchie, J.T. 1991. Specifications of the ideal model for predicting crop yields. In: Muchow, R.C and J.A. Bellamy(Eds.), Climatic Risk in Crop Production: Models and Management for the Semi-arid Tropics and SubtropicsProc. Intnl. Symposium, St. Lucia, Brisbane, Queensland, Australia. July 2–6, 1990. C.A.B. International, Wallingford, U.K., pp. 97–122.

Ritchie, J.T., Singh, U., Godwin, D.C and W.T. Bowen. 1998. Cereal growth, development and yield. In: Tsuji, G.Y., Hoogenboom, G and P.K. Thornton(Eds.), Understanding Options for Agricultural Production. Kluwer Academic publishers, Dordrecht, The Netherland, pp. 79–98.

Ritchie, J.T.1996. International consortium for agricultural systems applications (ICASA): establishment and purpose. Agricultural Systems, 49 (4):329–335.

Rosegrant, M.W., Fernandez, M., Sinha, A., Alder, J., Ahammad, H., de Fraiture, C., Eickhour, B., Fonseca, J., Huang, J., Koyama, O., Omezzine, A.M., Pingali, P., Ramirez, R., Ringler, C., Robinson, S., Thornton, P., van Vuuren, D and H. Yana-Shapiro. 2009. Looking into the future for agriculture and AKST. In: McIntyre, B.D., Herren, H.R., Wakhungu, J and R.T. Watson (Eds.), International Assessment of Agricultural Knowledge, Science and Technology for Development (IAASTD): Agriculture at a Crossroads, Global Report. Island Press, Washington, DC, USA, pp. 307–376.

Rosenzweig, C and D. Hillel. 1995. "Potential impacts of climate change on agriculture and food supply," Consequences, 1, Nr. 2.

Rosenzweig, C and M.L. Parry. 1994. Potential impact of climate change on world food supply. Nature 367: 133-138.

Rosenzweig, C., Elliott, J., Deryng, D., Ruane, A.C., Arneth, A., Boote, K.J., Folberth, C., Glotter, M., Müller, C., Neumann, K., Piontek, F., Pugh, T., Schmid, E., Stehfest, E and J.W. Jones. 2013b. Assessing agricultural risks of climate change in the 21st century in a global gridded crop model intercomparison. Proceedings of the National Academy of Sciences of the United States of America. doi.org/10.1073/pnas.1222463110.

Rosenzweig, C., Jones, J.W., Hatfield, J.L., Ruane, A.C., Boote, K.J., Thorburn, P., Antle, J.M., Nelson, G.C., Porter, C., Janssen, S., Asseng, S., Basso, B., Ewert, F., Wallach, D., Baigorria, G and J.M. Winter. 2013a. The agricultural model intercomparison and improvement project (AgMIP): protocols and pilot studies. Agricultural and Forest Meteorology, 170: 166–182. doi.org/10.1016/j.agrformet.2012.09.011.

Rotter, R.P., Hohn, J.G and S. Fronzek. 2012. Projections of climate change impacts on crop production: a global and a Nordic perspective. Acta Agriculturae Scandinavica, Section A — Animal Science, 62, 166e180.

Rousseau, C., Belin, E., Bove, E., Rousseau, D., Fabre, F., Berruyer, R., Guillaumes, J., Manceau, C., Jaques, M.A and T. Boureau. 2013. High throughput quantitative phenotyping of plant resistance using chlorophyll fluorescence image analysis. Plant Methods, 9:17.

Royal Society of London, Reaping the Benefits: Science and the Sustainable Intensification of Global Agriculture (Royal Society, London, 2009).

Rumpf, T., Mahlein, A.K., Steiner, U., Oerke, E.C., Dehne, H.W., and L. Plumer. 2010. Early detection and classification of plant diseases with Support Vector Machines based on hyperspectral reflectance. Computers and Electronics in Agriculture, 74:91-99.

Rupakumar, K. 2002. Regional Climate Scenarios. TERI Workshop on Climate Change: Policy Options for India, New Delhi.

Rupakumar, K., Kumar, K., Prasanna, V., Kamala, K., Desphnade, N.R., Patwardhan, S.K. and G.B. Pant. 2003. Future climate scenario, In: Climate Change and Indian Vulnerability Assessment and Adaptation. Universities Press (India) Pvt Ltd, Hyderabad, pp. 69–127.

Rupakumar, K., Sahai, A.K., Kumar, K.K., Patwardhan, S.K., Mishra, P.K., Revadekar, J.V., Kamala, K. and G.B. Pant. 2006. High-resolution climate changes scenarios for India for the 21st century. Curr Sci. 90:334–345.

Sabir, N., Singh, B., Hasan, M., Sumitha, R., Deka, S., Tanwar, R.K., Ahuja, D.B., Tomar, B.S., Bambawale, O.M and E.M. Khah. 2010. Good Agricultural Practices (GAP) for IPM in Protected Cultivation, Tech. Bull. No. 23, National Centre for Integrated Pest Management, New Delhi-110 012 India, July 2010, P. 16

Sahoo, S.K. 1999. Simulating growth and yield of maize in the different agro-climatic regions. M.Sc Thesis, Division of Environmental Sciences, IARI, New Delhi.

Salack, S., Sarr, B., Sangare, S.K., Ly, M., Sanda, I.S and H. Kunstmann. 2015. Crop–climate ensemble scenarios to improve risks assessment and resilience in the semi-arid regions of West Africa. Climate Research, 65: 107–121.

Sankar T., Ga.Dheebakaran, S.Panneerselvam, and S.Kokilavani. 2019. Astrometeorological relationship between planet azimuth and cyclone events in Bay of Bengal. International Journal of Agriculture Sciences Volume 11, Issue 11, pp.-8595-8598.

Sankaranarayanan, K., Nalayini, P., Sabesh, M., Usha Rani, S., Nachane, N.P and N. Gopalakrishnan. 2011. Low cost drip- cost effective and precision irrigation tool in Bt cotton. Technical Bulletin No. 1/2011.

Sarkar, R. P and B.C. Biswas. 1988. A new approach to climatic classification to find out crop potential. Mausam,39(4): 343-358.

Savary, S., Bregaglio, S., Willocquet, V., Gustafson, D., Mason D'Croz, D., Sparks, A., Castilla, N.,et al., 2017. Crop Health and Its Global Impacts on the Components of Food Security. Food Security, 9 (2): 311–27. doi.org/10.1007/s12571-017-0659-1.

Savary, S., Srivastava, R.K., Singh, H.M and F.A. Elazegui. 1997. A characterisation of rice pests and quantification of yield losses in the rice-wheat system of India. Crop Protection, 16, 387-398.

Savary, S., Teng, P.S., Willocquet, L and F.W. Nutter. 2006. Quantification and modeling of crop losses: a review of purposes. Annual Review of Phytopathology, 44:89–112.

Savin, R., Satorre, E.H., Hall, A.J and G.A. Slafer. 1995. Assessing strategies for wheat cropping in monsoonal climate of pampas using the CERES-wheat simulation model. Field Crops Research, 42:81–910.

Schumann,G.J.P.,Neal,J.C.,Voisin,N.,Andreadis,K.M.,Pappenberger,F.,Phanthuwongpakdee, N., Hall, A.C and P.D. Bates. 2013. A first large scale flood inundation forecasting model. Water Resources Research, 6248–6257.

Scofield, R.A and V.J. Oliver. 1977. Using satellite imagery to estimate rainfall from two types of convective systems. Proceedings of 10th Conference on Hurricanes and Tropical Meteorology, Miami Beach, Florida, AMS Boston, Massachusetts, 204-211.

Sehgal, J.L., Mandal, D.K., Mandal, C and S. Vadivel.1992. Agro-Ecological Regions of India, Second edition, Technical Bulletin, NBSS and LUP, Publ.24, P 130, NBSS&LUP, Nagpur, 440010, India.

Sejian, V., Bhatta, R and S.M. K. Naqvi.2017. Role of biometeorology in livestock. pp 39-53. IN. Livestock Meteorology (edts. G.S.L.H.V. Prasada Rao., Girish Varma and V. Beena.,) New India Publishing Agency, New Delhi.P.522.

Sejian, V., Krishnan, G., Bagath, M., Lipismita Samal., Soren, N.M., Malik, P.K. and R. Bhatta.2017. Weather extremes and livestock production under humid tropics. IN Livestock Meteorology (edits. G.S.L.H.V. Prasada Rao., Girish Varma and V. Beena.,) New India Publishing Agency, New Delhi.P.522.

Selcuk, R., Nisanci, R., Uzun, B., Yalcin, A., Inan, H and T. Yomralioglu. 2003. Monitoring land-use changes by GIS and remote sensing techniques: case study of Trabzon. http://www.fig.net/pub/morocco/proceedings/TS18/TS18_6_reis_el_ al.pdf 5.

Semenov, M.A and P. Stratonovitch. 2015. Adapting wheat ideotypes for climate change: accounting for uncertainties in CMIP5 climate projections. Climate Research, 65: 123–139

Shakeel, A., Khan, Sanjeev Kumar., Hussain, M.Z and N. Kalra. 2009. Climate Change, Climate Variability and Indian Agriculture: Impacts Vulnerability and Adaptation Strategies S.N. Singh (ed.), Climate Change and Crops, Environmental Science and Engineering, DOI 10.1007/978-3-540-88246-6 2, C .Springer-Verlag Berlin Heidelberg 2009.

Shamsudduha, M., Chandler, R. E., Taylor, R. G., & Ahmed, K. M. (2009). Recent trends in groundwater levels in a highly seasonal hydrological system: the Ganges-Brahmaputra-Meghna Delta. Hydrol Earth Syst Sc, 13(12), 2373-2385.

Sharma, R.C and G.P. Bhargava, G. P. 1988. Landsat imagery for mapping saline soils and wet lands in north-west India. International Journal of Remote Sensing, 9, 39-44.

Sharma, V.K., Pandey, R.N., Kumar, S., Chobhe, K.A and S. Chandra. 2016. Soil test crop response based fertilizer recommendations under integrated nutrient management for higher productivity of pearl millet (Pennisetumglaucum) and wheat (Triticumaestivum) under long term experiment. Indian Journal of Agricultural Sciences 86 (8): 1076–81.

Sheehy, J.E., Mitchell, P.L and A.B. Ferrer. 2006. Decline in rice grain yields with temperature: Models and correlations can give different estimates. Field Crop Res. 98: 151-156.

Shinde, S., Taneja, V.K and A. Singh1990. Association of climatic variables and production and reproduction traits in cross breds. Indian J. Animal Sci.60(1):81-85

Shiva Shankar, M., Ramanjaneyulu, A.V., Neelima, T.L and Anup Das. 2015. Sprinkler Irrigation–An Asset in Water Scarce and Undulating Areas. Rajkhowa, D.J., Das, Anup, Ngachan, S.V., Sikka, A.K. and Lyngdoh, M. (Eds.) Integrated Soil and Water Resource Management for Livelihood and Environmental Security. 2015. pp 259-283. ICAR Research Complex for NEH Region, Umiam–793 103, Meghalaya, India.

Shodhganga, INFLIBNET Centre https://shodhganga. inflibnet.ac.in/bitstream/10603/ 107846/ 13/13_chapter-v.pdf

Shweta Pokhariyal and N.R. Patel. 2020. Variation in energy fluxes over wheat ecosystem along the Gangetic plain. Paper presented in National Seminar on Agrometeorological Interventions for enhancing farmers income (Agmet-2020) 20-22 Jan. 2020. Kerala Agricultural University, Thrissur, Kerala. Abstract Compendium. Theme: 0.11.21. P 159.

Singh, P., Gupta, P and M. Singh. 2014. Hydrological inferences from watershed analysis for water resource management using remote sensing and GIS techniques. Egypt. J. Rem. Sens. Space Sci. 17, 111–121.

Singh, S.N., Singh, G.P., Rai J.P., and S. Prasad. 2016. Problem and Solution for the dryland agriculture in India. Indian Agriculture and Farmers. Pp - 88-91

Singh, S.V., Yogendra Kumar and Sunil Kumar,2019. Impact of temperature humidity index (THI) on physiological responses and milk yield of Tharparkar and Karan Fries cows exposed to controlled environment. Journal of Agrometeorology. 21(4): 405-410.

Sinha and Banik. 2009. Plant Disease Forecasting and monitoring: An imperative in precision Agriculture. Indian Farming. 59(8):46-50.

Sinha, S.K and M.S. Swaminathan. 1991. Deforestation, climate change and sustainable nutrition security. Clim Change 16:33–45.

Sivakumar, T., Thennarasu, A and J.S.I. Rajkumar, 2012.Effect of season on the incidence of infectious diseases of bovine in Tamil Nadu. Elixir Meteorology47,8874-8875.Online www. Elixirpublishers.com (Elixir International)

Sivaprakasam S and V.Kanakasabai. 2009. Traditional Almanac predicted rainfail - A case study, Indian Journal of Traditional Knowledge, Val. 8 (4). pp. 621-625

Sorooshian, S., Hsu, K.L., Gao, X., Gupta, H.V., Imam, B and D. Braithwaite. 2000. Evaluation of PERSIANN system satellite–based estimates of tropical rainfall. Bulletin of the American Meteorological Society, 81, 2035–2046.

Stapleton, H.N., Buxton, D.R., Watson, F.L., Notling, D.J and D.N. Baker. 1973. COTTON: a computer simulation of cotton growth. University of Arizona, agricultural experiment station. Tech. Bull. 206.

Steddom, K., Heidel, G., Jones, D and C.M. Rush. 2003. Remote detection of rhizomania in sugar beet. Phytopathology, 93:720-726.

Steduto, P., Hsiao, T.C., Raes, D and E. Fereres. 2009. AquaCrop-The FAO crop model to simulate yield response to water: I. Concepts and underlying principles. Agronomy Journal, 101, 426e437

Stern, N and N. H. Stern. 2007. The economics of climate change: the Stern review. cambridge University press.

Stuckens, J., Coppin, P.R and M.E. Bauer. 2000. Integrating contextual information with per-pixel classification for improved land cover classification. Remote Sensing of Environment, 71, 282-296.

Suraj, P.T and T. Siva Kumar.2013. Existing dairy housing systems and its suitability in different agro-climatic regions of Tamil Nadu. P.17-21. IN. Fundamentals of Livestock Meteorology. Vol. II (edts.G.S.L.H.V. Prasada Rao and G. Girish Varma). Centre for Animal Adaptation to Environment and Climate Change Studies, KVASU. Mannuthy, Thrissur, Kerala, India,295p.

Talsma, T. 1963. The Control of Saline Ground Water. , 1- 68. Med. Landb. Wageningen, 63 (10): 1–68. http://edepot.wur.nl/185086.

Tao, F., Rötter, R.P., Palosuo, T., Höhn, J., Peltonen-Sainio, P., Rajala, A and T. Salo. 2015. Assessing climate effects on wheat yield and water use in Finland using a superensemble-based probabilistic approach. Climate Research 65: 23–37

Tao, × and H. Xin.2003. Acute synergistic effects of air temperature, humidity and velocity on homeostasis of market-size broilers. Transactions of the ASAE.46(2): 49`-497.

Tennakoon, S.B., Murty, V.N and A. Eiumnoh. 1992. Estimation of cropped area and grain yield of rice using remote sensing. International Journal of Remote Sensing, 13, 427-439.

Thorburn, P., Boote, K., Hargreaves, J., Poulton, P and J. Jones. 2014. Crop systemsmodeling in AgMIP: a new protocol-driven approach for regional integrated assessments. In: Hillel, D and C. Rosenzweig (Eds.), Handbook of Climate Change and Agroecosystems: Agricultural Model Intercomparison and Improvement Project Integrated Crop and Economic Assessments ICP Series on Climate Change Impacts, Adaptation, and Mitigation vol. 3 & 4. Imperial College Press (In Press).

Thornthwaite, C. W and J. R Mather.1955. * The water balance. In climatology. VIII (1), Drexel Institute of Technology, New Jersey, USA.104.

Thornthwaite, C.W and J.R Mather.1955. The water balance. Pub. In climatology, C.W. Thornthwaite Associates, Centeron, New Jersey. 8(1): p.86.

Thornton, P.K., Jones, P.G, Owiyo, T., Kruska, R.L., Herrero, M., Kristjanson, P., Notenbaert, A., Bekele, N., Omolo, A., Orindi, V., Adwerah, A., Otiende, B., Bhadwal, S., Anantram, K., Nair, S., Kumar, V and U. Kelkar. 2006. Mapping climate vulnerability and poverty in Africa. Report to the Department for International Development, ILRI, Nairobi, Kenya, (198 pp).

Timsina, J and E. Humphreys. 2006. Performance of CERES-rice and CERES-wheat models in rice-wheat systems: a review. Agricultural Systems, 90:5–31.

Timsina, J., Godwin, D., Humphreys, E., Yadvinder-Singh Bijay-Singh Kukal, S.S and D. Smith. 2008. Evaluation of options for increasing yield and water productivity of wheat in Punjab: India using the DSSAT-CSM-CERES-wheat model. Agricultural Water Management, 95:1099–1110.

Timsina, J., Sing, U., Singh, Y and F.P. Lansigan. 1995. Addressing suitability of RW systems: testing and applications of CERES and SUCROS models. In: Proceedings of the International Rice Research Conference. 13-17 February 1995. IRRI, Los Banos, Philippines. pp. 656–663.

Tingem, M and M. Rivington. 2009. Adaptation for crop agriculture to climate change in Cameroon: Turning on the heat. Mitigation and Adaptation Strategies, 14, 153–168.

Tsuji, G., Uehara, G and S. Balas. 1998. DSSAT. Version 3 Vol. 1–3 University of Hawaii, Honolulu, HI.

Tsuji, G.Y., Hoogenboom, G and P.K. Thornton. 1998. Understanding Options for Agricultural Production. Springer (400 pp.).

Tubiello, F.N., Rosenzweig, C., Goldberg, R.A., Jagtap, S and J.W. Jones. 2002. Effects of climate change on U.S. crop production: simulation results using two different GCM scenarios. Part I: wheat, potato, maize, and citrus. Climate Research, 20 (3):259–270.

Uehara, G and G.Y. Tsuji. 1998. Overview of IBSNAT. In: Tsuji, G.Y., Hoogenboom, G., Thornton, P.K. (Eds.), Chapter 1, pp. 1–7. Understanding Options for Agricultural Production. Springer, The Netherlands.

UNESCO. 2012. Facts and figures from WWDR4. Paris. Available at; unesdoc.unesco.org/images/0021/002154/215492e.pdf.

UNFCCC. 2013: Reporting and accounting of LULUCF activities under the Kyoto Protocol. United Nations Framework Convention on Climatic Change (UNFCCC), Bonn, Germany. Retrieved from: http://unfccc.int/methods/lulucf/items/4129.php.

United Nations Framework Convention on Climate Change, New York, 9 May 1992, in force 21 March 1994. International Legal Materials (31, 1992).

Vaidya,V.B., Suvarna Dhabale, K.S.Damle1, L.D.Chimote and M.S.Kulshreshtha. 2019. Astro-Meteorological Rainfall Prediction and Validation for Monsoon 2018 in Gujarat, India. Int. J. Curr. Microbiol. App. Sci., 8(5): 2359-2370

Van Dyne, G.M and J.C. Anway. 1976. Research programfor and process of building and testing grassland ecosystem models. Journal of Range Management, 29, 114–122

Van Dyne, G.M. 1980. Grasslands, systems analysis, and man. In: Breymeyer, A.I. (Ed.), International Biological Program 19. Press. New York, Cambridge University 953 pp

Van Ittersum, M.K., Ewert, F., Heckelei, T., Wery, J., Alkan Olsson, J., Andersen, E., Bezlepkina, I., Brouwer, F., Donatelli,M., Flichman, G., Olsson, L., Rizzoli, A., Van derWal, T., Wien, J.E and J. Wolf. 2008. Integrated assessment of agricultural systems — a component based framework for the European Union (SEAMLESS). Agricultural Systems, 96:150–165.

Van Ittersum, M.K., Leffelaar, P.A., van Keulen, H., Kropff, M.J., Bastiaans, L and J. Goudriaan. 2003. On approaches and applications of the Wageningen crop models. European Journal of Agronomy, 18:201–234.

Van Ittersum, M.K., Rabbinge, R and H.C. Van Latesteijn. 1998. Exploratory land use studies and their role in strategic policy making. Agricultural Systems, 58:309–330.

Varshneya, M. C, Vaidya, V. B, Vyas Pandey, Shekh, A. M. and B. I. Karande. 2008. Validation of Astrometeorological Rainfall forecast for Gujarat. Journal of Agrometeorology, 10(2):345-348.

Varshneya, M.C. 2008. Blend of Ancient Wisdom and modern science of Meteorology for improving agriculture. In: Souvenir of International symposium on Agrometeorolrology and Food security (INSAFS),CRIDA, Hyderabad. p.15.

Vasi Raju Radhakrishna Murthy. 1996. Terminology in Agricultural Meteorology. Sri. Venkateswara Publisher, Hyderabad. P 250.

Veeraputhiran, R., Karthikeyan, R., Geethalakshmi, V., Selvaraju, R., Sundersingh, S.D and T.N Balasubramanian 2003. Crop planning -climate atlas. Principles.A.E. Publications. Coimbatore.41. Tamil Nadu. India. P157.

Venkataraman, S and A. Krishnan.1992. Crops and Weather. Publications and Information Division. Indian Council of Agricultural Research, Krishi Anusandhan Bhavan, Pusa, New Delhi. 110 012.P 586.

Verma, M.K., Singh, S. K., Srivastav M and Jai Prakash. 2013. Good agricultural practices for temperate fruit production in India. S.K. Singh and M.K. Verma (eds.) In: Good Agricultural Practices GAP in Production of Horticultural Crops, Edition: 1st, Chapter: Division of Fruits and Horticultural Technology, Indian Agricultural Research Institute, New Delhi 110 012, India, , pp.28-44.

Vimal Antony.2017. Heat stress management in poultry. PP.475-485. IN. Livestock Meteorology (edts. G.S.L.H.V. Prasada Rao., Girish Varma and V. Beena.,) New India Publishing Agency, New Delhi.P.522.

Vlassopoulos, C. A. 2012. Competing definition of climate change and the post-Kyoto negotiations. International journal of climate change strategies and management.

Waha, K., Muller, C., Bondeau, A., Dietrich, J.P., Kurukulasuriya, P., Heinke, J and H. Lotze-Campen. 2013. Adaptation to climate change through the choice of cropping system and sowing date in Sub-Saharan Africa. Global Environmental Change, 23 (1):130–143.

Wahabzada, M., Mahlein, A.K., Bauckhage, C., Steiner, M., Oerke, E.C and K. Kersting. 2015. Metro maps of plant disease dynamics-automated mining of differences using hyperspectral images. PLoS One,10:e0116902.

Wang, X., Zhang, M., Zhu, J and S. Geng. 2008. Spectral prediction of Phytophthora infestans infection on tomatoes using artificial neural network (ANN). International Journal of Remote Sensing, 29:1693-1706.

Weart, S. 2008. The carbon dioxide greenhouse effect. The Discovery of Global Warming. American Institute of Physics.

Weart, S. R. 2008. The discovery of global warming. Harvard University Press.

Weart, S. R. 2010. The idea of anthropogenic global climate change in the 20th century. Wiley Interdisciplinary Reviews: Climate Change, 1(1), 67-81.

Welch, S.M., Croft, B.A., Brunner, J.F and M.F. Michels. 1978. PETE: an extension phenology modeling system for management of multi-species pest complex. Environmental Entomology, 7, 487–494.

Whish, J.P.M., Herrmann, N.I., White, N.A., Moore, A.D and D.J. Kriticos. 2015. Integrating pest population models with biophysical cropmodels to better represent the farming system. Environmental Modelling & Software, 72, 418–425.

Wiegand, C., Anderson, G., Lingle, S and D. Escobar. 1996. Soil salinity effects on crop growth and yield - illustration of an analysis and mapping methodology for sugarcane. Journal of Plant Physiology, 148, 418-424.

Wijekoon, C.P., Goodwin, P.H and T. Hsiang. 2008. Quantifying fungal infection of plant leaves by digital image analysis using Scion Image software. Journal of Microbiological Methods, 74:94-101.

Wilkerson, G.G., Jones, J.W., Boote, K.J., Ingram, K.T and J.W. Mishoe. 1983. Modeling soybean growth for crop management. Transactions - American Society of Agricultural Engineers, 26:63–73.

Williams, J. R. 1995. The EPIC Model. Computer Models of Watershed Hydrology. V. P. Singh, Water Resources Publications, Highlands Ranch, Colorado: 909-1000

Williams, J.R., Jones, C.A., Kiniry, J.R and D.A. Spanel. 1989. The EPIC crop growth model. Trans. American Society of Agricultural and Biological Engineers, 32 (2):497–511.

Williams, J.R., Renard, K.G and P.T. Dyke. 1983. EPIC: a new method for assessing erosion's effect on soil productivity. Journal of Soil and Water Conservation, 38 (5):381–383.

World Meteorological Organisation,1992. International Meteorological Vocabulary (WMO. NO.182).Geneva.

World Meteorological Organisation,2008. Guide to Meteorological Instruments and Methods of Observations (WMO.NO.8) Geneva

Worthington, E.B. (Ed.). 1975. The Evolution of IBP. Cambridge University Press, Cambridge

Xiong, W., Asseng, S., Hoogenboom, F., Ochoa, I.H., Robertson, R., Sonder, K., Pequeno, D., Reynolds, M., and B. Gerard. 2019. Different uncertainty distribution between high and low latitudes in modelling warming impacts on wheat. Nature food, 1: 63–69

Xiong, Y., Guo, S., Chen, J., Deng, X., Sun, L., Zheng, × and W. Xu. 2020. Improved SRGAN for Remote Sensing Image Super-Resolution Across Locations and Sensors. Remote Sensing. 12, 1263; doi:10.3390/rs12081263.

Yadav. A.K. 2010. Organic Agriculture: Concept, Scenario, Principals and Practices. National Centre of Organic Farming, Department of Agriculture and Cooperation, Ministry of Agriculture, Govt of India, CGO-II, Kamla Nehru Nagar, Ghaziabad, 201 001, Uttar Pradesh.

You, L., Wood, S and U. Wood-Sichra. 2006. Generating Global Crop Maps: From Census to Grid. Selected paper, IAAE (International Association of Agricultural Economists) Annual Conference, Gold Coast, Australia.

Yu-liang, Q. 1996. An application of aerial remote sensing to monitor salinization at Xinding Basin. Advances in Space Research, 18, 133-139.

Zadoks, J.C. 1971. Systems analysis and the dynamics of epidemics. Phytopathology 61, 600-610.

Zhang, X., Wang, S., Sun, H., Chen, S., Sho, L and X. Liu. 2013. Contribution of cultivar, fertilizer and weather to yield variation of winter wheat over three decades: a case study in the North China plain. European Journal of Agronomy, 50, 52–59.

Zhao, G., Hoffmann, H., van Bussel, L.G.J., Enders, A., Specka, X., Sosa, C., Yeluripati, J., Tao, F., Constantin, J., Raynal, H., Teixeira, E., Grosz, B., Doro, L., Zhao, Z., Nende, C., Kiese, R. Eckersten, H., Haas, E., Vanuytrecht, E., Wang, E., Kuhnert, M., Trombi, G., Moriondo, M., Bindi, M., Lewan, E., Bach, M., Kersebaum, K.C., Rötter, R., Roggero, P.P. Wallach, D., Cammarano, D., Asseng, D., Krauss, G., Siebert, S., Gaiser, T and F. Ewert. 2015. Effect of weather data aggregation on regional crop simulation for different crops, production conditions and response variables. Climate Research, 65: 141–157.

Zhou, W.T and S. Yamamoto,1997. Effect of environmental temperature and heat production due to food intake on abdominal temperature, skin temperature and respiration rate of broilers. Bri.Poult. Sci. 8: 107–114.

Ziska, L.H., P.A. Manalo and R.A. Ordonez. 1996. Intraspecific variation in the response of rice (Oryza sativa L.) to increased CO2 and temperature: growth and yield response of 17 cultivars. J. Exp. Bot. 47:1353-1359.

Bibliography

Agricultural Meteorology, Kerala Agricultural University, Thrissur, Kerala, India. P 326 (Prasad Rao, G.S.L.H.V.2003)

Agricultural Meteorology. Sri. Venkateshwara Publishers, Hyderabad.P.203. (Vasiraju Radha Krishna Murthy. 1996.Terminology)

Measuring Vulnerability to Natural hazards: Towards disaster resilient societies. TERI.United Nations University Press. P 524 (Jorn Birkmann, 2006)

Text book of Agricultural Meteorology, ICAR, New Delhi.P221. (Varshneya, M.V. and P. Balakrishna Pillai 2004.)

Weathering the storm options for framing adaptation and development, World Resources Institute. (Heather MGGRAY; ANNE HAMMILL., Rob Bradley with E. Lisa Schipper., Jo. Ellen Parry, 2007)

Annexure I

Selected Questions and Answers

This Chapter gives answers for the questions created by the authors from the perception of reading clients. Only selected and relevant questions have been answered. This chapter is included in this book so as to understand the subject properly.

1. Introductory Knowledge on Atmosphere

1.1 What is atmosphere?

Among the three components of the earth, namely lithosphere (solid), hydrosphere (ocean) and atmosphere (gaseous), atmosphere is very important for the life of the organisms living in the earth.

It is the gaseous envelope that surrounds the earth. The presence of atmosphere in the earth planet regulates temperature and light to the earth. This atmosphere is the origin for the weather that prevails on the earth.

1.2 What is the composition of the atmosphere?

Atmosphere is the gaseous envelope of the earth, which consists of relatively a stable mixture of number of gases. Four gases *viz.* nitrogen, oxygen, argon and carbon-dioxide account for 99.98 per cent of the air by volume. In addition to these gases, atmosphere also contains water vapour and aerosol (sea salt, dust, organic matter and smoke and other pollutants).

Nitrogen forms about 78 per cent of the total volume of dry air, followed by oxygen (21%) and argon (0.93%). The volume of carbon-dioxide is 0.03 per cent and the remaining per cent is hold by neon, helium, methane, krypton and hydrogen.

1.3 Can you list out the properties of the atmosphere?

- It exerts pressure over the earth
- It adds weight
- It supports combustion
- It diffuses heat and gases
- It conducts sound
- It is denser to the surface of the earth and becomes thinner as it goes up to a height of 5.25 km from the earth
- The temperature of the atmosphere decreases by about 6.5°C/km in altitude. But this fall in temperatures does not take place beyond a certain limit.

1.4 What is the relationship between earth and atmosphere, hydrosphere?

The lithosphere(earth) and atmosphere are closely linked together by the gravitational force of the earth. When earth rotates around the sun, it does along with the atmosphere. The radiation from the sun reaches to the earth through space and atmosphere.

Hydrological cycle is the result of interaction between lithosphere(earth)–hydrosphere–atmosphere. There is no head and tail for this processes.

1.5 How atmosphere is scientifically stratified?

The atmosphere is divided vertically into four layers on the basis of temperature. In addition, there is one last layer above the four layers and this is called as Exosphere. This exosphere extends up to 10,000 km and this layer is above the thermosphere. Beyond this layer only space exist.

1.6 What are these four layers and list out their physical and chemical properties?

(a) ***Troposphere***: This is the lower most layer of the atmosphere touching the ground level. Its elevation is 16 km at the equator and 8 km at the poles.

It is the zone, where weather phenomena and atmospheric turbulence are most marked and contains 75 per cent of the total gaseous mass of the atmosphere and virtually all the water vapour and aerosols exist. Throughout this layer, there is general decrease in temperature with height at an average rate of 6.5°C/km.

The whole layer is capped in most places by temperature inversion layer (layer of relatively warm air above a colder one). This inversion layer is called as tropopause. Variations in the altitude of troposphere exist between different latitudes.

(b) ***Stratosphere***: This is the second layer, which extends upwards from the tropopause to about 50 km. In the stratosphere, the temperature at first remains nearly constant to the height of 20 km and then it begins to increase until the height of about 50 km above the earth's surface. The top of the layer, where maximum temperature is attained and is called as stratopause. Higher temperature occurs in the stratosphere because of absorption of ultraviolet radiation by ozone.

(c) ***Mesosphere***: Above the stratopause, average temperature decreases to minimum of about (–) 90°C at around 80 km. Above 80 km temperature again begin rising with height and this inversion is referred to as mesopause.

(d) ***Thermosphere***: Above the mesopause, atmospheric densities are extremely low. The lower portion of the thermosphere is composed mainly of nitrogen and oxygen in molecular (O_2) and atomic (O) forms, whereas above 200 km, atomic oxygen predominates over nitrogen (N_2 and N). Temperature rise with height owing to the absorption of ultraviolet radiation by atomic oxygen.

2. Introductory Knowledge on Weather Elements

2.1 Rainfall

2.1.1 What is rainfall?

Precipitation is the deposition of atmospheric moisture on the earth in the form of rainfall, drizzle, snowfall, sleet and hail.

Rainfall is the most predominant form of precipitation in which water droplets of size greater than 0.5 mm fall on the earth surface. It is defined as the quantity of water falling on unit area.

If the daily rainfall is more than 2.5 mm in a day, and that day is called as rainy day. The other word rain day means the received rainfall is between 0.1 and 2.49 mm/day. The day without rainfall is called as rainless day.

2.1.2 What is the mean annual rainfall of India and Tamil Nadu?

Mean annual rainfall of India	:	117 cm (1170 mm)
Mean daily rainfall in India	:	2 cm (20 mm)
Mean annual rainfall of Tamil Nadu	:	98 cm (980 mm)

2.1.3 What is the distribution of mean annual rainfall in Tamil Nadu in terms of percentage?

Cold weather period (January–February)	4 per cent
Summer season (March–May)	16 per cent
Southwest monsoon season (June–September)	32 per cent
Northeast monsoon season (October–December)	48 per cent

2.1.4 What are the weather systems in India that bring rainfall?

There are three systems that bring rainfall in India and those are;

(a) Trade wind (moist and warm air) with orographic rainfall processes: Southwest monsoon season (June–September)

(b) Low pressure/cyclone rainfall processes: Northeast monsoon season (October–December)

(c) Western disturbance, jet stream of low pressure: November–March

2.1.5 What is the classification given for rainfall quantity description as per India Meteorological Department (IMD) for planning purpose?

Normal rainfall	:	±19 per cent rainfall from the mean rainfall
Wet	:	+19 per cent to +59 per cent above the mean
Flood/Excess	:	>59 per cent above the mean
Deficit rainfall	:	(–)19 to (–)59 per cent down from the mean
Drought	:	> (–)59 per cent down from the mean

2.1.6 *Can you say, what is the relationship between quality statement and quantity statement of rain to describe its intensity?*

Very heavy rainfall	:	> 12.5 cm/day
Heavy rainfall	:	6.5 to 12.5 cm/day
Rather heavy	:	3.5 to 6.5 cm/day
Moderate	:	0.8 to 3.5 cm/day
Light rain	:	0.3 to 0.7 cm/day
Very light rain	:	up to 0.2 cm/day

2.1.7 *What is the instrument used to measure rainfall in India?*

The name of the instrument is rain gauge. There are three types:

Ordinary rain gauge (100 and 200CM^2)

Self-recording rain gauge

Dipping bucket rain gauge (only for automatic weather station)

2.1.8 *If 10 mm of rainfall is received over one hectare of land, what is the quantity of water added to that one hectare in terms of litres.*

Rainfall received	:	10 mm or 0.001 m
1 ha.	:	10,000 m^2 Quantity of water added in one ha (cm^3) 10,000 × 0.001 = 100 cm^3
1 cm^3 = 1000 litres	:	100 cm^3 = 100 × 1,000 = 1,00,000 litres of water

2.2 Hail storm

2.2.1 *What is hail storm?*

Precipitation of small pieces of ice with diameter ranging from 5 to 50 mm or more is known as hail storms and are frequent in the tropics. In India the period from March to May offers the ideal condition for hail storm occurrence. It is most dangerous and destructive form of precipitation produced in thunder storms or by cumulonimbus clouds.

Three forms of hail are recognized:

(a) **Graupel or soft hail:** It is composed of loosely compacted ice crystals and is roughly spherical with tendency to fracture upon striking the ground. Diameter is less than 5 mm.

(b) **Small hail:** It is semi-transparent. When it strikes hard surface, often in conjunction with rain, the hail storm remains intact.

(c) **True hail or severe hail:** It is composed of hail storms greater than 5 mm that often causes extensive damage.

Hail is formed from the ice embryos form, as liquid water droplets, from the warmer and lower portion of a cloud, all carried aloft by strong vertical currents and subsequently freeze upon encountering the sub-freezing temperature associated with upper portion of the cloud.

2.3 Air temperature

2.3.1 What is air temperature and the source for air temperature may be described?

The hotness of the air is called air temperature and the source for air temperature is radiation received from the sun. Differential heating of air mass would lead to variation in atmospheric pressure.

Air temperature is one among the variable weather elements. Growth and development of all organisms are controlled by air temperature. It affects the plant and animal life at all stages of their growth.

2.3.2 How the air temperature is quantified?

The atmospheric temperature can be quantified in many ways and one among them is as follows:

(a) *Maximum temperature* (°C)–T_{max}: It is the highest temperature attained by the atmosphere in diurnal variation (this will occur around 2.30 pm).
(b) *Minimum temperature* (°C)–T_{min}: It is the lowest temperature attained by the atmosphere in diurnal variation (this will occur between 5 and 5.30 am).
(c) *Mean daily temperature*: This is the average of both T_{max} and T_{min} (T_{max} + T_{min}/2).
(d) *Base temperature*: This is the temperature below which the plant does not grow (growth arrested)

2.3.3 What are the scales used to measure air temperature?

Scale	Ice point	Steam point
(a) Celsius (°C)	0°	100°
(b) Fahrenheit (F)	32°	212°
(c) Kelvin (°K)	273°	373°

(d) What are the conversion scales used to convert from one value to another one?

$C = (F-32) \times 5/9$

$F = (C + 32) \times 9/5$

$K = C + 273$

°K = Absolute (°A) temperature

2.3.4 What are the criteria for declaring heat wave?

When the normal maximum temperature of the area is 40°C or less, the following Table value may be used to declare heat wave.

Category	Deviation from normal
Normal	–1°C to +1°C
Above normal	+2°C
Appreciably above normal	+3°C to 4°C
Moderate heat wave	+5°C to 6°C
Severe heat wave	+7°C and above

When the normal maximum temperature of the area is more than 40°C, the following Table value may be used to declare heat wave.

Category	**Deviation from normal**
Normal	–1°C to +1°C
Above normal	+2°C
Heat wave	+3°C to 4°C
Severe heat wave	+5°C and above

2.3.5 What are the criteria for declaring cold wave?

When the normal minimum temperature of the area is less than 10°C, the following Table value may be used to declare cold wave.

Category	**Departure from normal**
Normal	–1°C to1°C
Below normal	–2°C
Cold wave	–3°C to–4°C
Severe cold wave	–5°C and less

When the normal minimum temperature of the area is 10°C or more, the following Table value may be used to declare cold wave.

Category	**Departure from normal**
Normal	1°C to –1°C
Below normal	–2°C
Appreciably below normal	–3°C to–4°C
Moderate cold wave	–5°C to–6°C
Severe cold wave	–7°C and less

2.4 Wind

2.4.1 What is wind?

When air is in motion with difference in atmospheric pressure as a result of radiation, then the moving air is called as wind or the horizontal movement of air is also known as wind.

2.4.2 Can you name the monsoon winds, which bring rainfall to India?

(a) Southwest monsoon wind comes from south-west direction of India. This wind is always moist and warmer and gives rain to 80 per cent geographical area of India. It blows from June to September for four months.

(b) Northeast monsoon wind comes from north-east direction to India.

The southwest monsoon wind after a stay of three months from July 15th to September 15th over Rajasthan, gets back to Bay of Bengal after withdrawal from Rajasthan, gains energy from Bay of Bengal and forms as low or depression or cyclone and blows to India from north-east direction. This wind benefits entire Tamil Nadu, coastal parts of Andhra Pradesh, Karnataka and Kerala States. This monsoon rainfall occurs from mid-October to mid-December.

2.4.3 What are the units used commonly to express wind speed?

Wind speed is given in kilometer per hour. It is also being expressed in different units like knots, m/second, miles/hour and feet/second and scales vary with different countries. The conversion Table is given hereunder:

Knots	m/second	m/hour	Km/hour	Ft/second
1	0.515	1.152	1.853	1.689
1.943	1	2.237	3.600	3.281
0.868	0.447	1	1.609	1.467
0.540	0.278	0.621	1	0.911
0.592	0.305	0.682	1.097	1

m: meter; Ft: Feet

2.4.4 How wind speed is measured?

Wind speed is measured by using cup anemometer at surface observatories. Sensors are also being used to measure wind speed at different altitudes of the atmosphere by using different carriers. In the case of Automatic Weather Station also, sensor is used.

In the absence of an anemometer, wind can be estimated reliably by its effect on objects on land and the state of the sea.

2.4.5 What is wind direction and different terminologies used?

There are two terminologies used. One is Windward direction and another one is Leeward direction. Windward direction indicates, where from the wind comes, while the Leeward direction indicates where the wind goes.

2.4.6 What is the instrument used to measure wind direction and unit used?

Wind vane is used to measure wind direction at the surface observatories. Wind direction is expressed in degrees. For east it is 90°, while it is 180° for south, 270° for west, and 360°/0 for north. Totally 16 wind directions are used in meteorological science. Some examples are as follows:

Northerly wind : Wind from 337.6 to 22.5 degrees
North easterly wind : Wind from 22.6 to 67.5 degrees
Easterly wind : Wind from 67.6 to 112.5 degrees
South easterly wind : Wind from 112.6 to 157.5 degrees
Southerly wind : Wind from 157.6 to 202.5 degrees
South westerly wind : Wind from 202.6 to 247.5 degrees
Westerly wind : Wind from 247.6 to 292.5 degrees
North westerly wind : Wind from 292.6 to 337.5 degrees

2.4.7 What is wind rose?

A wind rose is a pictorial representation of wind direction and wind speed. It provides an overall idea of the distribution of wind direction and wind speed at a given location for given period of time. It is very useful to design wind break and shelter belt.

2.5 Relative humidity

2.5.1 What is humidity and how many types of humidity are available in the meteorological science?

The actual moisture content of atmosphere is called as humidity. Humidity is expressed in different forms and those are absolute humidity, specific humidity, relative humidity, mixing ratio, vapour pressure, and dew point temperature.

Among them in agriculture, relative humidity form is used.

2.5.2 What is relative humidity and how it is measured?

Relative humidity is defined as the ratio of the amount of water vapour that is actually present in air (actual vapour pressure) to the amount of water vapour that air can hold at its maximum capacity (saturation vapour pressure) at a given temperature. It is expressed in per cent.

Relative humidity is measured at 07 IST at morning and is called as maximum relative humidity. Similarly, it is measured at 014 IST in the evening and this measurement is called as minimum relative humidity.

2.5.3 How relative humidity is measured?

Station hygrometer, Hygrometer and Hygrograph are used to record relative humidity. Among them, in the agricultural meteorological observatories, station hygrometer is largely used.

Both dry bulb and wet bulb thermometers are fixed in the Stevenson Screen wooden box. Dry bulb thermometer will exhibit the real time temperature at any point of time. In the wet bulb thermometer, the bulb of the thermometer is covered with moistened cloth so as to indicate the temperature at moisture saturation point.

Readings from both dry and wet bulb thermometers are taken simultaneously both at 07 and 014 IST and the recorded readings will be referred in hygrometric table. From this table, relative humidity value, dew point temperature and vapour pressure could be picked up against the concerned dry and wet bulb readings.

2.5.4 What are the properties of the relative humidity?

It is clear that the relative humidity must change whenever the amount of water vapours in the air changes. The relative humidity varies inversely with the temperature. A decrease in temperature causes corresponding capacity decrease, if the capacity decreases the relative humidity increases as the air is brought nearer the saturation point.

2.6 Dew

2.6.1 What is dew?

The temperature of air and earth surface decreases to such an extent that water condenses over the surface and this is called as dew. It occurs in the early morning hours. It generally forms at night due to radiational cooling (clear, cool nights) which causes the temperature of the air to fall below dew point (dew point is the temperature at which saturation of the air occurs).

Dew is often the only form of moisture available to plants and animals in extreme deserts.

2.7 Fog

2.7.1 What is fog?

The temperature of air decreased to such an extent that water droplets remain suspended in air, but do not deposit over surface is called as fog.

Fog is a stratus cloud that lies on or very close to the surface of the earth. The horizontal visibility in fog is reduced to less than 1 km according to International definition.

2.8 Mist

2.8.1 What mist?

Mist consists of an aggregate of microscopic sized droplets producing a thin grayish-veil over a landscape, reducing visibility to a lesser extent than fog.

2.9 Frost

2.9.1 Can we define frost?

The temperature of the air and earth surface decreases to such an extent that water condenses and freezes over the surface of deposition is called as frost. It is a solid phase of water and hence damages the crops to 100 per cent.

Frost occurs under the following condition:

(a) Surfaces on which the frost form must be at 0°C or below
(b) The surrounding air is saturated at 0°C or slightly below
(c) Nuclei are present, so that the process of sublimation can take place

2.10 Evaporation

2.10.1 What is evaporation?

It is a physical process of converting water molecules into water vapour and goes to the atmosphere with energy involving process and this takes place from water and land surfaces to the atmosphere.

2.10.2 How evaporation is measured?

Evaporation is measured in agricultural meteorological observatories thorough open pan evaporimeter.

2.10.3 What is the rate of evaporation per day in different seasons?

(a) During cold weather period (January–February) the evaporation per day would be around 2 to 3 mm/day.

(b) During hot weather period (March–May) the evaporation per day would be around 5 to 8 mm/day.

(c) During southwest monsoon season (June–September), the evaporation per day would be around 4 to 6 mm/day.

(d) During northeast monsoon season (October–December) the evaporation per day would be around 4 to 6 mm/day.

(e) The annual evaporation would be1800 mm–2100 mm and this value differs with location. In terms of litre of water per hectare, the annual loss would be 180 lakh litres to 210 lakh litres of water.

2.10.4 What is Evapo-Transpiration (ET)?

Evaporation from soil surface and transpiration from plant is called as evapo-transpiration. It can be also described as water loss to the atmosphere from the cropped land.

2.10.5 Is there any relation between crop yield and evapo-transpiration?

Yes. Relationship exists. If ET is more, the yield would be more (*e.g.*, Rice ET is 1200–1500 m and the yield is 5000 to 6000 kg/ha.). If ET is lesser, the yield would be lower (*e.g.*, Pulses ET is 200–250 mm and the yield is 300–500 kg/ha.).

3. Elementary Knowledge on Agro-met Observatory, Meteorological Instruments and their Maintenance, Data recording and Keeping Data Registers

3.1 What is agro-met observatory?

Meteorological observatory, which is established at agricultural research and service stations would provide meteorological data on rainfall, maximum and minimum temperature, wind speed, wind direction, bright sun shine hours, evaporation for farm decision making to reduce the crop production risks.

Its uses are many and among them the following are important.

(a) Data are useful to develop weather forecast in coordination with India Meteorological Department

(b) To take farm decisions for prevailing weather

(c) Developing weather-based technologies and testing

(d) To undertake response farming

(e) To manage pest and disease

(f) To enhance input efficiency

(g) Understanding climate and weather of the particular block/taluk/district/State/Country

3.2 What are the types of agro-met observatory?

Under agricultural meteorological observatory, there are three types and those are:

(a) Principal or A class observatory

Size: 36 × 55 m

Essential instruments:

- Maximum and minimum thermometers
- Wet and dry bulb thermometers
- Soil thermometers (5, 15, 30 and 60 cm depth)
- Grass minimum thermometer
- Self and ordinary rain gauges
- Wind vane and cup anemometer
- USD open pan evaporimeter
- Sunshine recorder
- Assman psychrometer
- Dew gauge
- Thermo and hydro graph
- Soil moisture equipment
- Solar radiation instruments

Optional instruments:

- Lysimeter
- Thermopile sensing elements for short and long wave radiation
- Automatic Weather Station (AWS)

At this observatory, both automatic and manual instruments of the listed items above would be installed.

This observatory will be established at State and districts headquarters of both India Meteorological Department and Agricultural Universities.

(b) Ordinary observatory or B class observatory

Size: 15 × 31 m

Essential instruments: All the instruments listed out in A class observatory except optional instruments.

Optional instruments:

- Sunshine recorder
- Dew gauge
- Self-recording rain gauge

- Thermo- hygrograph

This observatory will be established at all Agricultural Research Stations and 90 per cent of instruments are manual.

(c) Auxiliary or village level observatory

Size: 8 × 3 m

Essential instruments:

- Anemometer
- Non-recording rain gauge
- Single Stevenson Screen to house maximum, minimum, dry and wet bulb thermometers

This observatory will be established at village level and all the instruments are only manual.

3.3 Can we say something on the rules and regulations to be observed for the establishment of agro-met observatory?

- All instruments must be purchased with IMD specification number
- To be purchased instruments must be certified by the IMD office at Pune
- The time of observation for rainfall and evaporation would be at 8.30 am (IST) and others at 7.00 am IST and 2.00 pm IST(convert to local time)

3.4 What are the criteria to be followed for selecting a site for establishing agro-met observatory?

- It must be in the open field without any obstructions like trees or buildings or at a distance of at least 10 times the height of the obstruction
- It must be at the ridge point to drain the rain water against water stagnation
- It must be free in east-west direction in order to let in clear sunrise and sunset
- Must be easily accessible
- Away from trespassing and well protected area

3.5 How to maintain the observatory?

(a) All the instruments must be cleaned once in a week against dust (dust free)

(b) Anemometer and wind vane must be lubricated with 3 in 1 oil once in 15 days in a month

(c) Observatory must be painted with white once in a year including instruments

(d) Open pan evaporimeter must be checked against leakage

(e) Around the observatory, there must be fence to a height of 5 feet against animal entry and trespassing

(f) No trees are allowed to come up with in the observatory

(g) Termite protection must be given to the wooden posts and Stevenson's Screen box

(h) Ground surface of the observatory must be table plain against any water stagnation

(i) Once in every month, herbicide may be sprayed with in the observatory to control abnormal weed growth

(j) Wet bulb thermometer must be kept cool with salt free water

3.6 What are the rules governing data recording?

(a) Meteorological observations must be taken at the appropriate time stipulated

(b) Must be recorded in pocket register first with pencil against wetting by rain or drizzling

(c) Immediately reaching the office, it must be entered in the daily register and communicated to immediate authorities for action as per daily requirement

(d) Extreme values of the day must be highlighted in the register for further reference at latter times

3.7 What are the meteorological registers to be maintained?

(a) Daily meteorological data register

(b) Weekly meteorological data register

(c) Monthly meteorological data register

(d) Seasonal meteorological data register

(e) Annual meteorological data register

(f) Register for recording extreme weather events with date, name of the disaster

(g) Pest and disease outbreak information

(h) Pocket register for initial recording

(i) Registers may be maintained in long size 200 pages notebook and also keep the data in the computer (both soft and hard copy).

4. Introductory Knowledge on Monsoon Rainfall and Its Management at Village Level

4.1 What is monsoon?

The word monsoon is derived from the Arabic word "*mausim*" and this means season. Season is defined as prevalence of homogenous weather elements especially rainfall over a period of more than three months. Season permits crop cultivation in dry land. Cultivated crops get their evapo-transpiration requirement from rainfall. Filling up water bodies like dam, tanks, ponds etc., occurs during the season along with recharging of ground water. Hence under Indian condition, monsoon is described as rainy season.

4.2 How to assess the strength of a monsoon over land area?

Weak monsoon: Rainfall is less than the 50 per cent of the normal seasonal rainfall

Strong monsoon: Rainfall is 150–400 per cent of the normal seasonal rainfall

Vigorous monsoon: Rainfall is more than 400 per cent of the normal seasonal rainfall

4.3 How many monsoons are seen in India?

Two monsoon seasons have been identified in India based on wind direction *viz*, southwest monsoon season or summer monsoon (June–September) and northeast monsoon season (October–December). In addition, from November to March Central India experiences chillness with drizzling as a result of western disturbance system.

4.4 What are the properties of Southwest monsoon?

Warm and moist trade winds travel from east Pacific Ocean and Indian Ocean enters India through Kerala during last week of May on every year. It prevails over India from June to September. It gives rainfall to 80 per cent geographical area of India except Tamil Nadu and Jammu and Kashmir and the mean rainfall amount is 88 cm (880 mm). Except Tamil Nadu, all Indian States receive 80 per cent of their annual rainfall from this monsoon.

The rainfall during this season is mainly due to orographic effect (mountain hit by the moist and warm trade wind). On-set of southwest monsoon is last week of May or first week of June. It's with drawl occurs during first week of October.

4.5 What are the properties of northeast monsoon?

It is also called as retreating monsoon. After withdrawal of southwest monsoon from Rajasthan by September 15th, it crosses India to Bay of Bengal and gains energy and through formation of low/depression/cyclone, it gives rainfall to all districts of Tamil Nadu, coastal parts of Andhra Pradesh, Karnataka and Kerala. The average rainfall of this season is 42 cm or 420 mm. Its on-set is by mid-October and gets withdrawn by second week of December.

4.6 What is western disturbance?

It is originated from Afghanistan and middle-west as low-pressure Jet stream and crosses central India from November to April. Its average rainfall amount is from 50–100 mm (5–10 cm).

4.7 What is length of growing period?

The length of growing period is the safest period, where in the crop gets enough moisture from the soil as a result of monsoon rain to meet its evapo-transpiration.

Normally the number of days or months between on-set and withdrawal of a monsoon rain is considered as length of growing period irrespective of distribution of monsoon rainfall in general.

4.8 Who is monsoon manager?

At each village, the trained Climate Risk Manager also can be called as Monsoon Manager. He/She understands in advance the amount of rainfall to be received from the monsoon season from the long-range forecast or seasonal climate forecast and accordingly in discussion with majority of the farmers of the village he/she plans agriculture at the village level.

4.9 What are advantages that accrue when monsoon rainfall is managed?

(a) Water resource at the village level could be improved
(b) Well organized and efficient cropping pattern
(c) Proper use of natural resources without any degradation
(d) Minimizing climate risks in crop production
(e) Reducing the cost of cultivation
(f) Enhancing the input efficiency

4.10 How to manage monsoon disaster?

Step I Understand the long-range weather forecast or seasonal climate forecast in advance from different websites (www.imd.gov.in; www.tnau.ac.in). Review past monsoon disaster that happened in the village and document the activities done to manage the situation through PRA and personal interview with aged people and village level teachers.

Step II Discuss with the fellow farmers on the weather codes to be anticipated (normal season rainfall, below seasonal rainfall or abnormal rainfall *i.e.*, normal season, drought or floods).

Step III Prepare contingency plan for floods and drought situation at the village level including area to be cultivated, crops to be selected with relevant technologies.

Step IV Plan water management strategies in response to normal weather or flood weather or drought weather at the village level.

Step V Second and third discussion with the villagers may be conducted at the mid-season for mid correction and also after monsoon season to document learnt experiences.

5. Introductory Knowledge on Relationship Between Crop and Weather and Between Animal and Weather

5.1. What is the science behind relationship between weather and crop and between weather and animals?

The life of plants and animals mainly depend on prevailing weather especially for their growth and multiplicity. The benevolent (favorable) weather gives benefit to the crops/animals, while malevolent (unfavorable) weather would insert negative impact on them.

In meteorological literature, it is recorded that the climate of a particular region selects its own organisms to come up for life, while the weather that prevails within that climate controls the development and multiplicity of the organisms already permitted by that climate to come up. (*e.g.* Apple in temperate climate, banana in tropical climate).

It is also further added that the physiology especially the activity of hormone and enzyme of both crops and animals have direct link with the prevailing weather. Weather gives both biotic (Pest and diseases, weed) and abiotic (drought) stresses to the crops and animals on the negative side.

5.2. Can we list some examples for crops and weather and between animals and weather relationship?

- Artificial insemination is successful in big animals when thermal humidity index in lesser than 72
- When maximum temperature is greater than 35 °C with night temperature greater than 25 °C, the milk production in dairy animals would get reduced.
- During summer especially from March to April, there would be sudden rise in day temperature from 2 to 3 °C of the prevailing temperature of 34 to 36 °C and this hot environment development would kill the poultry birds en-mass.
- Bright sun shine hours are required to rice after flowering to have higher yield.
- Cloudy weather and high relative humidity would induce micro nutrient deficiency in turmeric during its vegetative stage.
- Wind free period is required for banana after its bunching against lodging.
- Rainfall is required for groundnut during its harvest.
- Minimum temperature must be above 18 °C during pegging stage and also during pod development stage of groundnut to have higher productivity.
- If the rice crop is planted in such a way that to make the rice crop to come for flowering towards full moon phase and this management would enhance rice yield to1.5–2 times more than the normal yield. (Nursery sowing has to be adjusted for this management).
- Wet spells (Continuous rain of more than 3 days) are detrimental to rice crop during its flowering stage.
- The optimum temperature appears to be near 30 °C for the maximum temperature and near 20 °C for the night minimum temperature in respect of rice crop for its higher productivity.
- During rice pollination, strong wind may induce sterility and increases the number of abortive endosperms
- Button shedding in coconut appears to be high during rainy season
- There is a lag period between the influence of weather and crop yield as the initiation of primordium in coconut takes 44 months before the final harvest. However, the duration of prolonged dry spell from November to May under south Indian condition may affect the coconut production in the following year under field condition.
- The maximum temperature has a positive impact on coconut yield under profuse irrigated condition, while the nut development is adversely affected if the maximum temperature is high during the second phase of nut development.

- High relative humidity together with moderate air temperature (both maximum and minimum) may be advantageous for both nut size and its quality in coconut
- Relatively dry atmosphere within winter season may be conductive for better flowering in cashew.

5.3. Can we reduce the impact from negative relationship between weather and crops and between weather and animals? If so how?

Yes, it can be done to a level of best. Response farming based on long range and seasonal climate weather forecasts would be a strategical decision to reduce/minimize the risks in the farm management of forthcoming season. For medium range and short-range weather forecast, tactical decision has to be taken based on the agro advisories given. This may reduce climate/weather-based risks in crops and animals' management programmes.

Some **examples** are given here under;

- To reduce the high maximum temperature beyond 34 °C, for sustainable crop production, irrigate the crop to field capacity immediately.
- Prop banana of more than six months old against lodging by heavy wind and rainfall.
- If heavy rain is anticipated/forecasted at the time of rice harvest, postpone rice harvest one or two days later along with providing drainage channel to drain excess rain water at the time of harvest.
- With the receipt of 25–30 mm of rain fall in one or two continuous day, dry land sowing may be undertaken to have uniform crop germination.
- Spraying and dusting of chemicals to crops towards leeward wind direction.
- If the evening relative humidity is more than 60 per cent and if it is continued for more than a week, preventive plant protection measures can be undertaken against the initiation of crop diseases.
- Providing wetted gunnies around the poultry shed during high temperature time would protect the birds from death. Similarly, during high wind period, cover the shed to reduce the wind speed against drifting of poultry feeds
- Bathing animals once or twice when thermal humidity index is more than 72 for effective artificial insemination and milking.

6. Introductory Knowledge on Contingency Plan Preparation for Different Weather Codes

6.1. What is contingency plan?

The contingency plan is a process that prepares action plan to respond coherently to aberrant weather situation like drought or flood or to prepare agricultural plan to an emergency situation to sustain productivity.

In literature it is indicated that contingency plan is a plan devised for an outcome other than in the usual plan. It is often used for risk management.

The crop management activities meant for normal weather situation cannot be practiced either during floods or droughts situation. Hence technologies suitable for crop production to manage floods and droughts must be tailored in advance to sowing season for each village in consultation with farming community and agricultural extension personnel. Past experiences also could be considered for this process.

6.2. What is weather code and how many types could be seen?

Any one of these three weather codes namely normal season (normal code), abnormal season (floods code) and subnormal season (drought code) would prevail in nature in rotation based on seasonal rainfall amount received.

Under climate variability situation which is triggered by natural balance, these weather codes would come in rhythm. This means that out of five years in Tamil Nadu, one year would be with drought code, one year with flood code and three years with normal weather code. But under "Climate change" scenario, the frequency occurrence of both flood and drought codes would be more and they may not follow any fixed rotation as noticed under climate variability scenario.

Hence proactive measures in terms of contingency plan have to be developed and executed.

6.3. When to prepare the contingency plan?

Preparation of contingency plan mainly depends on forecasted long range weather forecast or seasonal climate forecast information and under Indian condition, these forecast on rainfall for more than three months are available 15 days in advance of seasonal sowing. India Meteorological Department provides south west monsoon rainfall forecast (June–Sept) in advance. Similarly, the Agro climate Research Centre of TNAU provides seasonal climate forecast for south west monsoon and as well as for north east monsoon seasonal rainfall for different districts of Tamil Nadu in advance. From the weather forecast, we can infer that whether the coming season is droughty season or flood season or normal season. Accordingly, contingency plan can be prepared for droughty and flood weather codes. It is not necessary to prepare plan for normal weather code.

6.4. Is it necessary to prepare the plan at the village level or block level?

It is suggested to prepare the plan for a village rather than for a block level, since variability is getting enlarged at block level. Precious result could be obtained at the village level.

6.5. What are the components of village level contingency plan?

- Preparation of contingency plan for agricultural activities
- Preparation of contingency plan for animal activity including fodder management
- Preparation of plan for drinking water and other natural resources management

6.6. What are the steps to be followed to prepare the contingency plan?

- Let us understand what is long range weather forecast or seasonal climate forecast information for the ensuing crops season (Whether it is droughty season or flood season or normal season)
- Arrange pre seasonal village level meeting at village level and inform the anticipated weather code and discuss the options to be implemented with the villagers.
- Prepare a detailed plan covering crop and animal management and discuss with selected elderly farmers and get their views and refine. The indigenous knowledge's that are already available may also be considered for inclusion as best bet technologies.
- Procurement of inputs and fixing responsibilities with the villagers on the division of work to be attended
- Mass communication of the plan prepared with in the village and keep ready for implementation.

6.7. What are the village level activities to be considered for preparing the contingency plan?

(a) Cropping pattern/cropping system
(b) Ground/surface water availability
(c) Forage/fodder requirement
(d) Crops to be cultivated
(e) Variety to be selected for the selected crop
(f) Technology to be selected including best bet technology
(g) Input requirement and procurement
(h) Pest and disease load at the village level
(i) Labour availability
(J) Normal agricultural practices of the village
(k) Farmer's previous experience at the village.

7. Weather Forecast And Its Application In Agriculture Including Preparation of Agro-Advlsories

7.1. What is weather forecast?

As indicated already elsewhere in this book, weather is defined as the day to day change in the atmosphere. The anticipated change in the weather for three to five days in advance in terms of change in rainfall, temperature, wind speed, wind direction, cloud amount and relative humidity will be communicated as weather forecast under medium range weather forecast. Prediction of change in weather elements is done by employing Global circulation model or by using Regional climate model as the case may be. This forecast will be communicated to the client farmers for making weather-based farm decisions. Especially in agriculture weather forecast information would be very useful to change or shift the proposed farm operations to reduce crop management risks. Some examples are as follows:

- If heavy rain is forecasted on the proposed day of rice harvest, post- pone the rice harvests.
- If rain more than 25 mm is received during dry land sowing season, sowing of dry land crops can be done successfully.
- Anticipating rain, proposed irrigation schedule can be postponed to 10–15 days later based on the soil water holding capacity

Considering the importance of weather forecast in biological life including agriculture, the following jargon has been developed by the senior author of this book.

"BE weather wise–otherwise–Not wise".

7.2. Can you explain what is weather based farm decision?

Altering the proposed farm operations/decisions to reduce climate and weather-related risks based on weather forecast received. This is called as weather-based farm decisions.

7.3. How many types of weather-based farm decision are available in the field?

There are two types of weather-based farm decisions and those are.

(a) Farm decisions based on the weather forecast
(b) Farm decisions based on the prevailing weather

7.4. Can you explain how farm decisions are taken based on weather forecast?

In practice four types of weather forecast are available and they are:

- Short range weather forecast
- Medium range weather forecast
- Long range weather forecast
- Seasonal climate forecast.

The details on these forecasts are given in the Table.

Name of the weather forecast	Issuing organization	Period of forecasting	Preciousness of the forecast (%)
Short range weather forecast	India Meteorological Department	1–2 days	75–85
Medium range weather forecast	India Meteorological Department	3–10 days	60–70
Long range weather forecast	India Meteorological Department	10 days–3 months	50–60

Name of the weather forecast	Issuing organization	Period of forecasting	Preciousness of the forecast (%)
Seasonal climate forecast	Agro-climate Research Centre, TNAU, Coimbatore	South west monsoon and North east monsoon seasons for Tamil Nadu alone	60

7.5. What is short range weather forecast and how it is useful to agricultural farm decisions?

Precious forecast on rainfall, temperature and wind speed will be given for next 24 hours with an outlook for another 24 hours and this is called as short-range weather forecast.

Based on the weather forecast, weeding, irrigation, plant protection and harvesting can be rescheduled or adjusted to reduce climate-based risks.

Example:

Weather forecast: Heavy rain is anticipated in 24 hours

Crops raised: Rice

Stage of the crop: To be harvested

Weather based farm decisions to be taken: Postpone rice harvest to later days and provide drainage channel to drain flood water.

7.6. What is medium range weather forecast and how it is useful in agricultural farm decisions?

This weather forecast carries information on anticipated change in rainfall amount, wind speed, wind direction, cloud cover, relative humidity, maximum and minimum temperature for the next five days. This is being communicated through radio, dailies, Television channels, mobile based SMS etc.

The forecasted information can be used effectively from land preparation to marketing.

Example:

Weather forecast: With daily rainfall of 20mm/day in the coming five days 100 mm of rainfall is anticipated.

Crops to be raised in dry land: Cotton

Stages of the field: Land has been prepared for sowing already

Weather based farm decisions to be taken

(a) Sow the cotton seeds @ 5 cm depth in the soil after basal application of fertilser under optimum soil moisture for uniform germination with the receipt of rainfall.

(b) Provide drainage channel to drain excess rain water with subsequent heavy rains.

7.7. What is long range weather forecast and how this will be useful in agriculture

Presently the India Meteorological Department gives rainfall forecast for southwest monsoon (June–September) separately for east India, West India, North India and South India. This forecast gives only the information on total seasonal rainfall anticipated. By utilizing the information, farm decision on the area to be cultivated, crop and variety to be selected, identification of appropriate technology etc., could be decided in advance to sowing season. This type of decision making always minimizes the crop production risks to a satisfactory level. This we will call it as response farming and dominantly seen in Australia country.

Example:

Weather forecast: In the coming north east monsoon season (Oct–Dec) there would be 800 mm of rainfall against normal seasonal rainfall of 420 mm

Farm area: 5 acres

Normal cropping system: 3 acres Maize, 2 acres rice

Weather based farm decision to be taken: Cultivate rice in all the five acres, since Maize can't tolerate water stagnation as a result of excess rainfall.

7.8. What is seasonal climate forecast and how it is useful?

Rainfall forecast, (total for the season, and as well as month wise breakup for each village/block/district) will be given 15 days in advance to start of the cropping season. This forecast carries information on rainfall covering both temporal and spatial dimensions.

(Refer the example given for question No 7.7)

7.9. Is it possible to integrate all the four weather forecasts in a single window system for taking effective farm decisions?

Yes. It is possible. First consider long range/seasonal climate forecast information and accordingly decide the major activities of the farm and executive them as given in Q7.7. Once the crop is sown, propose farm operations based on the medium range weather forecast information given for five future days and fine tune to short range weather forecast information on day to day basis. This type of integration reduces climate risk, cost of cultivation and enhance input use efficiency.

7.10. What is farm decision based on prevailing weather?

Based on the weather event or prevailing weather as felt by the farmer reschedule the proposed farm operations and take new weather-based farm decisions.

Example

Weather event/prevailing weather	Farm decision is to be taken.
Receipt of 50mm of rainfall	Stop irrigation, drain the stagnated water.
Receipt of 10–15 mm rainfall	Advised for fertilizer application

Receipt of 10–15 mm rainfall	Advised for hand weeding and hoeing.
Receipt of 5–10 mm rainfall	Advised for plant protection.
Wind < 5 km/hour	Plant protection can be taken.
Wind > 5 km/hour	Not advised for plant protection.
Wind 25 km/hour	Advised for propping in banana of more than 6 months and sugarcane of more than eight months.
Maximum temperature 32 °C for a week	Advised for plant protection against sucking pest.
Minimum temperature 20 °C for a week	Advised for plant protection against disease.
Cloudy environment with high RH for a week	Brown plant hopper may infest rice crop. Micro nutrient deficiency would appear in turmeric at its vegetative stage.

7.11. What is agro advisory and what for this is given?

Advisory in terms of changed farm operations with reference to changing weather information (weather forecast) is termed as agro advisory.

7.12. What are the points to be considered for preparing agro advisory?

- Previous week weather and their impact on standing crops at village.
- Changing weather in terms of rainfall, maximum temperature, minimum temperature, wind speed, wind direction, cloud amount, relative humidity for the next five days.
- Crops cultivated in the village and their stages, pest and disease incidence
- List out normal farm operations to be carried out at this stage.
- Considering sensitiveness of crop stages to changing weather for the crops cultivated in the village, select best weather-based technology for the present stage of the crop and communicate as agro advisory to the farmers.

7.13. Is it possible to prepare agro advisory for a village?

Yes. The example is as follows;

Name of the village: Perumpalayam.

Season: North east monsoon.

Month: November

Crops cultivated and their stages: Rain fed groundnut: 45 days, Rainfed black gram: 30 days. Irrigated maize: 55 days, Rice (medium duration): 35 days from transplanting, Banana: 8 months old, No of milch animals in the village: 100.

Medium range forecast received for five days

Date	Maximum temperature (°C)	Minimum Tempera-ture (°C)	Rainfall (mm)	Relative humidity (Morning) in per cent	Relative humidity (Evening) in per cent	Cloud cover (octa)	Wind speed (kmph)	Wind direction (degrees)
5.11.14	31	24	0	85	60	6	50	90
6.11.14	31	23	20	90	62	7	40	91
7.11.14	32	22	15	94	67	7	50	90
8.11.14	32	21	30	88	65	8	40	92
9.11.14	31	20	05	80	70	3	40	90

Preparation of agro advisories:

a. Observation made on the weather forecast received.

- Optimum maximum temperature is anticipated in the next five days
- Minimum temperature found reducing over days
- Total of 70 mm rainfall is expected.
- Both morning and evening relative humidity found to be on the higher side.
- Sky is heavily covered with clouds except on 9.11.14.
- High wind speed is expected, may be due to cyclonic effect.
- Wind direction is from North east direction.

b. Sensitiveness of crops and their stages to changing weather has to be considered.
Rain fed groundnut: 45 days. Gypsum has to be applied on 45th day, and hence rainfall is required to dissolve applied gypsum so as make available to the crop.

Rain fed black gram: 30 days. It does not tolerate water stagnation in the field.

Irrigated maize: 55 days. The crop would be under tasseling and silking stages. Since the crop is cross pollinated, it does not tolerate water stagnation.

Rice: Boot leaf stage

Banana: 8 months old. Bunches have come; it adds weight to the plant.

c. Selection of technologies is to be done.
Ground nut: Gypsum has to be applied.

Drainage is to be provided because of continuous rain.

Rain fed black gram: Drainage is to be given.

Irrigated maize: Cross pollination is sensitive to rain. Drainage is to be given.

Rice: Drainage is to be given.

Banana: Support is to be given against lodging with wind, since banana has come for bunching.

d. Agro advisories to be given.

Expecting rainfall, gypsum application to rain fed ground is to be given.

Provide drainage channel to drain excess rain storm water from groundnut field.

Provide drainage to standing black gram, since the crop does not tolerate water stagnation from rain.

Since irrigated maize does not tolerate water stagnation, provide drainage channel to drain rain water.

Don't top dress 'N' to rice though rice is at boot leaf stage. Anticipating heavy rainfall provide drainage facility to rice crop.

Since the wind speed is more and the banana is with growing weight from bunch, provide bamboo support to banana against lodging.

8. Introductory Knowledge on Weather Relationship for Pest and Disease Epidemic.

8.1. Do pests and diseases depend on weather?

Yes. Pest and disease depend on weather input (both directly and indirectly) for their initiation, development and multiplication. Pests are related to maximum temperature, diurnal variation, wind speed and wind direction, while the disease occurrence depends on relative humidity, wind speed, wind direction and minimum temperature.

Pest and disease outbreak are mainly triggered by weather elements.

8.2. Is there any pest and disease weather dependent model?

Yes. Three types of model are available and they are:

(a) Thumb rules developed by local people

(b) Statistical model (Step down regression model)

(c) Dynamic Simulation model.

Similar to weather forecast, pest and disease outbreak in relation to weather can be given as forecast in two frames namely short range and medium range pest and disease forecast. These forecasts have been generated from thumb rules.

In the case of statistical model, equations have been developed from the long past data set of both dependable and independent variables. In science, this type of statistical analysis is termed as postmortem analysis, that means it cannot be used further. But this limitation is eliminated now and statistically equation can be used effectively for future prediction.

In the case of dynamic model, many soft wares are available like Oryzae for rice blast disease, but this area needs further improvement.

8.3. Can we use those models to forecast pest and diseases out break?

Yes. But they need some scientific refinement except thumb rules.

8.4. Can you provide some examples for these models?

Thumb rules

- Sowing groundnut very late than the normal sowing would result in heavy rust disease incidence
- Drop in mean temperature (T max + T min/2) (< 28 °C) in association with continuous light rain was found to be congenial for the outbreak of rice stem borer.
- Rainfall of 20–30 mm did reduce thrips population in black gram up to 51 per cent.
- Tea mosquito bug population in cashew would be higher, when the night temperature is between 15 and 22 °C.

Statistical model

- Under normal date of sowing for groundnut, the rust disease was noticed only after 70 days of sowing, while it was 61 and 51 days after sowing for 15 days and 30 days delayed sowing from the normal date of sowing.
- High maximum and minimum temperatures of pre monsoon period with low relative humidity followed by continuous rains coupled with high maximum and minimum temperature followed by monsoon break for a week or two, favored the outbreak of groundnut leaf miner.
- These models will be given under statistical equation

Dynamic model

- Oryzae model for rice blast

9. Introductory Knowledge on Climate Change and its Impact

9.1. What is climate?

Climate is defined as long term average of the atmospheric variables (Rainfall, At. pressure, Wind, Relative humidity) over larger area or the composite of the day to day values of weather elements over a long period of a given place or region or aggregates of weather over longer period of 30 years covering larger domain.

9.2. What is the importance of climate?

Climate is the basic requirement for any living organism. The importance of climate and weather in bio-sphere is that the climate of a particular geographical area dictates the selection of a particular organism to come up, while the growth and the productivity of the selected organism is controlled by the prevailing weather of that climate.

9.3. List out the type of climates and how they are classified

There are six climate types and those are "arid, semiarid, sub -humid (dry and wet), humid and per humid".

In a simple way based on annual rainfall and evaporation, climate types can be classified.

If the ratio between annual rainfall and evaporation is lesser than 0.3, it is called as **arid climate**. In other words, maximum of 30 per cent of annual evaporation is met by rainfall and 70 per cent is rainfall deficit.

If the ratio between annual rainfall and evaporation is between 0.31–0. 50, it is called as **semi-arid climate**, where in 50–69 per cent of annual evaporation is met by rainfall and deficit is from 31–50 per cent.

If the ratio between annual rainfall and evaporation is between 0.80–0.99, then the climate is **sub humid**, where with temperature variation, there are two sub types:

(a) Sub humid climate–wet: The temperature is mild.

(b) Sub humid climate–dry: The temperature is very hot.

If the ratio between annual rainfall and evaporation is around one, the climate is called as **humid** where in the annual evaporation demand (Breakeven point) is completely met by the rainfall received.

If the ratio between annual rainfall and evaporation is more than one, then the climate is called as **per humid climate,** where the rainfall is more than annual evaporation demand.

Among these climates, agricultural production could be enhanced heavily in semi-arid climate, when crops are irrigated.

9.4. What is climate variability?

This is a natural phenomenon. Out of five continuous years in a region, drought will come in one year followed by one-year flood and three years normal rainfall. This cycle is getting repeated in sequence. This is mainly due to natural balance. This climate variability does not add greater risk to crop production, since farmers are already adapted.

9.5. What is global warming?

Greenhouse gases namely carbon dioxide, nitrous oxide, methane and chlorofluro carbon have been released to the atmosphere after industrial period heavily without any control. This happened freely in all developed countries. These gases have long shelf life and stays in the stratosphere and spreads like a blanket over the earth. This green gas blanket permits incoming solar radiation to the earth from the sun, but does not permit the outgoing radiation back to the space which is normally happening. Hence the arrested and accumulated infra-red rays (hotness) between the greenhouse gas blanket and earth surface, gets rotated/articulated and increases the earth air temperature. In a short way, global warming is nothing but increase in global surface temperature over a long period of time with greenhouse gas accumulation in the atmosphere.

In the past of 150 years, global mean temperature did increase by almost 0.8 °C. If the present rise in temperature is continued without any attempt to check (mitigation), it is estimated that the global temperature may rise up to 5.8 °C by the year 2100 with an average value of 3 °C. But the IPCC has requested all countries to reduce greenhouse gas emission to stabilize rise in temperature to 2 °C.

9.6. What is irradiative index?

This index is used to measure the heat imbibing capacity of a molecule of greenhouse gases or heat holding capacity of the greenhouse gases.

9.7. If the radioactive index of carbon dioxide is one, what would be the value for other greenhouse gases?

Methane: 30

Nitrous oxide: 150

Chlorofluro carbon: 10000

9.8. Does global warming lead to climate change? Then what is climate change?

The answer is yes. When surface air temperature gets increased, with the accumulation of greenhouse gases, the atmospheric condition gets changed and there is atmospheric confusion. This may lead to permanent shift in climate from one to other (Arid becomes semi-arid and vice versa)

Example: The State Rajasthan climate would shift to humid climate from arid climate, while the State Assam comes to arid climate from humid climate as a result of climate change, which is triggered by global warming.

9.9. Can you list out what will happen if climate is changed?

- Agricultural potential will be lost.
- With melting of ice glacier, perennial river would become seasonal river.
- With sea level rise, land–ocean ratio will be altered.
- Human health will be affected including rise in infertility cases.
- Water resource will be affected with shift in hydrological cycle.
- Energy resource would be affected.
- Bio diversity will be lost.
- Livelihoods of rural mass will be affected
- Frequent occurrence of extreme weather events.
- Fish population and fish breeding will be affected.
- Marine wealth will be lost.
- The positive interaction between different eco system would become negative.

- Animal production would be affected
- Meteorologically frequent occurrence of strong lightning and dangerous thunders
- Conflict between countries or chance of getting economical war or cold war between countries

10. Introductory knowledge on coping mechanisms and adaptation strategies to meet climate change impact.

10.1. What is adaptation?

Having understood climate change, we must understand the word "adaptation".

It is nothing but adjustment in natural or human systems to a new or changing environment.

Adaptation to climate change refers to adjustment in natural or human systems in response to actual or expected climatic stimuli or their effects, which moderates harm or exploits beneficial opportunities.

10.2. What are the three types of adaptation available in the literature against climate change?

Any effective development and planning process will need to take climate adaptation into account and conversely adaptation efforts themselves will often require development interventions to succeed.

The three types of adaptations are:

(a) Serendipitous adaptation: Activities undertaken to achieve development objective incidentally achieve adaptation objectives.

(b) Climate Proofing of ongoing development: Activities added to an ongoing development initiative to ensure its success under a changing climate.

(c) Discrete adaptation: Activities under taken specifically to achieve climate adaptation objectives.

10.3. How to improve the adaptation capacity of the villagers against climate change?

The best way is to enhance meteorological knowledge of the community and skill development. Some ways to enhance meteorological knowledge are:

- Running climate school at the village level during evening/night once in two to three days in a week and exchange meteorological knowledge.
- Preparing climate atlas for the village that includes information on mean values and extreme values of rainfall, temperature, relative humidity, wind speed, evaporation and details on flood and drought occurrence.
- Conducting pre-season, mid-season and post-season village level crop–weather link workshops.
- Empowering the farmers to use weather-based farm technologies for the forecasted weather information.

- Empowering the farmers to watch and monitor weather change during cropping season.

10.4. What are steps to be taken to introduce adaptation processes successfully?

(a) Addressing the drivers of vulnerability of the village to climate change impact.

(b) Building response capacity at the village level (Capacity building to farmers and establishing infrastructure)

(c) Managing climate risk

10.5. What is coping capacity?

This can be defined in two ways.

(a) The ability to cope with threats includes the ability to absorb impacts by guarding against or adapting to them. It also includes provisions made in advance to pay for potential damage, for instances by mobilizing insurance repayments, saving or contingency reserves.

(b) The manner in which people and organisations use existing resources to achieve various beneficial ends during unusual, abnormal and adverse conditions of a climate disaster event. The strengthening of coping capacities usually builds resilience to withstand the effects of natural and other hazards.

10.6. What is difference between adaptation and coping capacity against impact from climate change?

When an organization supports the adaptation capacity of the people, it is called as coping capacity, while action of the people by themselves without any support from any organization and infrastructure and then it becomes adaptation.

10.7. Can we say some example for coping infrastructures at the village level?

(a) Village level mini agro–met observatory

(b) Village Knowledge Centre.

10.8. What is mitigation?

Action taken to reduce the release of greenhouse gases from industries and others to the atmosphere or to minimize the release of greenhouse gases to the atmosphere.

It can also be explained as structural and nonstructural measures' undertaken to limit the adverse impact from climate change, environmental degradation and technological hazards.

Some **examples** are given below.

(a) Planting trees to act as sink to released carbon dioxide.

(b) Soil carbon sequestration.

(c) Fixing threshold level to release green houses to the atmosphere by the Government

(d) Levying tax on per unit of carbon release to the atmosphere.
(e) Carbon emission trading
(f) Clean development mechanism.

11. Vulnerability assessment of a village to climate change impact.

11.1. What is vulnerability of a village?

The degree to which a village is susceptible to or unable to cope with adverse effects from climate change, including climate variability and extremes.

Vulnerability of a village to climate change is a function of the character, magnitude and rate of climate variation to which the villages is exposed, its sensitivity and its adaptive capacity.

11.2. What is the necessity for assessing village vulnerability to climate change?

Strength and weakness of the village in terms of coping and adaptive capacities have to be assessed in advance to achieve satisfactory resilience capacity.

11.3. How to assess a village whether it is vulnerable to climate change or not.

Many methods including models are available. Simple method is given here under.

Step I

Select the parameters, which make the village vulnerable to climate change.

For example, one village "A "has been selected. Ten parameters have been selected for assessing the village for its vulnerability.

1. Population of the village/km^2
2. Village forest area (per cent)
3. Irrigated area in the village (per cent)
4. Water source for irrigation
   Tank/Wells/Canals
5. Ground water depth (mean value in m)
6. Poverty per cent at the village
7. Literacy per cent at the village
8. Per capita/day in come from the livelihood
9. Per capita food production at the village
10. Coping infrastructure facility available at the village

These parameters need not be considered for all the villages. Parameters can be changed based on the assessment of village for example, if a village is very nearer to sea, parameter on sea level rise, area to be submerged. salinization of agricultural lands, fish catching etc., can be considered.

Step II

Fix threshold level for each selected parameter. State level or National level mean values can be considered for fixing threshold level.

1. Population of the village: The optimum is 300/km^2
2. Village level forest area: Optimum is 33% of geographical area.
3. Irrigated area in the village: 30% of cultivated area in the village.
4. Water source for irrigation: Optimum is any two sources viz Well/tank/canal etc.
5. Ground water depth: 30–50m depth from the ground level (Optimum)
6. Poverty per cent at the village: < 20 % (Optimum)
7. Literacy per cent at the village: > 70% (Optimum)
8. Per capita/day in come: ₹500 (Optimum)
9. Per capita food production/day: 500g (Optimum)
10. Coping infrastructure facility available at the village: Village knowledge centre/ Agro met observatory/etc. (Any one)

Step III

Fixing marks for threshold level and also fixing marks for below and above threshold.

Selected Parameter	Mark to be given for threshold	Marks to be given for below and above threshold level	
Population of the village/km^2	0 mark for the population of 300/km^2	Population Marks to be given Density	
		301–400	10
		401–500	30
		> 500	60
Village level forest area	0 mark for 33%	Forest area	
		< 10%	60
		10.1–20.9%	30
		21–32%	10
Irrigated area in the village	0 mark for 30%	Irrigated area	
		< 10%	60
		10.1-19.9%	30
		20–29%	10
Water source for irrigation	0 mark for any two sources	Water source	
		canal alone	60
		Tank alone	30
		Well alone	10
Ground water depth	0 mark for up to 50 m	Ground water depth	
		51–75m	10
		76–100 m	30
		> 100 m	60

Poverty per cent at the village	0 mark for < 20 %	Poverty per cent 20–30 % 10 30–40% 30 > 40% 60
Literacy per cent at the village	0 mark for > 70%	Literacy level 60–69% 10 51–59% 30 < 51% 60
Per capita/day income (Rupees):	0 mark if the income is ₹500/	Per capita income 400–499 10 301–399 30 < 301 60
Per capita food production at the village level/day (g):	0 mark if the per capita food production is 500 g/day.	Per capita food production 400–499 10 301–399 30 < 301 60
Availability of coping infrastructure facility at the village	0 mark for optimum. (any one)	None 60

Step IV

Evaluate the vulnerability of the village for the following values collected from the village in comparison with threshold value for awarding marks

Parameters	Values collected from the village	Marks given	Vulnerability index
Population of the village/km^2	500/km^2	60/100	0.6
Village level forest area	15%	30/100	0.3
Irrigated area in the village	20%	10/100	0.1
Water source for irrigation	Tank and wells	0/100	0
Ground water depth	70m	10/100	0.1
Poverty per cent at the village	60%	60/100	0.6
Literacy per cent at the village	40%	60/100	0.6
Per capita/day in come:	₹300/-	60/100	0.6
Per capita food production at the village level/day:	₹450g/day	10/100	0.1

Availability of coping infrastructure facility available at the village	None	60/100	0.6
Total marks	—	360/1000	0.36

V Step

Assessing the village for its vulnerability to climate change
Scale has to be developed and used. One scale is developed and given below for this exercise.

Scale to be used:	**Vulnerability grade**
Not Vulnerable	
0.1–300	Slightly vulnerable to climate change
300.1–600	Moderately vulnerable to climate change
> 600.1	Highly vulnerable to climate change

In the evaluation process, the village did record 360 marks and hence this village is moderately vulnerable to climate change and accordingly adaptation skill can be given. This is for an example. Similar to this you can assess the vulnerability of your village by selecting parameters purposefully.

12. Developing knowledge to keep records on climate information to be maintained at the village level.

12.1. What are the records on climate information to be maintained at the village level?

- Performance of monsoon rainfall in terms of its on- set, with drawl and number of dry spells occurred within a season
- Extreme rainfall information over years
- Annual and seasonal mean values of maximum temperature, minimum temperature, relative humidity, wind speed and wind direction over years (wind rose)
- Pest and disease outbreak in relation to weather
- Natural disaster like earthquake, Tsunami and agricultural losses in detail due to cyclone
- Severe floods and droughts occurrence and agricultural losses met
- ITK on crop wealth relationship
- Ruling proverbs on climate and weather at the village
- Crops raised and yield over seasons of different years
- Best bet technologies

12.2. Who will maintain the registers?

Climate risk manager will maintain both hard and soft copies of the information given above (12.1).

13. Resilience in village level agriculture

13.1. What is resilience?

The ability to resist downward pressures and to recover from a shock.

It can also be defined as "a measure of how quickly a system (village) recovers from failures. Some more definitions are given hereunder.

- Resilience is a measure of the recovery time of a system.
- The capacity of a group or organization to withstand loss or damage or to recover from the impact of an emergency or disaster. The higher the resilience, the faster and more effective recovery is likely to be.
- A resilience system/village is not sensitive to climate variability and change and have the capacity to adapt.

13.2. Is there any difference between coping capacity and resilience?

In real life, the damage caused to a village by climate change not only depends on its intensity, vulnerability of the village to climate change impact and exposure of the village to climate change in temporal scale, but also on coping capacity and the resilience capacity of the village at risk.

There is always overlapping between coping capacity and resilience.

Coping capacity encompasses those strategies and measures that act directly upon the impact as caused by climate change by alleviating or containing the damage or by bringing about efficient relief, as well as those adaptive strategies that modify behavior or activities in order to circumvent or avoid damaging effects.

Resilience is all of these things plus the capability to remain functional during the period of climate change and to completely recover from it. Hence, resilience includes coping capacity also.

13.3. How to enhance the resillence capacity of the village against the anticipated impact from climate change?

- Creating awareness with the villages particularly farmers on the cause and effect of climate change.
- Enhance the forest area in the village by planting trees in waste lands etc.
- Enhancing recharge capacity of ground water at the village by providing percolation ponds and check dam across streams.
- Provide in situ moisture conservation practices in rain fed/dry land areas.
- Practicing response farming.
- Water harvesting at houses/offices/institutions/common buildings.

- Establish village knowledge centre and mini agro met observatory.
- Capacity building to the villages including farmers on the need of adaptation to climate change.
- Demonstration on weather-based farm technologies at the village
- Creating infra-structure facilities

Annexure II

Practical Tools (Computations and Calculations)

I. General Rules and Regulations of Meteorological Data Set Use

1.1 Period of Calculation

Under the Technical Regulations of WMO, NO. 49, climatological standard normal is the average of climatological data computed for a data set of a minimum of thirty years. This is called as a block of 30 years. For getting further precision of the results, a minimum of three blocks of past data have to be taken for analysis. This will reduce the standard deviation and CV of data set to be used. Blocking of 30 years may be taken as given below;

1st January 1901 to 31st December 1930 and 1st January 1931 to 31st December 1960, and so forth.

The most recent period for climatological standard normal is from 1961 to 1990, and the next period for the calculation of climatological standard normal will be *1991 to 2020.*

1.2 Missing data

Normal calculated from incomplete data sets can be biased.

As there is often considerable autocorrelation in climatological data, consecutive missing observations can have more impact on normal than the same number of missing observations scattered randomly through the period in question.

As a guide, normal or period average should be calculated only where values are available for at least 80 per cent of the years of record, with not more than three consecutive missing years.

No missing monthly normal is permitted in the calculation of annual normal. It is recommended that a monthly value should not be calculated if more than ten daily values are missing or five or more consecutive daily values are missing.

II Rainfall analysis

2.1 Extreme weather events study

To assess the Extreme weather events - Return period can be analyzed

(Veeraputhiran *et al.,* 2003).

Step 1: Arranging the data in the ascending order.

Step 2: Giving the serial number as order number to the arranged data of step 1.

Step 3: Computing cumulative frequency (CF). CF can be computed as per the following model.

$$CF = \text{Number of order/Total order} + 1$$

Step 4: Find out the value for 1-CF.

Step 5: Compute uncorrected return period with the following model. Uncorrected RP = 1/1 – CF

Step 6: Find out a multiplier which makes the uncorrected return period values as a whole number (product).

Step 7: Find out the ratio between the multiplier and the product.

2.2 Coefficient of Variation (CV in per cent)

$$CV = \frac{SD}{\overline{X}} \times 100$$

where in;

CV: Coefficient of Variation

SD: Standard deviation of the data set

$\overline{X}$: Mean of the data set

By our experience We feel that the following threshold values of CV are acceptable.

Yearly rainfall data set = < 25%

Seasonal rainfall data set = < 50%

Monthly rainfall data set = < 100%

Weekly rainfall data set = < 150%

Daily rainfall data set = < 250%

2.3 Initial Probability (IP)

$$IP = \frac{n \times p}{100}$$

Where in; IP: initial probability in per cent

n: sample size (number of observations used for analysis)

p: Probability to be obtained (whether the required result is for 50 or 60 or 75 per cent probability level)

Steps to be followed

Arrange the data set in descending order

Compute the probability to be required based on the number of samples used. For example, it is 30 years rainfall data and arrange in descending order as per the rule and we require 75 per cent of rainfall amount, then 30 × 75/100 = 22.5. Select the 23rd data arranged in descending order and this is the 75 per cent rainfall amount to be anticipated. This can be worked for daily, weekly, monthly and yearly data set of any meteorological weather variables.

2.4 Conditional Probability (CP)

$$CP = \frac{\bar{X} - x}{SD}$$

Where in CP: conditional probability

$\bar{X}$: = Mean of the data set

X: Required quantity of rainfall for carrying out selected farm operation like weeding, sowing etc.,

SD: Standard deviation

Since the resultant value does not fall under normal distribution, it has to be referred to Z Table (annexed) for normal distribution and multiplied by 100 to find out the actual probability in per cent. If the probability is more than 60 per cent, it can be considered for weather-based farm decision.

At this context two situations have to be solved for arriving probability.

Situation 1: If the resultant value from the above formulae used for computation, is positive, the values thus obtained from the formulae can be directly referred in Z Table and multiplied by 100 to arrive probability in per cent.

Situation 2: If the resultant value from the above formulae used for computation, is negative, the obtained value from the formulae may be referred in Z table and the obtained value from Z table may be deducted from 1 before multiplying with 100 to arrive probability in per cent.

2.5 Monsoon Rainfall Onset and Withdrawal

The knowledge on probable date of commencement and with drawl of rainy season is very important to find out the length of growing period and accordingly cropping pattern/cropping system can be decided. This minimizes crop risks to malevolent weather. There are two methods and one is Raman (1974) method and the other one is IMD method.

(a) Raman (1974) Method.

Raman has identified the commencement of sowing rain for black cotton soils of Maharashtra State. He had taken criteria of a seven days wet spell with a total rainfall of 25 mm with not less than five days of one mm or more rainfall in each day as the commencement of sowing rain. The major lacuna of this criterion is that it has no variable factor incorporated in it to account the daily evaporation. Another defect is one mm of rainfall was too meager to call for a rainy day.

Subsequently three criteria have been fixed to define the date of onset of monsoon rain and those were;

- The first day rain in the seven days wet spell must be not less than 'e' mm (e is the mean daily evaporation in mm)
- The total rain during the seven-day spell must be not less than 5e+10mm
- At least four out of these seven days must have rainfall > 2.5mm

(b) IMD Method

The date of either onset or withdrawal of monsoon rainfall is fixed based on the plotting of five days total normal rainfall for the selected monthly window of the mean of the past 30 years.

Five days total of past 30 years (example; 1961–1990 or 1991-–2020) of rainfall can be taken. For computing SWM onset, five days total rainfall may be taken from May 20th to June first fortnight and for NEM five days total rainfall may be taken from September 15th to October 30th. Similarly, for the with drawl of SWM monsoon, five days total rainfall from September 15th to October 30th and from December first to December 30th can be taken for computing NEM with drawl. These data can be plotted in graph sheet separately for on set and with drawl and as well as for each season. The five-day total in which there is sudden spurt in rainfall compared to earlier five days total with sustained rainfall thereafter and this five day can be taken as onset and the mid date of the five days total can be taken as onset day. Similarly, in respect of with drawl, sudden decline in rainfall occurs in a five days total and there after no rain and the mid date of the five days total may be taken as with drawl of monsoon rainfall. The number of between onset and with drawl would be length of growing period. But the deficiency is to find out separately the distribution of rainfall within the growing season identified. The mean of five days potential evapo-transpiration corresponding to the five days rainfall was also taken in to consideration for the purpose of more précised assessment.

2.6 Rainfall forecast verification

2.6.1. Forecast Accuracy (ACC) or Ratio score or Hit score

$$ACC = YY + NN/YY + NN + YN + NY$$

First letter in the pair: Forecasted one

Second letter in the pair: Events occurred

Y = yes; N = No

2.6.2. Heidke Skill Score (HSS)

$HSS = ZH- FM/\{(Z+M) (M+ H) + (Z+F) (F+H)\}/2$

Z = No of correct predictions of no rain (neither predicted nor observed)

F = No of false alarms (predicted but not observed)

M + No. of misses (observed but not predicted)

H = No of hits (predicted and observed)

2.6.3. Root Mean Square Error (RMSE)

$RMSE = [\ 1/n \sum (f_1 - 0_1)^2]^{1/2}$

F_1 = Forecast value

0_1 = Observed value

N = Total number of observations.

Correlation also can be used for validation between forecasted value and observed value. For this a minimum of 30 pairs must be there for analysis. Chi square test also can be used for this purpose.

2.7 Markov- Chain Analysis

Markov-chain analysis is a universal mathematical tool used for rainfall analysis especially for wet and dry spell and its four combinations for rainfall data series. It is continuous pair analysis in which the consecutive rainfall data are compared for wet and dry spells and the process is continued up to 52 meteorological standard weeks for weekly data series. This method is useful for crop planning especially for rainfed/ dry land situation. For planning purpose, it is important to know the persistence or sequence of dry and wet periods (Veeraputhiran *et al.*, 2003).

There are two probabilities in this analysis. One is initial and the other one is conditional. In the initial probability analysis, the activity is to identify the probability of a particular week is dry or wet. The formula/equation is as follows.

$$P(W) = F(W)/F(W) + F(D) \times 100$$

$$P(D) = F(D)/F(D) + F(W) \times 100$$

Where,

P(W): Probability of the week being wet (%)

P(D): Probability of the week being dry (%)

F(W): Frequency of wet weeks in the set data (number of weeks)

F(D): Frequency of dry weeks in the set data (number of dry weeks)

For doing analysis 30 years meteorological standard weeks data were analysed for weekly mean, which is required as input. The other action is considering the weekly Potential Evapo Transpiration (PET), this is assumed to be 3mm/day and for a week it is 21mm. The weekly rainfall quantity is to be fixed for wet spell, which must be above the weekly PET. Hence, considering 21 mm as weekly PET, 30 mm rainfall is fixed as wet spell week and lesser than $<$ 30 mm is taken as dry spell week. The initial probability was computed for each week by using mean weekly rainfall of 30 years.

The conditional probability of dry week preceded by dry week or wet week and vice versa are computed by using the following formulae

$$P(W_2/D_1) = F(W_2D_1)/F(D_1)$$

$$P(D_2/D_1) = F(D_2D_1)/F(D_1)$$

$$P(W_2/W_1) = F(W_2W_1)/F(W_1)$$

$$P(D_2/W_1) = F(D_2W_1)/F(W_1)$$

Note: subscript to W and D indicates the number of weeks, where,

$P(W_2/D_1)$ = Probability of second week being wet with preceding week dry

$P(D_2/D_1)$ = Probability of second week being dry with preceding week dry

$P(W_2/W_1)$ = Probability of second week being wet with preceding week wet

$P(D_2/W_1)$ = Probability of second week being dry with preceding week wet

$F(W_2/D_1)$ Frequency of wet week preceded by dry week.

F (D_2/D_1) Frequency of dry week preceded by dry week.
F (W_2/W_1) Frequency of wet week preceded by wet week.
F (D_2/W_1) Frequency of dry week preceded by wet week.
This conditional probability may be worked for the 52 weeks.

2.8 Length of Growing Period (LGP)

The length of growing period is the duration of crop growing period, where in crop gets sufficient soil moisture for its growth and yield. The LGP can be worked out by many methods and those are;

Weekly moisture availability period of FAO model (Higgins and Kassam,1981), Weekly R/PE ratio of Jeevananda Reddy (1983) and Weekly moisture Availability Index of Sarkar and Biswas (1988) have been presented here under.

2.8.1 Sarkar and Biswas (1988) Method

This is based on weekly Moisture Availability Index (MAI).

MAI = Assured weekly rainfall(mm)/Weekly PET (mm).

To avoid risks, in getting précised MAI, the assured weekly rainfall is considered against mean weekly rainfall. The dependable weekly rainfall or assured or guaranteed rainfall amount is 50 per cent when annual rainfall of the region is more than 400 mm. If the annual rainfall is lesser than 400 mm, 30 per cent assured weekly rainfall may be taken for the computation of MAI. Based on the number of weeks fall under different MAI values, the classification of LGP may be done as follows.

Zone	Weekly MAI		Remarks
	≥ 0.3	≥ 0.7	
D	<10 weeks	< one week	Crop production needs irrigation support
E	≥ 10 weeks	≥ one week	Short and medium duration crops may be raised
F	≥ 11 weeks	≥ 4 weeks	Medium duration crop may be raised
G	≥ 14 weeks	≥ 7 weeks	A crop of 13 to 18 weeks may be raised
H	≥ 18 weeks	≥ 9 weeks	Medium duration crop is recommended
I	≥ 20 weeks	≥ 10 weeks	One crop followed by fallow crop to be raised under stored soil moisture of 4 to 6 weeks duration
J	≥ 24 weeks	≥ 12 weeks	Two crops can be raised

2.8.2 FAO Water Balance Method (Higgins and Kassam,1981)

In this model the growing season starts when the monthly/weekly rainfall exceeds 0.5 of concerned monthly/weekly PET and ends with the utilization of assumed quantum of stored soil moisture of 100 mm after the concerned monthly/weekly rainfall falls below 0.5 PET of concerned monthly/weekly rainfall.

2.8.2 Jeevananda Reddy Model (Jeevananda Reddy,1983)

Jeevananda Reddy used the method as a simple tool to compute length of growing period for SAT based on mean weekly rainfall and weekly PET. He has suggested 14 weeks moving average based on the output of weekly rainfall and PET ratio(R/PET). The 14 weeks have been suggested since the duration of dry land crop is around 100 days.

He has identified six climatic variables from his calculations and those are;

i. Available Effective Rainy Period(G).

The number of consecutive weeks in which the 14 weeks moving average of R/PET is > 0.75, the initial week of this effective rainy period should have a simple R/PET ratio of ≥ 0.50

ii. Sowing rain(S)

The week before the beginning of the available effective rainy period is taken as the week of commencement of sowing rain Here the simple weekly R/PET ratio should be ≥ 0.50

iii. Pre-sowing cultivation (Ps)

Pre -sowing cultivation or seed bed preparation is started when the 14 weeks moving average of R/PET crosses 0.50 limit and when the particular week has weekly simple R/PET ratio of ≥ 0.25

iv. Wet spell(W)

Within the G if the simple weekly R/PET is ≥ 1.5, that week is called as wet spell week

v. Dry spell(D)

Within the growing period if the simple weekly R/PET is < 0.5, that week is said to be a dry week

vi. Crop failure(A)

If the G period is ≤ five weeks, the growing of dry land crop is limited.

Abstract of climatic variable

S. No	Climate variable	14 weeks moving average of R/PET	Simple weekly R/PET
1	G	≥ 0.75	≥ 0.50
2	S	--	≥ 0.50
3	Ps	≥ 0.50	≥ 0.25
4	W	-	≥ 1.50
5	D	-	≤ 0.50

Example:

Coimbatore weekly rainfall data have been collected and mean for each week and used for this analysis and the model is presented in Table A 3.1.

Table A 3.1 Computing R/PET ratio

Std. week	Mean weekly rainfall (mm)	Mean weekly PET (mm)	Weekly simple R/PET ratio
1	3.96	23.4	0.17
2	1.59	22.7	0.07
3	2.03	24.3	0.08
4	0.16	26.7	0.01
5	0.66	28.7	0.02
6	0.86	29.6	0.03
7	2.97	29.8	0.10
8	3.74	33.0	0.11
9	1.65	34.1	0.05
10	9.35	33.8	0.21
11	3.67	34.5	0.11
12	1.66	34.9	0.05
13	3.40	36.0	0.09
14	5.05	35.8	0.14
15	3.19	35.2	0.09
16	20.02	36.6	0.55
17	12.41	35.7	0.35
18	16.26	34.3	0.47
19	19.07	34.0	0.56
20	18.20	33.7	0.54
21	13.45	34.4	0.39
22	7.90	29.6	0.27
23	6.03	27.9	0.22
24	5.99	25.9	0.23
25	8.56	28.2	0.15
26	12.78	27.6	0.46
27	15.56	29.6	0.53
28	10.45	26.6	0.39
29	7.46	25.6	0.29
30	13.55	25.3	0.54
31	6.14	26.5	0.23
32	8.39	25.0	0.34
33	8.57	28.3	0.30
34	9.07	28.0	0.33
35	5.78	30.0	0.19

Std. week	Mean weekly rainfall (mm)	Mean weekly PET (mm)	Weekly simple R/PET ratio
36	6.13	28.2	0.22
37	16.34	25.9	0.63
38	18.91	26.7	0.71
39	19.34	26.6	0.73
40	24.67	25.6	0.96
41	33.10	26.3	1.26
42	41.01	26.3	1.56
43	43.70	22.2	1.97
44	39.81	20.9	1.90
45	32.39	19.4	1.67
46	30.90	24.3	1.27
47	25.90	24.0	1.08
48	12.07	22.5	0.54
49	27.20	19.8	1.37
50	11.23	21.6	0.52
51	4.43	21.9	0.20
52	7.55	22.0	0.34

The computation of 14 weeks moving average is presented in Table A 3.2

Table A 3.2 Computation of 14 weeks moving average

Standard week (1)	14 weeks moving total (2)	14 weeks moving average (3)	Reference average (4)	Reference standard week (5)
1–14	1.31	0.09	0.09*	7
2–15	1.23	0.09	0.11*	8
3–16	1.71	0.12	0.14	9
4–17	2.05	0.15	0.17	10
5–18	2.51	0.18	0.20	11
6–19	3.05	0.22	0.24	12
7–20	3.56	0.25	0.27	13
8–21	3.85	0.28	0.28	14
9–22	3.94	0.28	0.29	15
10–23	4.11	0.29	0.29	16
11–24	4.06	0.29	0.29	17
12–25	4.08	0.29	0.31	18
13–26	4.51	0.32	0.34	19
14–27	4.95	0.35	0.36	20

Standard week (1)	14 weeks moving total (2)	14 weeks moving average (3)	Reference average (4)	Reference standard week (5)
15–28	5.20	0.37	0.38	21
16–29	5.40	0.39	0.39	22
17.30	5.39	0.39	0.39	23
18–31	5.27	0.38	0.38	24
19–32	5.14	0.37	0.36	25
20–33	4.88	0.35	0.34	26
21–34	4.67	0.33	0.33	27
22–35	4.47	0.32	0.32	28
23–36	4.42	0.32	0.34	29
24–37	4.83	0.35	0.37	30
25–38	5.31	0.38	0.40	31
26–39	5.89	0.42	0.44	32
27–40	6.39	0.46	0.49	33
28–41	7.12	0.51	0.55	34
29–42	8.29	0.59	0.65	35
30–43	9.97	0.71	0.76	36
31–44	11.33	0.81	0.86	37
32–45	12.77	0.91	0.95	38
33–46	13.70	0.98	1.01	39
34–47	14.48	1.03	1.04	40
35–48	14.67	1.05	1.09	41
36–49	15.87	1.13	1.15	42
37–50	16.17	1.16	1.14	43
38–51	15.74	1.12	1.11	44
39–52	15.37	1.10	1.08	45
40–1	14.81	1.06	1.03	46
41–2	13.92	0.99	0.95	47
42–3	12.74	0.91	0.86	48
43–4	11.19	0.80	0.73	49
44–5	9.24	0.66	.60	50
45–6	7.37	0.53	0.47	51
46–7	5.80	0.41	0.37	52
47–8	4.64	0.33	0.30	1
48–9	3.61	0.26	0.25	2
49–10	3.35	0.24	0.20	3

Standard week (1)	14 weeks moving total (2)	14 weeks moving average (3)	Reference average (4)	Reference standard week (5)
50–11	2.09	0.15	0.14	4
51–12	1.62	0.12	0.12	5
52–13	1.51	0.11	0.10	6

From the value of column 3, compute for column 4 as given below.

* = 0.09 + 0.09 = 0.18/2 = 0.09

* 0.09 + 0.12 = 0.21/2 = 0.11

*0.12 + 0.15 = 0.27/2 = 0.14 and so on and fill up the

Draw the graph for the values given in column 4 for the standard week given in column 5 and find out climatic variables as discussed elsewhere.

3. Heat Unit Concepts

3.1 Growing Degree Days (GDD)

The GDD is an arithmetic accumulation of daily mean temperature above base temperature of a particular crop. This decides the phenological development in crops (Iwata, 1984).

GDD = Daily Tmax°C + daily T.min°C/2 - base temperature

A base temperature of 10°C was adopted for rice. Similarly, for other crops it is available

3.2 Photo Thermal Units (PTU)

PTU = GDD × Mean day length (N) (to be obtained from Table for a particular location based on coordinate)

3.3 Helio Thermal Units (HTU)

Helio Thermal Units (HTU) at different stages of crop growth were computed by using the formula of Rajput (1980).

HTU = GDD × Bright sunshine hours (n) (from sunshine cards)

3.4 Heat Unit Efficiency (HUE)- *Kiniry et al., 1992*

HUE = Dry matter production (g)/GDD

3.5 Relative Temperature Disparity (RTD)

The Relative Temperature Disparity (RTD) for the cropping period was calculated by using the formula as suggested by Kaladevi (1998). If the RTD in a season is greater, the productivity would be greater.

RTD = T max (°C)–Tmin(°C)/Tmax (°C) × 100

3.6 Relative Humidity Disparity (RHD)

RHD = Max.RH (%)–Min RH (%)/Max RH (%) × 100

3.7 Radiation Use Efficiency (RUE)

RUE = Dry matter Production(gm^2)/Cumulative light absorbed (MJm^{-2})

4. Others

4.1 Relative Spread Index (RSI) and Relative Yield Index (RYI)

Relative Spread Index (RSI) and Relative Yield Index (RYI) as given by Kanwar (1972) are presented.

$$RSI = \frac{\text{Area of the crop expressed as \% of total cultivable area in a district}}{\text{Area of the crop expressed as \% of total cultivable area in the State}} \times 100$$

$$RYI = \frac{\text{Mean yield of a particular crop in a district}}{\text{Mean yield of particular crop in the State}} \times 100$$

4.2 Relative water content (Barrs and Weatherly,1962)

$$RWC = \frac{\text{Fresh weight of leaves} - \text{Dry weight of leaves}}{\text{Turgid weight of leaves} - \text{Dry weight of leaves}} \times 100$$

(expressed in per cent)

4.3 Water balance allied (Thornthwaite and Mather,1955)

1. AET = PET, when P > PET
2. AET = P + ΔS, when P < PET
3. WS = P-AET
4. WD = PET- AET
5. Humidity Index (Ih) = Water surplus/PETx100
6. Aridity Index (Ia) = Water deficit/PET $\times$ 100
7. Moisture Index (Im) = P-PET/PET $\times$ 100

4.4 Water Requirement Satisfaction Index (WRSI)

WRSI = 100–Deficit/Total water requirement

AET: Actual Evapotranspiration; PET: Potential evapotranspiration;

P: Precipitation; ΔS: Soil moisture; WS: Water surplus; WD: Water deficit

4.5 Light interception (LI)

LI (%) = (1–I b/I o) $\times$ 100 (when ACCU PAR Ceptometer, model-LP-80 was used)

LI; Light interception by the crop canopy

Io = PAR measured at above canopy

Ib = PAR measured at below canopy

4.6 Percentage of Light Intercepted (PLI)

PLI = LI–LT/LI $\times$100

LI; Light intercepted at above crop canopy

LT; Light intercepted at below crop canopy

4.7 Thermal Humidity Index (THI)- Tao and Zin, (2003)-Poultry

THI = 0.85 Tdb + 0.15Twb

T db: dry bulb temperature; wb: wet bulb temperature

4.8 Thermal humidity Index (U.S. Weather Bureau)-animals

$THI = 0.72(C_{db}+C_{wb}) + 40.6$
db.: dry bulb temperature(°C)
wb: wet bulb temperature (°C)

4.9 Cloud cover estimation (Balasubramanian et al., 2009; Geethalakshmi, et al., 2000.)

Normally at the surface observatory cloud cover is estimated through observer's experience. The first author of this book met Dr.A.S.R.A.S. Sastry, the Head of the Department of Agricultural meteorology, Indira Gandhi Agricultural University, Raipur, during mid 1990's and discussed about scientific measurement of cloud cover. During discussion, Dr. A.S.R.A.S. Sastry indicated the availability of the following equation and this is more scientific.

$CC = 1 - (n/N) \times 8$
Where CC = Cloud cover in octa
N = Normal day length for concerned latitude (Table from FAO Tech. bulletin, 24)
n = Number of bright sun shine hours of the concerned latitude recorded at the observatory
8 = Overcast sky sale in octa

Unit code of measurement from the result in OCTA

Unit octa	Cloud amount
0	clear sky
1	Few clouds
2 to 6	Increasing cloud cover
7	Cloud covering almost the entire sky with at least one opening through which sky is seen
8	Completely overcast sky

NB: One situation where in, the amount of clouds could not be estimated due to fog, dust etc., presence in the atmosphere and hence the code 9 is used.

Example:
N = 11 hours
n = 04
Then $CC = 1-(4/11) \times 8$
$1 - 0.36 \times 8$
0.64×8
5.12 octa

Z Table- Probability of normal distribution(home.ubalt.edu/ntsbarsh/Business-Stat/Statistical Tables.pdf)

Z	0.00	0.01	0.02	0.03	0.04	0.05	0.06	0.07	0.08	0.09
0.0	0.5000	0.5040	0.5080	0.5120	0.5160	0.5199	0.5239	0.5279	0.5319	0.5359
0.1	0.5398	0.5438	0.5478	0.5517	0.5557	0.5596	0.5636	0.5675	0.5714	0.5753
0.2	0.5793	0.5832	0.5871	0.5910	0.5948	0.5987	0.6026	0.6064	0.6103	0.6141
0.3	0.6179	0.6217	0.6255	0.6293	0.6331	0.6368	0.6406	0.6443	0.6480	0.6517
0.4	0.6554	0.6591	0.6628	0.6664	0.6700	0.6736	0.6772	0.6808	0.6844	0.6879
0.5	0.6915	0.6950	0.6985	0.7019	0.7054	0.7088	0.7123	0.7157	0.7190	0.7224
0.6	0.7257	0.7291	0.7324	0.7357	0.7389	0.7422	0.7454	0.7486	0.7517	0.7549
0.7	0.7580	0.7611	0.7642	0.7673	0.7704	0.7734	0.7764	0.7794	0.7283	0.7852
0.8	0.7881	0.7910	0.7939	0.7967	0.7995	0.8023	0.8051	0.8078	0.8106	0.8133
0.9	0.8159	0.8186	0.8212	0.8238	0.8264	0.8289	0.8315	0.8304	0.8365	0.8389
1.0	0.8413	0.8438	0.8461	0.8485	0.8508	0.8531	0.8554	0.8577	0.8599	0.8261
1.1	0.8643	0.8665	0.8686	0.8708	0.8729	0.8749	0.8770	0.8790	0.8810	0.8830
1.2	0.8849	0.8869	0.8888	0.8907	0.8925	0.8944	0.8962	0.8980	0.8997	0.9015
1.3	0.9302	0.9049	0.9066	0.9082	0.9099	0.9115	0.9131	0.9147	0.9162	0.9177
1.4	0.9192	0.9207	0.9222	0.9236	0.9251	0.9265	0.9279	0.9292	0.9306	0.9319
1.5	0.9332	0.9345	0.9357	0.9370	0.9382	0.9394	0.9406	0.9418	0.9429	0.9441
1.6	0.9452	0.9463	0.9474	.9484	0.9495	0.9505	0.9515	0.92525	0.9535	0.9545
1.7	0.9554	0.9564	0.9573	0.9582	0.9591	0.9599	0.9608	0.9616	0.9625	0.9633
1.8	0.9641	0.9649	0.9656	0.9664	0.9671	0.9678	0.9686	0.9693	0.9699	0.9706
1.9	0.9713	0.9719	0.9726	0.973	0.9738	0.9744	0.9750	0.9756	0.9761	0.9767
2.0	0.9772	0.9778	0.9783	0.9788	0.9793	0.9798	0.9803	0.9808	0.9812	0.9817
2.1	0.9821	0.9826	0.9830	0.9834	0.9838	0.9842	0.9846	0.9850	0.9854	0.9857
2.2	0.9861	0.9864	0.9868	0.9871	0.9875	0.9878	0.9881	0.9884	0.9887	0.9890
2.3	0.9893	0.9896	0.9898	0.9901	0.9904	0.9906	0.9909	0.9911	0.9913	0.9916
2.4	0.9918	0.9920	0.9922	0.9925	0.9927	0.9929	0.9931	0.9932	0.9934	0.9936
2.5	0.9938	0.9940	0.9941	0.9943	0.9945	0.9946	0.9948	0.9949	0.9951	0.9952
2.6	0.9953	0.9955	0.9956	0.9957	0.9959	0.9960	0.9961	0.9962	0.9963	0.9964
2.7	0.9965	0.9966	0.9967	0.9968	0.9969	0.9970	0.9971	0.9972	0.9973	0.9974
2.8	0.9974	0.9975	0.9976	0.9977	0.9977	0.9978	0.9979	0.9979	0.9980	0.9981
2.9	0.9981	0.9982	0.9982	0.9983	0.9984	0.9984	0.9985	0.9985	0.9986	0.9986
3.0	0.9987	0.9987	0.9987	0.9988	0.9988	0.9989	0.9989	0.9989	0.9990	0.9990
3.1	0.9990	0.9991	0.9991	0.9991	0.9992	0.9992	0.9992	0.9992	0.9993	0.9993
3.2	0.9993	0.9993	0.9994	0.9994	0.9994	0.9994	0.9994	0.9995	0.9995	0.9995
3.3	0.9995	0.9995	0.9995	0.9996	0.9996	0.9996	0.9996	0.9996	0.9996	0.9997
3.4	0.9997	0.9997	0.9997	0.9997	0.9997	0.9997	0.9997	0.9997	0.9997	0.9998
3.5	0.9998	0.9998	0.9998	0.9998	0.9998	0.9998	0.9998	0.9998	0.9998	0.9998
3.6	0.9998	0.9998	0.9999	0.9999	0.9999	0.9999	0.9999	0.9999	0.9999	0.9999
3.7	0.9999	0.9999	0.9999	0.9999	0.9999	0.9999	0.9999	0.9999	0.9999	0.9999
3.8	0.9999	0.9999	0.9999	0.9999	0.9999	0.9999	0.9999	0.9999	0.9999	0.9999

Index

E

F

G

H

I

J

L

M

N

O

P

R

S

T

V

W

Z

Printed in the United States
by Baker & Taylor Publisher Services